T0342386

Tabula Raza

Tabula Raza

MAPPING RACE AND HUMAN DIVERSITY
IN AMERICAN GENOME SCIENCE

Duana Fullwiley

UNIVERSITY OF CALIFORNIA PRESS

University of California Press
Oakland, California

© 2024 by Duana Fullwiley

Library of Congress Cataloging-in-Publication Data

Names: Fullwiley, Duana, author.
Title: Tabula raza: mapping race and human diversity in American
    genome science / Duana Fullwiley.
Other titles: Atelier (Oakland, Calif.) ; 14.
Description: Oakland, California : University of California Press,
    [2024] | Series: Atelier : ethnographic inquiry in the twenty-first
    century ; 14 | Includes bibliographical references and index.
Identifiers: LCCN 2023040930 (print) | LCCN 2023040931 (ebook) |
    ISBN 9780520401167 (cloth) | ISBN 9780520401174 (paperback) |
    ISBN 9780520401181 (ebook)
Subjects: LCSH: Genomics—United States. | Human gene mapping—
    United States. | Human beings—Variation—United States. | Human
    genetics—Variation—United States. | Scientific racism.
Classification: LCC QH447 .F84 2024 (print) | LCC QH447 (ebook) |
    DDC 572.8/60973—dc23/eng/20231222
LC record available at https://lccn.loc.gov/2023040930
LC ebook record available at https://lccn.loc.gov/2023040931

33 32 31 30 29 28 27 26 25 24
10 9 8 7 6 5 4 3 2 1

For
Aunt Re
Zoe
and Amalie

Contents

Preface

SKIN AND CODE

It is the twenty-first century and humans are grappling with which features of their identities matter most, and when. Digital tools that fragment aspects of the person are now adopted to better express a limited part of what can, or cannot, be exposed. Technology platforms and algorithms aggregate social and DNA data, helping people create untold possibilities—and new vulnerabilities. Personal privacy is facing the threat of extinction. Yet people's stories and experiences remain essential to how humans understand the world and how best to navigate it. Science and technology have now been recognized as highly malleable products, shaped by human desires.

In the future that is now, the field of anthropology has updated its interest in the concept of culture to focus on power. Meanwhile, the violence of race keeps reinventing itself, and anthropologists keep calling it a social construct. Some begin to see how the construct can include biology and genetics and that we need better language to describe how this new race is made. In the preface to a book on these issues, an anthropologist archives a scene where the human names of study subjects—at all levels—are replaced with code to disguise them. That their identities are abstracted and possibly effaced does not negate what the codes attempt to obscure.

· · · · ·

Head scientist, geneticist 00XPI.PHD, burst into the lab with a Santa smile bearing gifts. "I got somethin' for y'all." In his arms were several cellophane-wrapped white lab coats for all of the members of his team. He handed one to each of the two researchers in this section of the cold workspace as he made the rounds. They carefully removed the glassy packaging.

"Wow," 00QNF exclaimed, her eyes wide. "My very own lab coat." She and 00BKE, both beaming, raced to put them on. 00QNF held out her arms, then tugged on the lapels. Their eyes darted back and forth admiring themselves and each other. As they wiggled the coats over their clothes and tried to get the fit right, it was obvious that the arms were too long, especially on 00QNF. Even the size S dwarfed her. It was only then that I realized that they hadn't been wearing the powerfully symbolic robes before.

"Wow, these are some long sleeves," 00QNF laughed. "It's okay, though." She was flying high in the super skin that she had just donned. She rolled up the cuffs with two folds. 00BKE did the same, taking himself in from left to right, asking, "Where's a mirror?"

"It's gonna shrink anyway when you wash it. And I'll grow into it," 00QNF offered. She whispered a few words to herself about having enough room to layer sweaters under it during winter. They both stood tall, proud, and all together new in the white vestments.

This was during the first of my three field stays in the lab, on a warm Chicago morning in June of 2007. But 00QNF's words took me back to the winters of my childhood—those Christmases when we sometimes got gifts. My struggling parents always bought us clothes that were a size too large so that we could wear them for a few years. But, of course, 00QNF and 00BKE were already full-grown adults. Having a childhood like 00QNF did, where there were lots of kids in the house and resources were tight, primed her affection for the research team and the principal investigator (PI) who had some renown. They were her new family, a group of mostly Black geneticists buzzing in a world of whiteness, who felt they had to stick together. I told them about my always-sized-up-clothes. 00QNF laughed, nodding, "Yep, yep!"

As the coat commotion calmed, and people started to get back to work, I asked OOBKE if I could shadow him and learn more about what he did in the lab. He played down his role and said that he wasn't working on any big ideas or papers at the time but was mostly just organizing and managing the data. He showed me his computer screen, where rows and rows and columns and columns contained little bits of genotyping information on hundreds of African Americans.

OOBKE scrolled through a little as I asked him about the ancestry genetic markers that were featured as small numeric codes with decimal points to indicate at what level each person in the data possessed each allele in their genomes. The data was anonymized, neatly standardized in a light-colored script on a black background. It made a din of visual white noise. In the blurred scroll of it all, it dawned on me that what we were looking at was no longer physically just about people. I pointed to the computer's square face—or more precisely to one line of information.

"Who is that?" I asked.

OOBKE smiled slowly, turned his head away then back and laughed. "I have no idea. Not anymore . . . ," he trailed off. "At this point," he continued, "we have all of the ID codes somewhere so we could call them, and find out who they are, if we need to. But we have to anonymize the data."

I told him why all of those tables and cells spoke out to me. "I know," I said. "I just wonder if you ever think about who these people are when you're sorting through and organizing these bits of information about them?"

He shook his head, smiling again. "It's funny—no, I don't." He paused. "But maybe I should. I'm mostly just focused on the data at this point. But okay, I see what you mean."

For a moment I thought about whether I would ever write this scene. And if I did, what the effect would be if I gave my scientific "subjects" lines of code as identifiers. On screen the codes were the surface digital skins that enveloped the lives stored beyond these skins' numeric ability to mask. To give codes to living, breathing scientists who have voice, language, and emotion, however, seals away very little of their lively selves. I stored the thought away until years later.

Most scientists in these pages agreed to be named in the writings that resulted from my fieldwork with them. As a human practice of familial belonging and recognition, personal names resonate for us and allow us to connect to people once we associate the feeling of their unique humanity to a memory we make of them in our minds. For the anthropologist of science, participants in an ethnographic study also publicly author the material cultures of science they produce in publications of all sorts— products that are key to their everyday training, aspirations, and professional identities. The experiment of giving them alphanumeric codes could only go so far, even if some degree of masking of their identities would be possible through pseudonyms.

The abstract nature of a more sterile code, however, provides an opening for another means of relating to people through what I am experimenting here with *as a conscious abstraction*—to see the world of scientists slightly differently.[1] This approach is expanded upon in chapter 1, but for now the reader should file away within their memory that this book will relay real-world ethnography to describe what I call its "sci non-fi" elements. These are the scientific nonfictional happenings in labs that might have once been imagined as futuristic mash-ups of our recognizable past and present thrust forward in technological time but that in actuality are all too real.

I looked back at OOBKE's simultaneously lifeless yet spirited screen and then up at him. I noticed that he was now practicing conscious abstraction. He seemed to be thinking about the lines of *people* now ordered on his screen. He drew back the curtain on his routine way of seeing and experienced it from a different vantage. I took a mental note and realized for the first time that such an idea should be developed.

Pulling ourselves out of our busy work lives, or busyness in general, to notice the details in the machinery that we drive on autopilot throughout the course of a day is not an easy reflex. I knew that OOBKE and many of the scientists I observed cared deeply about the people in their studies. In OOXPI.PHD's lab the data was mostly from Black people, and it was clear how much the researchers cared about Black health.

I knew that they did not just reduce people to data, even if the person could mentally fade from daily view and the data was what lived on in code, on screen, in the tables and cells of what I call, later in this book, the *tabula raza.* Yet I was grappling with how easy the work that the scientists before me did with ancestry informative markers (AIMs) could quickly become racialized, and even used against Black people in the realm of biased police work through various applications of forensics that utilize the same DNA. This could make the scientists frustrated or uncomfortable. These were moments when they tried to work with a largely inherited troubled genetic science that—like the ill-fitting lab coats—was a size or two off, but was what they had.

The care that 00XPI.PHD and his research team showed for bettering the plight of people who lived and sometimes thrived despite generations of discrimination was part of what connected us. But beyond that care and bond was the power of the machine interface and the larger societal impulse to digitize human data that allow for various levels of more unwitting abstraction that can mask, blur, and displace from view how that data is captured, circulated, mined, and exploited. In many instances in my fieldwork, in this lab as well as others I write about in this book, I witnessed what seemed to be well-meaning geneticists faced with career structures and academic enticements to digitize Black and other lives that are racialized in such "genetic ancestry" datasets. On this particular day the rows of cells on the computer screen of a young scientist had been far removed from, then easily and intentionally rejoined to, the lives of the people who gave of their DNA.

In *The Immortal Life of Henrietta Lacks*, writer Rebecca Skloot includes a compelling scene where the twenty-one-year-old research assistant to Dr. George Guy, the scientist who first took Henrietta's cervical cancer cells that would go onto be the first immortalized cell line, came back to collect more tissue when Lacks died. The year was 1952. The assistant, Mary Kubicek, recounted decades later that when she went into the autopsy room she noticed Henrietta's red painted toenails. Seeing that human detail, she gasped and thought, "Oh jeez, she's a person," and nearly fainted.[2]

Readers of Skloot's book are brought into a visceral scene where the human body of a Black woman with an aggressive cancer who died too soon is made into a commodity through her cells, her genetic material, and the technologies that her cells helped bring into being. The scientific immortalization, storage, production, sharing, circulation, and reproduction of her cells—now called HeLa—made them one of the most utilized living therapeutic substances that have spawned countless medical innovations and improved human lives ever since. These advances range from the crucial role of HeLa in the development of the polio vaccine to the production of drugs that inhibit HIV/AIDS, from techniques now known as in vitro fertilization (IVF) to those needed to understand cellular aging, as well as the effects of x-ray technology on the body. Drugs for multiple myeloma, sickle cell anemia, vaccines for COVID-19, and countless others were all brought to market with the help of Henrietta's cells.[3]

There was also a problem, however, in that these discoveries and the wealth and recognition that they brought to scientists and scientific industries were possible because of Henrietta's place in America's racial order. She was an uneducated tobacco farmer who grew up in poverty, who had no health insurance, and who received a second-class form of medical care during Jim Crow. When she and her family migrated to the Baltimore area for work in the 1940s, Henrietta was treated with inferior care in the colored ward of Johns Hopkins hospital. It was there that her cervical cancer cells were taken for research purposes without her consent. At the time, gaining medical consent from people for research was not enshrined in medical ethics codes, but it was still practiced and often perceived as the correct thing to do. After Henrietta's death her family continued to live with poor health-care options. Most inequitable, they did not have medical insurance even decades later, while many in the scientific world made enormous profits that would tally into the billions by the time Skloot's book appeared in 2010. Since then, the wealth from Henrietta's unique biology has continued to accumulate.

I tell the story of Henrietta Lacks and HeLa to emphasize the everyday poverty that still affects so many Black Americans as well as other racialized minorities in the United States and around the globe in contrast to the wealth that can be generated from their biological material and DNA. This structural chasm should figure front and center in conversations on when and how human DNA might be collected, researched, and economically exploited in the world we now live in. This is a world where research and data gathering have been intensely amplified since 1952. The storage, sharing, and circulation of DNA is not only more common but also an expected feature of investment-worthy projects. Large databases that have collected thousands of samples—such as the 1000 Genomes Project and the UK Biobank as well as the *All of Us* initiative in which the US government's goal is to collect data on one million Americans—also feature the immortalization of cell lines from those who have given their DNA. This DNA will live on, be reproduced, and mail-ordered for a broad range of future research applications that could in no way be specified to donors in advance.

According to a 2021 report from the American Society of Human Genetics, the estimated wealth generated from the human genetics and genomics industry in the United States alone exceeded $108 billion in 2019 and generated more than $265 billion across the US economy.[4] The questions that the history of science, as it intersects with the racial history of the United States and its founding, should pose for us at the outset are: How will this wealth garnered from people's genetic material—that is so easily reproduced and amplified for technological pursuits—involve American underrepresented groups, the poor, and those targeted for research in the name of diversity, in truly inclusive ways? More specifically, how can we create a world where the medical blood draw or cheek scrape that leads to DNA extraction does not necessarily translate into the extraction of capital from poor bodies into the pockets of wealthier ones? And how are underrepresented scientists being increasingly brought into these endeavors with a potential promise to diversify what has been reported over and again as an overwhelming overrepresentation of DNA from European white populations in genetic and genomic studies throughout the world?

In all of this, we must also track how a pattern of cultural forms where Western European colonizers and their descendants control social and scientific economies, including those of modern genetics, translates to older outlines of empire and enslavement—and where they diverge.

THE ANCESTORS WATCH IN DISBELIEF

In May of 2016 a celebrated Black American writer invited me to her house to have tea and to discuss what I described as my "troubled book." I was documenting how several influential scientists and academics, some of whom hold incredible sway in the media and popular culture, might be overselling the precision of genetic ancestry technology to reconstruct the past. In their positive promise of self-discovery, of tools to search for "roots," of scientifically pinpointing geographic origins, and in some cases of giving people speculative information about disease risk, they also downplayed how curious test takers lose control over the fate of their biological material.

Powerful academics, who now had their own genes and money in the ancestry game—through DNA testing companies where they had financial and personal investments—were also doing what they could to undermine my work and reputation. Colleagues working in genetics and anthropology who knew the players involved told me that their actions were meant to "bully" me into silence. The stress and professional negotiations were not trivial. The ethical question for me was how to publicly tell a story about a troubled science of race and genetics in an era where privacy is often treated as a relic of the past and the rise of big data makes its possibility appear a lost cause from the outset. At the same time, an increased awareness of differential racial outcomes in medicine, of racial bias in policing, and a simplified rendition of ancestry in commercial genealogical pursuits have amplified the need for scientists and social scientists to confront these issues head on.

In most instances, the spirit of the work I do has been in a curious and collaborative mode. I have learned an incredible amount from those who have spent time with me and allowed me into their work and thinking spaces. In other instances, when I published a piece in 2014 showing that when people gave their DNA for ancestry studies some of it ended up in the hands of forensic scientists for the policing of Black Americans, the geneticist responsible asked to withdraw from my project. Since the evidence for the connections were in the scientific literature and in the public domain, I asked where in the work he felt that I had misrepresented him. I wanted to make sure that was not the case, and if there was an error, to correct my mistake. I got no response.

A similar issue had come up a few years earlier regarding my analysis of a research team working on genetic ancestry and a common disease where the lab head was sure that "African ancestry" would be associated with more severe cases. His dogged certainty did not immediately play out, but the narrative that he had woven for his lab primed their expectations, which were confusing to many in the group when the data did not sort along the racial lines that they had been led to believe were "real." When I was about to publish on these lab dynamics, including the point when their expectations shifted, I emailed the PI the piece in 2008 before it was published. He enthusiastically responded and said he would let me know his thoughts. He never did. In 2011 he wrote to the journal editor and tried to have the piece retracted. His claims, however, that he had not signed consent and name attribution forms, did not stand once I reminded him of them and provided copies for him and the journal editor.

In 2013, after a few years of silence, this same researcher and I were both attending a large multidisciplinary conference on ancestry inference with the goal of finding some standardization for genetic ancestry tools. Several other geneticists who had consented to be in my ethnography were also in attendance, as were many key players in both of our fields. After I broke the icy awkwardness at the coffee and pastries table in the hallway before the morning session, the researcher and I exchanged small talk throughout the day. That night a group of social scientists and geneticists were going to dinner. We walked behind and made a digression in the hotel to talk. I asked him about his earlier reaction, this time in person.

"So what exactly in my article upset you?" I asked.

"You called me a racist," he said as he shifted his backpack on straight. "I let you into my lab. I trusted you because you're a kid from the ghetto, like me."

When I was observing his team in the early 2000s, we had discussed all kinds of issues about race, our biracial families, and both of us having struggled as first-generation college students. We shared stories of our parents laboring in the fields. His mother was a farm worker and my father had traveled from Memphis to pick cotton and peas in rural Mississippi when he was a child. Although I didn't ever describe or think of myself as "from the ghetto," I didn't contest and just listened before asking him to reread the paper that he was upset about. "I actually said you were not a racist, right in the abstract," I told him. "I did my best to describe your background and you as a person."

He never attempted to point out where I had allegedly made the racist call. That said, I did understand that he felt hurt about my close reading of his lab's highly racial hypotheses at the time. I then remembered when he would introduce me to colleagues as "my anthropologist" who was "writing a book about me." I'd had to remind him about my role. I was not a PR agent, and I was studying multiple labs. In these exchanges it became clear that I was confronting a fundamental occupational hazard of being an anthropologist of science. At issue in these two emotional requests was a perhaps predictable and inevitable clash: where my care and protection of people working in genetics—as "human research subjects" in my anthropological study who often wanted to be described in heroic and flattering terms—had run up against my own academic freedom to study, interpret, and write up my findings and a larger commitment to the public more broadly. If people, patients, donors, or students gave their DNA for research purposes, or for a pedagogical classroom exercise on genetic ancestry, I felt that it was important to describe the culture and economy of science where their DNA could end up in a highly racializing scientific article, or, as in the first instance discussed above, in a forensics database used for police work that at the time was designed to target Black faces based on "ancestry" genetic data.

As with so many other moments in this research, as in the field that is human genetics writ large, at stake are overlapping and simultaneous hopes. Hopes for freedom and liberation through better science, especially as it might bear on improving health, and hopes for social justice commitments to address if not cure societal ills tied to race and power. These hopes mark the political pledge that often serves as an irreproachable wedge on the part of some geneticists to close the gap in racial health disparities. Real differences in disease burdens are cited as the core reason that genetic pursuits for causation should continue, despite admittedly paltry findings on the genetic sources of common conditions after decades of research and billions of taxpayer dollars spent.[5] When people speak out or critique the reductive emphasis on genetic causation, they are often accused of being against minority health: *Who could be against saving lives?* At stake is how these politics are fed by the capitalistic excesses that sustain their virtuous endeavors when "inclusion" at all costs too readily becomes corrupt.

.   .   .   .   .

Before we got very far into my story, at my first mention of "ancestry through DNA," the writer put her head down and said, "Oh, dear." She let me know that we had more in common than I probably suspected. First, she clearly wanted to get on the same page and get me up to speed on why she had this nagging feeling about the whole industry. She asked: "Can you please explain what genetic ancestry testing can actually do?"

Unbeknownst to me, she had already begun to doubt the promises. She told me how she had submitted her DNA for a documentary film about tracing ancestry a few years prior, but it was never used. Now, she wanted it back. She had looked high and low, but she couldn't find her consent form. She wanted to read its details again, which she couldn't remember— that is, she admitted, if she had ever really taken them in at all. She wanted to know how her DNA was being handled and stored. "Who owns it now?" she asked.

I told her that people's genetic material collected for such films is analyzed by the companies that the films feature. "Most likely it's in their hands."

She paused in a meditative way, then admitted that she knew at the time she gave her DNA that she shouldn't have. Admonishing herself, she said, "I knew better. I felt it when I was doing it." She paused again. "I knew the ancestors weren't happy. But somehow I didn't listen."

Collapsing both then and now, cupping her tea for comfort, she mulled over the lesson. "We need to pay deeper attention to what we sense, better attention to ourselves." She straightened up in her seat and seemed to take a broader view beyond her own immediate regret. She thought back to those who were telling her something that day she gave her DNA. "Our poor ancestors," she lamented. "With this genetic ancestry business, they are being sold all over again."

GENETIC GOLD

Over time it became increasingly clear that DNA, especially if it comes from Black people, was seen as a scarce and lucrative commodity. Geneticist Rick Kittles, who is Black, once told me: "You have to realize that Black DNA—it's like gold." As he saw it, very few people could build large datasets with the consent of significant numbers of Black people because, in his experience, African Americans still hold widespread fears of experimentation, continue to feel socially denigrated and devalued, and share a general sentiment that scientists might think that Black people are interesting as specimens, but they lack a sincere interest in African American well-being more generally. Kittles and other geneticists would often cite the value of African genetic diversity for understanding how disease variants found in populations outside of Africa actually functioned, especially since candidate markers found in US cohorts for common diseases did not seem to automatically trigger illness in the first Africans whose genomes were fully sequenced in 2010.

Black Americans can in no way stand in for the genetic diversity of all of Africa, yet it is common knowledge that they possess a subset of it. Like the writer cited above, Kittles could not separate out the emotional lever of genetic ancestry from the will to know as a scientific and political quest. Yet it wasn't always clear how ancestry by DNA as an advantageous tool to reconstruct some version of the past could easily backslide, becoming a complicated force of disadvantage, on the fulcrum of supposed "objective" science. But unlike the writer worried about the resale of her enslaved ancestors in today's marketplace, Kittles has had substantial financial stakes in the industry. In fact, concerning DNA ancestry tests for Black people, he helped create them. Kittles, who has spent his career research-ing the genetic and lifestyle contributors to prostate cancer as an academic scientist, was the first person to start a company that specifically promised to trace DNA for Black Americans in order to reconnect them to the con-tinent of Africa. The enterprise, called African Ancestry, launched in 2003. He made it clear to me, however, that he did not sell or share the data with inquirers who asked for it, such as consumer assessors and mar-ket analysts at companies like American Express. Nor did he share the data with law enforcement. Kittles took pride in being able to say that he was not selling his people out. Still, today, African Ancestry's website boasts: "We DO NOT sell or share your genetic information."[6]

After an initial few years of bold market success with his private enter-prise, Kittles found himself struggling with whether to share this business and cultural platform with celebrity academic Henry Louis Gates Jr., who in the mid-2000s wanted to buy into the company. When Kittles and his business partner, Gina Paige, refused, the Harvard professor started his own outfit. Gates's venture launched with a glowing article in the *Wall Street Journal* about his superior product.[7] The journalist covering the story recounted in vivid detail Gates's disappointment in his erroneous maternal DNA result that he received from Kittles early on in 2000, before he founded African Ancestry. The journalist, through the professor, cited the many ways that Kittles's company was failing consumers like Gates, while lauding Gates for bringing historians on board to interpret DNA matches for his newly launched subsidiary of the company Family Tree DNA. In an interview with the journalist, Kittles admitted his error, in which he placed Gates's maternal line in Egypt. He emphasized the difficulty of providing exact matches since most people have multiple matches to different places. At base, Kittles argued, the attractive pull and cultural limitation of his needed database was that it focused specifically on African populations. (Gates's maternal DNA markers were eventually localized to Europe.) The coverage on the whole, however, was damning for Kittles despite his transparency.

In my own experience at conferences with both Kittles and Gates at different points during that decade, historians, sociologists, other geneticists, and anthropologists discussed these critiques in detail with both of them. Yet in the *Wall Street Journal* article there was no mention of our larger concerns: that genetic ancestry tests, more generally, simply do not have a gold standard. There was no mention that test "results" are only as good as the reference samples are—that they are merely reflections, *as possibilities*, of a circumscribed data pool. There often remains a lack of transparency concerning how accurate, exclusive, or reliable ancestry matches are because the details of who exactly has been sampled for population databases (and how extensively any scientific team has included vast, or not so vast, reference samples) is rarely addressed. The latter were often treated as company trade secrets. That the professor was on the board of the *Wall Street Journal* and also launched his own television series focused on DNA ancestry (that had featured Kittles quite favorably up until that point) did not make print.

The tension in their relationship was hard on Kittles, who felt embarrassed and slighted, as he conveyed to me during one of my subsequent field stays. In 2011, a few years later, he was faced with whether to participate in a commercialized effort to get large numbers of Black people to give their DNA to the direct-to-consumer (DTC) company 23andMe. At the time, the relatively nascent yet growing company was preparing to launch a highly mediatized DNA recruitment effort called Roots into the Future at the National Urban League's 2011 meeting—a collaboration the company had enthusiastically formed with Gates. Kittles still had the respect of many, however. If members of the public didn't know him by name, they knew of his African Ancestry and his highly publicized work with the lower Manhattan African Burial Ground project, where Kittles first matched the DNA of human remains from a cemetery of enslaved Africans in New York City to various peoples and places in Africa.[8]

Getting Kittles's buy-in for the 23andMe campaign to recruit more Black clients would greatly benefit 23andMe's efforts. The goal was to expand the 23andMe database, which was—and to this day remains— largely "European-American." In a scheme of win-win capitalism, Roots into the Future entailed a major seemingly cool giveaway of ten thousand so-called "free" ancestry tests to "Black" or "African-descended people." The catch was that these Black folks would also be asked to consent that their DNA could be used in the company's health and pharmaceutical drug discovery research—23andMe's primary lucrative plan for its con- sumers' data from the beginning.[9] At the time of this writing, after multi- ple deals with Genentec, Pfizer, GlaxoSmithKline (now GSK), Lundbeck, Janssen, Biogen, and Alynlam Pharmaceuticals, the company has gone public. Its worth hovers around $3 billion.

Another seam in the proverbial gold mine of African peoples' DNA branched out a year earlier in 2010 as the US National Institutes of Health and the UK's Wellcome Trust inaugurated a massive effort that would map and store genetic materials from different groups throughout Africa. At the press release unveiling the effort, called the Human Health and Heredity Project (H^3 Africa), NIH director Francis Collins told the world: "There are unique aspects of the African population that empower [the] ability to track down genetic contributions to common diseases." He went on to say that hundreds of potential disease variants had been identified more generally in other populations, but that scientists were still left ques- tioning which exact letters "of the DNA code [are] responsible for the risk." He ended with a dramatic takeaway: *Africa will answer the question.*"[10]

Collins prefaced his singular hope of harnessing African genomes on a racialized idea of genetic diversity on the continent, saying that "the African population is older than that of Europe and Asia" and that its patterns of genetic variation are more compact, making the search for functional variants theoretically easier. Although this comparison may be true as well for people beyond Europe and Asia, the triangulation of global diversity as "Europe," "Asia," and Africa," almost as a reflex, expressed a familiar fallback on continental race thinking that simultaneously vindicated the importance, utility, and richness of Africa (or its genomes) to provide clues to better health for humanity the world over. It also relegated Africa to a place for extracting raw materials—again. The "world" in question was positioned as "Out of Africa" historically, while at present those living in Europe and Asia and elsewhere were now in need of the continent's valuable pieces to the genomic health puzzle. "Africa" would hopefully comply since the initiative was simultaneously one to move beyond a prior "colonial mode" (which was called out by African scientists in attendance) and to "include" African geneticists and health researchers in the genomic revolution going forward.

SCIENCE UNDER THE MICROSCOPE

In this book I develop a further exploration of how geneticists, as people, approach and conduct science that does not happen in a vacuum. For many, the very idea that science, molecular genetics in this case, is not strictly biological constitutes an uncomfortable reveal. That someone might pull back the curtain on its cultural and political aspects can bring feelings of discomfort for those who are used to operating within relatively enclosed esoteric spaces. The labs I studied are undramatically flooded with florescent light. They are full of material cultural artifacts, such as scientists' white coats, companion computers and sequencing machinery, that combine to convey a hermetically distinct space filtered of airborne microbes and often, but not always, filtered of ambient explicit politics.

What I and others in my subfield (the anthropology of science) have witnessed in such presumed purified spaces is not alien. Scientists' daily work is quite terrestrial and of public concern, even if the space of the lab might contain life forms that are reassembled, abstracted, and relatable only through growth media and assessed not directly but through precise measurements and binary codes. Despite these initial markers of a cold out-of-the-way place, the scientists within these spaces are everyday people theorizing and hypothesizing about how their science will answer questions about their own and others' needs. In this way, any notion of the vacuum is an imaginary that conceals the human hand and its work that integrates elements of what legal scholar Richard Ford has called America's larger "racial culture" into a politics of truth bound up with the twists and turns of DNA.[11]

Regarding US genetic racial culture, this is not a straightforward concept. Within it we witness how individuals and collectives of geneticists simultaneously work to redress past harms of scientific racism, medical neglect, and the denigration of bodies belonging to people of color all the while many of these same professionals participate in science as an institution that has yet to figure out how to get past a reliance on reductive racial categories that perform all sorts of double dealings. They can foster the seeds for racism and antiracism as well as contribute to the racialized body's hypervisibility and invisibility that, in Ralph Ellison's words, "keep the Negroes running—but in their same old place."[12] Put otherwise, a discipline of genetics that has attempted to "map" human populations since the human genome was published in 2003—with all of its ingenuity and granular modeling—somehow keeps returning to that most familiar starting point: the eternal recurrence of a Western imperialist race-making that relies on a vision of division as a way of seeing that splits some humans from others and also divvies the body from within by fractionating the genomic self in percentage terms.[13]

Much of this book essentially documents the telling fits and starts of genetic science in these specific contemporary dynamics. I am concerned with the myriad efforts to both upend racial harm and to redefine the race concept's limitations by shrouding it in the language of ancestry. Ancestry has now been updated from the older notion of genealogical source within family lines to ancestry as a statistical phenomenon that does not actually need ancient family lines to function. Its "source" materials are contemporary humans who are "conceptually cast as past."[14] Ancestry in this field roots itself in the rhetorical use of percentages that numerically transform persons into convincing probabilities of relative relatedness, pegged to geographical origins most commonly rendered in the continental terms of "Africa," "Asia," "Europe," and "Pre-Columbian Native America." All of this movement in the form of ideas, all this cognitive dynamism, are expressions of the long and persistent constraints that paradoxically demarcate a broad cultural stasis. By "stasis," I mean a stuck place, which is the more or less persistent belief that people of different sociological racial groups are indeed—biologically *and* genetically—distinct, one from the other.

On a broader level, I tell this story through several principal protagonists— some human, some conceptual, some biochemical. They all bear the birthmarks of a troubled dynamic that we might think of as captive inclusion.[15] Inclusion in the political sense, I argue, has a captive property. It also has ancestral ties to other forms of captivity and labor that have always already been racialized. At base, the primary reason one would ever need "including" rests on the fundamental reality that one has structurally been excluded. The familiar, historical arteries that run through the body of the United States as a maturing (or at least growing) being have circulating within them blood and resource wealth based on racial exclusions. These are some of the founding dynamics of the United States, and as such they guide, influence, and shape the new race of genetics at every turn, even when scientists are trying to make a break or clearly pivot to create a space beyond it.

Captive inclusion is a two-headed hydra of sorts. But as writer Ocean Vuong reminds us, such figures as monsters are not strictly and definitively terrible things. Vuong writes: "To be a monster is to be a hybrid signal, a lighthouse: both shelter and warning at once."[16] Now, instead of explicitly hierarchizing peoples, with labels of black, brown, red, yellow, below "the European," the new genetics of race premises itself on a hopeful cultural technology of "inclusion" tied to ideas of embracing "diversity." The only problem is the diversity is not so diverse, and the inclusion comes with the social contract to be forever joined to all-consuming racial groupings that for most, but especially for African-descended people, locks individuals into what literary scholar Hortense Spillers has called a racial grammar grounded in schemas of simplification and potentially recurrent vulnerability.[17]

Attention, inclusion, and vindication—the importance of presence for once marginalized people—meet the pragmatic goals of more democratic health that, when racialized, force researchers to reach backward. This backward bend to redress past wrongs while extending inclusion as a way forward begs the question to what extent a future-grasping politics of racial science backslides onto old ground of harmful racial categories, now in seemingly objective biogenetic terms. And now we turn to a twenty-first-century ethnography of these layered issues. Our story could have many points of origin, but I have chosen to begin with the very namesake of "America" itself, and the explorer whose name and legacy quickly replaced, and attempted to erase, the land and people of this hemisphere.

OODNAFUL
Stanford, California
June 5, 2023

Abbreviations

AIM   ancestry informative marker
ASHG   American Society of Human Genetics
CEPH   Centre d'Étude du Polymorphisme Humain (Human
            Polymorphism Study Center)
DTC   direct-to-consumer
GWAS   Genome Wide Association Studies
HapMap   International Haplotype Map Project
HGP   Human Genome Project
NHGRI   National Human Genome Research Institute
NIH   National Institutes of Health
PG-ED   Personal Genomics Education Project
PGP   Personal Genome Project
PMT   pharmacogenetics of membrane transporters
SNP   single nucleotide polymorphism
SOPHIE   Study of Pharmacogenetics in Ethnically Diverse
            Populations

Introduction

AMERICA AND THE TABULA RAZA

After his third voyage to the Western Hemisphere, at a time when renowned sailors like Christopher Columbus assumed they were sailing to Asia, Amerigo Vespucci wrote a letter to his former patron and family friend Lorenzo Pietro di Medici of Florence. The year was 1503. In it he eagerly described a new world—a *Mundus Novus*—of which, in his words, "our ancestors" had no knowledge. The Atlantic did not in fact stretch out across the globe back to the land masses of Africa, Asia, and Europe with nothing in its wake as "the ancients" had assumed. The vessel carrying Vespucci anchored under constellations of stars that had yet to be penned in Latin, the literary lingua franca that facilitated such communications between seafarers in Italy, Spain, and Portugal who all wanted a foot in new lands.

In his dispatch to Medici, Vespucci set about detailing the skies and territories of this near "terrestrial paradise"—its many people, languages, animals, trees, resins, pearls, and precious metals in abundance—that his various sponsors, the Spanish and then the Portuguese crowns, imagined as their claim. At the beginning of the so-called Age of Exploration, records and letters from Europeans forced to sail to Asia by traveling west instead of east by maritime restrictions put in place by the Ottoman Empire were prolix in their descriptions that the New World was vibrantly inhabited. In his 1503 letter, Amerigo, after whom "America" would be named, described those he met as the first order of his report: "We found in those parts such a multitude of people that no one could enumerate, a race I say gentle and amenable." He then quite quickly went on to dehumanize them. For Vespucci and his European readers (as this particular letter would soon become a bestseller when reprinted), the humans who welcomed him were "monstrous" wonders: "comely" yet "ugly"; "dignified" yet "libidinous"; "communal" yet "cruel."[1]

Reassuring Medici that he was offering only the briefest detail because there was too much to tell in a mere correspondence and that a great "geographical and cosmological book" of his voyages, from 1497 to 1503, would be forthcoming, Amerigo recounted dramas of cannibalism, the people's propensity for unfeeling violence, their pervasive lawlessness, wanton incest, primeval sociality, lack of property, and an unconscious ignorance as to the value of gold. The claim to rich lands and resources in need of a civilized exploitation and the beginning of a simultaneous racialized discounting and eventual erasure of those "reddish" people of excellent "bodily structure," "with hair plentiful and black," was one and the same gesture. With the very spying of the "New World" by European eyes, we are invited to witness one of the first instances of what I describe in these pages as the *tabula raza*.

My concept of the tabula raza is an attempt to understand tangled and persistent habits of power. These begin with the ways that people have gone about building versions of the world that approach the human with a lens focused on differences. The ways that such visions have helped to amass material goods that yield profit and status have long characterized "discovery," or "science," in the so-called New World. Raw materials from foreign and distant lands have often funded scientific expeditions at different scales. Belief in the correctness of such missions has often justified them. A simultaneously racializing and self-proclaimed beneficent handling of peoples and resources has allowed those with state, higher institutional, and economic powers in different eras to enlist topographies, landscapes, and—when scientists now assemble biomarkers to learn something about human diversity—putative ancestral "genescapes" into conceptual edifices of racial worldviews.

These behaviors and missions do not stem from agentless acts of wizardry. They require humans with a vision who sometimes unwittingly wade deeper into these long legacies and vexed inheritances than they would like. As we will see, for some scientists who wanted to combat health disparities, it became an accepted practice in human genetics to sort the world's people into what were essentially race-based criteria for various ends. There was, however, also a forward-looking follow-up in the early days when the human genome was drafted that this ordering system would merely be temporary. The thinking went that at some point, human genetics as a field would achieve the much anticipated technological advance of easily accessible genotyping for diseases and individual pharmaceutical prescription regimes, known as "precision medicine." Beyond this, other scientists envisioned affordable full genome sequences for everyone where information about inter-individual genetic variation would far exceed race as a "proxy" for population-based patterns of allelic difference. (See table 1.)

In its essence, however, the tabula raza can only be made sense of by its underbelly—that is the more familiar tabula rasa—with an "s." This is, of course, that long-standing notion within the history of Western thought that from birth humans possess a mind that is a "blank slate"—or quite literally, in Latin, a "razed tableau." From Aristotle to John Locke, philosophers have emphasized the human soul, the mind, and the capacity for thought as containing potential, yet born bare of social custom, knowledge, and higher-order thinking of classification.[2] The tabula raza, as I argue, consists of a scientific, observational, taxonomic, and unrelenting entrepreneurial impulse, at times violent, at other times gentler, to imprint race—*raza*—as an order of value for the observer-discoverer onto bodies and land that require abstracting life in some way to create value anew. In this, I want to highlight the ways that *rasa* and *raza* work together. At issue is how racial objectification becomes imprinted on material, bodies, and life forms once these are rendered redefinable and conceptually detached from their prior state in some way. I borrow the Spanish *raza* to inflect racial ascriptions that can also efface and displace previous contextual understandings and relationships to highlight the long-lived influence of Spanish conquest in the Americas.[3]

For its part, the emphasis on the "blank slate" of the tabula rasa has not only come to connote an abstract state but also a pristine one. Vespucci and other Europeans' fascination and simultaneous dismissals of Native populations in their early accounts joined these various tabulas as

Table 1 A Literal Tabula Raza

Exon	SNP #	Exon Position	Nucleotide Change	Amino Acid Position	Amino Acid Change	Total Freq	AA Freq	CA Freq	AS Freq	ME Freq	PA Freq
1	1	(−38)	C⇒T	—	—	**0.023**	0.005	**0.053**	0.000	0.000	0.000
						n = 494	n = 200	n = 200	n = 60	n = 20	n = 14
1	2	(−23)	C⇒T	—	—	**0.119**	**0.053**	**0.218**	**0.050**	**0.100**	0.000
						n = 472	n = 190	n = 188	n = 60	n = 20	n = 14
1	3	9	C⇒T	3	Synonymous change	**0.042**	**0.105**	0.000	0.000	0.000	0.000
						n = 472	n = 190	n = 188	n = 60	n = 20	n = 14
1	4	38	C⇒G	13	Ser ⇒ Cys	0.004	**0.011**	0.000	0.000	0.000	0.000
						n = 474	n = 190	n = 190	n = 60	n = 20	n = 14

NOTE: This table reproduces a screenshot of how data was organized in a lab where I conducted fieldwork with scientists who work on questions of differential drug metabolism in US groups. The first half of the table shows a series of DNA base pair changes in the coding region of a drug transporter gene. The first six columns are concerned with the details and location of the variants (SNP = single nucleotide polymorphism). The second half inscribes these nucleotide changes within a grid organized by US census categories of racial groups. AA = African American, CA = Caucasian American, AS = Asian American, ME = Mexican American, and PA = Pacific Islander. "Freq" denotes the frequency of each variant in question, which gets interpreted through the lines of these racialized cells, reinforcing perceptions of biological race. Other instances of the tabula raza concern panels of Ancestry Informative Markers (AIMs), detailed in chapter 2.

Figure 1. An instance of virgin plains, an example of a putatively pristine tabula, in American popular art. Notably, two Indians look on as settlers cross the plains. They are separated by a cut in the land that divides by race at the outset. *The Rocky Mountains, Emigrants Crossing the Plains*, Frances Flora Bond Palmer, 1866. Printed by N. Currier and J. M. Ives. Yale University Art Gallery, Mabel Brady Garvan Collection.

concerned people, minds, souls, and lands. As he sailed and disembarked along the shores of what we now know as southern Brazil, Amerigo artfully constructed the people as part of a landscape empty of civilization, religion, and commonsense notions of wealth or individual property.[4] Thus raw material elements extracted from the life around them, from colorful birds to lustrous pearls, were collected for shipment to Europe. In some cases, Indigenous people were corralled and shipped as well.

Over the next few centuries, as Europeans settled South and North America en masse, the lucrative idea of a terra nullius inhabited by groups who related to the land in terms other than "property" continued to overlay a belief that the people themselves were slates without worthy content, or tableaux to be razed. In refusals to see the inhabitants in their complexity, in manipulations of territories recast as "discoveries," replete with "legal" and "empirical" explanations as to why the acreage Europeans dutifully defined and then measured, belonged to them and not to the various tribes, Northern European emigrants seized what they fashioned as possessable terrain.[5]

Some of these older habits of power have paved the way for newer ones in the late twentieth and early twenty-first century. These include approaching human DNA with a certain inheritance of vision—that is, of seeing the world's lands and many of its peoples through a racialized lens that often collapses geography, simplified lines of descent, and contemporary naming conventions of population groups that are customarily stripped of their historical complexities.

Digging even deeper, the tabula raza's spiritual force relationally draws from the planet's very material geography. In this process territory itself has often been razed physically and conceptually. Native names were habitually effaced or dissociated from their origins, places were then reimagined to represent the nations and markers of provenance held dear to those who conquered with racial domination as their scythe. For centuries the continental geographic slate was divided through ordered taxonomies. In Europe and North America some eighteenth- and nineteenth-century scientists used fine artistic detail to graph human comparisons, separated by rows and columns amid environmental sketches of fauna and flora. The associated "types of mankind" in each setting were drawn or painted such that aesthetic judgments could be read onto peoples' physiognomic attributions.

In colonial Mexico, called New Spain, different dynamics of taxonomic thinking appeared in what were known as *casta* paintings, which represented diverse humans through relationships that led to newly racialized offspring. In this genre the viewer is summoned into the divisions of the taxa, into literal *escenas de mestizaje* (scenes of mixing) regarding an abundance of human types. These are comprised of mixtures upon mixtures of people seen as originally from Africa, Europe, and Indigenous America. Also tabulated on a broad grid, each square portrays—in miniature—a man, a woman, and their resultant (ad)mixed progeny brushed in different skin tones, painstakingly labeled. Such visual modalities of organizing "*la naturaleza humana de los habitantes de las tierras americanas*" still inform one of the primary expressions of the tabula raza

I deal with herein—the data-filled tables and charts of racialized DNA minutiae.[6] (Looking ahead, table 3 in chapter 2 provides a clear example of the racialized "mixture" featured in certain articulations of the tabula.)

Stripped of their racist overtones, today's graphs are now digitally expressed and ordered via shared computer interfaces where lines of information about human allelic differences are communicated. Imagined continental essences nonetheless persist. Finally, the tabula raza concerns instances of when scientists' very own subjectivities undergo erasures—such as when their cultural or other intersecting identities (e.g., immigrant status, group affiliation, or gender) are minimized, devalued, or altogether ignored within their own labs, in some instances, or within the larger world of their science, in others.

.   .   .   .   .

The Age of Exploration, alternatively called the Age of Discovery, cannot merely be cast as the violence and witlessness of the past. Part of my hope in writing this book is to trouble the conceptual divisions that allow for the schemas that sustain the tabula raza. These call for a questioning of what theoretical physicist and feminist science studies scholar Karen Barad calls the ideational "cuts" that humans have made—cuts that entrench difference in our perceptions that naturalize borders, land masses, historical periods, and groups of peoples as somehow inevitably marked by inherent distinctions.[7] The fact that unfathomable numbers of Indigenous people in the Americas, together with the millions of Africans who were brought here as slaves, experienced ongoing displacement, cultural extermination, language loss, dispossession, and in some cases, slower forms of genocide makes any division of then severed from now as violent a cut as any.

In light of these losses, many descendants from these groups are still struggling and in certain cases, suffering psychically, physically, and biologically. Well-documented health disparities, poverty, drug and alcohol addiction, fatalism, and high rates of suicide and homicide necessitate a long overdue process of social reordering. Calls for reparations and "Land Back" demands have usually fallen on deaf ears in the United States, where the state itself has a difficult time accepting blame or even admitting that its America was ever anything other than an earnest successful democracy for the world to model. In a word, no. When official history and political will continuously fail to acknowledge the need for redress, smaller efforts have attempted to fill in the breach. As for such efforts in the field of genetics, there is no singular approach to address the many iterations of loss cited above stemming from what anthropologist Elizabeth Povinelli has aptly termed the "ancestral catastrophe" of colonialism and enslavement.[8] The field of genetics can be a source of ire, especially for the many Indigenous groups who have pointed out its potential to further damage their hard-won sovereignty where it exists (detailed further in chapter 8).

Nonetheless there are myriad flexible political claims and attempted uses for genetic ancestry products—as genealogical sleuthing tools, as medical technologies that aid in case-control studies, as forensic apparatuses that narrow down suspect pools, as keys to open doors of no return, and as ties to lost generations that might restore obliterated or stolen histories. Across the span of these applications, the tools themselves are emotional, political, and technological devices that have folded into the working structures of their models and algorithms aspects of colonization, slavery, and the ways these forced people to "admix."[9] In interviews and field research I carried out with American geneticists working with ancestry concepts, several leaders in the field cited the arrival of Europeans in what they in their contemporary lab spaces—in model building and hypothesis testing—called the "New World." One scientist who was pivotal in the routinization of what is referred to as Ancestry Informative Marker (AIMs) technology for admixture studies told me flatly that his models to reconstruct ancestry take the date 1492 as their starting point.

This inherited concept of the *Mundus Novus* was often mentioned by scientists along with the trans-Atlantic slave trade, most often discussed as "slavery" or "the slave trade," to acknowledge enslaved Africans who were forced to develop colonial wealth. Finally, Indigenous groups were most often designated as "pre-Columbian" to signify the fact that they were here before Europeans, yet their identities as entities were inescapably yoked to conquest. These markers of history, to my mind, often mimicked the selective and atomized ways that genetic markers of "ancestry" were made to function within the everyday science I witnessed. The DNA ancestry indicators were said to convey continental origin, relatedness, and some notion of race, while blotting from view the ways these base pair changes (possible versions of alleles at specific locations) had environmental or physiological adaptive functions that surely entailed more complicated notions and concepts of biological inheritance.

Taken together, the signposts flagging historical flashpoints and the genetic variants that showed frequency differences, when compared between groups, comprised the scaffolding of the scientific schema that was fundamental for scientists to reproduce a genetic depiction of what some called "the colonial encounter." Their science was largely focused on the supposed newness of people "mixing" in the New World (the possibility of humans from different regions mixing in the "Old World" did not usually figure into their theories of genetics and "admixture"). When I interviewed yet another specialist about New World ancestral assumptions similar to those laid out above, he told me flatly: "All models are wrong, but some models are useful." The uses of these narratives vividly express how scientists reconstruct scenes of the past in their attempts to reverse-engineer New World human mixing by parsing the genome in present-day Latinos and African Americans. On a broader scale, the markers of time and DNA precision that some geneticists used to build their models allowed them to reimagine what one called an attempt at "nation building." This was in reference to building a database of people of color for healthy inclusion in "the genetic revolution," which I take to be a form of genomic world building through restructuring scientific priorities to focus on aspects of diverse genomes as valuable materials and participatory investment.

As the power of narrative and the social uses of genetics becomes clear, I now invite the reader to ponder the broader "kin" relations within the very family tree of genetic ancestry technology itself. On one branch a contemporary offshoot that we might imagine to be this technology's first cousin constitutes racialized medicine; another line that shares this DNA in some contexts constitutes the forensics clan; and yet another faction bears the only somewhat estranged *enfants terribles* who politicize genetic variants like one that codes for lactose tolerance found in some Northern European populations. Lactase biomarkers have now been adopted into racist bonding rituals on the part of white supremacists who gather to collectively "chug milk."[10] Despite the fact that several lactose digestion variants have been found in multiple groups in Africa, the ideological grip of "milk genes" for bare-chested white men participating in communal cow juice–guzzling exhibits their embrace of a particular take on how the science of the "code of life" itself sets them apart from people whom they take to be lesser humans—humans from places other than their ideal Europe (which is read as somehow always already "white," despite the historical record to the contrary).[11]

Anyone familiar with the larger scholarship on race and genetics will be acquainted with a much gentler kind of primal scene: that of the Clinton White House ceremony announcing that the human genome draft showed our common humanity to be free of racial divisions at the genetic level. My recounting of this now oft-cited event relates back to my earlier discussion of the razed slate, only now in the scientific historical moment of the year 2000. In yet another instantiation of the tabula rasa, the genome draft map itself was said to be birthed free of racial social constructs, a kind of unwritten tabula reminiscent of Aristotle's newborn soul.

At the Clinton White House the US president was flanked by two of the genomics field's most famous scientists associated with the initial Human Genome Project, Francis Collins and J. Craig Venter. They themselves were in a once feisty competition to see who would be the first to achieve the then highly coveted "holy grail" of the first draft map of our species' genome. The stakes were high because of the billions of dollars invested in the project, the genetic determinism that ran through the importance of cracking the "code" for major diseases and also because of the fact that one effort was governmental and "public" while the other was private. This raised the specter that human genes might be patented in a capitalist system where many felt that life and health should be protected from all that drives commodification. In the end, both men led teams to map the genome from a few individual study subjects who were Americans of different racial backgrounds. Despite the scientific sprint to be the first to unveil the map, in the end Collins and Venter aligned scientifically and politically regarding human variation and a lack of evidence in the data for traditional racial divisions.

This genetic blank slate, the tabula rasa of our supposed raceless biology, was a short-lived reality, however. When the actual map was filled in beyond the draft in 2003, the "no race" pronouncements made headlines again. This time a war of scientific words ensued. Geneticists, epidemiologists, physician-researchers, and bioinformaticians wrote letters, opinions, and debate articles in flagship journals such as *Science*, *Nature*, and the *New England Journal of Medicine* to contest or defend the idea that race could, or could not, be found in the genome. Some were obviously incensed. In one of the most illustrative sets of articles, the no-race side argued that genetic variation is continuous and discordant with race, systematic variation according to continent is very limited, and there is no evidence that the units of interest for medical genetics correspond to what we call "races."[12]

The pro-race side cited medical trials in cardiology and genetic studies on population differences where people of social racial groups seemed to cluster by continent. This, they maintained, meant that even if humans were 99.9 percent the same genetically, the 0.1 percent of difference could in fact make all the difference. Some aspects of race, they argued, could indeed be excavated from the As, Ts, Cs, and Gs. These patterns, they pleaded, would furthermore be informative for making sense of the genetic bases for the gross health disparities that disproportionately burden US minorities.[13] Although some of those in the debates on the pro-race side were prominent white American men, others involved in these discussions were professionals from underrepresented groups. This quickly complicated any simple attempt to label them scientific racists or to see this new vocal movement as scientific racism.

A new set of articles covering much of the same ground appeared in 2021. This time some of the scientists involved evoked their racial identities and minority statuses to make arguments about the social justice need to address health disparities.[14] Yet it was with the gestures of scientists in the immediate post-map era, who saw value in researching genetic racial difference, that the tabula rasa of our raceless genome was first etched upon. Its letters, spelling out human groups and DNA sequence bits, as well as its numbers, portraying probabilistic percentages, created the literal tabula raza. It is this tabula that would live on to perceptibly shape the data of many genetic studies for decades to come.

With regard to US scientists' collection and sorting practices of human biological material, many saw their work as logical, needed, and even socially responsible. Nearly all of the professionals I encountered during my fieldwork were well-meaning on these counts. Yet there were times when their personal ambitions inside the space of the laboratory met with a sense of ambivalence about how their research might live outside of it. As I spoke with lab heads as well as lower-level scientists who understood themselves to be the direct descendants of a whole range of people who mixed through the violence of conquest, slavery, and dispossession, they were sensitive to the importance of studying specifically diverse genomes—the genetic legacies that people carry in the here and now as a result of the violent human toll that it took to build the Americas.

At the same time, scientific professionals could be very much aware of the racializing brushstrokes it takes to conjure an imagined past when putatively pure humans existed, which was implied in their methods of describing New World bodies in terms of combinations of Old World "genetic admixture." The language of admixture itself emerged from a particular line of scientists as well as animal breeders who were concerned with "pure types," contamination, and racial stocks.[15] On its face, however, this language was more neutrally used in the labs where I conducted fieldwork. Nonetheless, it still evoked the possibility that at some point, in several places in the world, there were (and possibly still are) pure human kinds that preceded the "admixed" varieties of people who inhabit this time and place of the Americas.

Today, the idea that contemporary Latinos and African Americans contain amalgams of specific genetic ancestry commonly referred to as "African," "European," and "Pre-Columbian Native American" reveals this inheritance of thought.[16] In several chapters I chronicle the work-lives of scientists who were often caught in a back-and-forth process of naming populations in terms that resembled continental racial thinking when they viewed samples of people's DNA and their patterns of genetic similarity at a group level. Key here is that this observed similarity was based on small allelic frequency differences that some individuals from the demarcated groups in question possessed.[17] Through statistical modeling that relied on geographical and racialized labels assigned to participants' DNA, these patterns of genetic similarity were then generalized to the broader groups of people to which these individuals' samples were associated in continental terms. The framing through which to interpret these artificially homogenized groups—which scientists created through what were referred to as "selection and validation criteria" (detailed in chapter 2)—has been referred to as genetic ancestry at a continental level, which was almost always visualized through racializing proclivities.

GENEALOGIES OF THE CONCEPTUAL PRESENT

In prior work, I have emphasized the ways that social constructs of race may now include references to such statistical frequencies of genetic variants found in and between specific populations, which are often assessed through optics of centuries-old ideas about continentally based human difference. In this I hoped to expand and complicate the notion that race is simply a "social" construct. I wanted to emphasize that separating out the domain of "the social" creates a sense that the biological sciences remain outside of cultural anthropologists' purview for analytic engagement. More important, my point was that social scientists must also get clear on the ways that carefully curated aspects of DNA figure into an increasingly pervasive "biologistical construction of race."[18] But no matter what our fields are, we can all learn from the ways that genetic science draws from and simultaneously bolsters seemingly fixed societal constructs, and vice versa. On a deeper register still, I wanted to underscore that DNA markers that are routinely interpreted through frameworks called "ancestry" could often just as easily be attributed to other associated catchall terms—for instance, "environment" or "ecology." Such nomenclature would likewise reflect scientific interests and priorities aligned with, or inflected by, still other social orientations and historical inheritances.[19] In all of these possibilities, conceptual delineations that are constructed as bounded phenomena that people create to describe the world reveal the very real and particular ways that they themselves are active participants in human renderings of what we call "nature."[20]

Recently, similar observations have been explicitly stressed by some geneticists as well. In a reflexive and clarifying tone, several have suggested that the field adopt alternative terminology with fewer built-in assumptions than the familial intimacy of direct relatedness that the word "ancestry" possesses. This is in part to avoid confusion about what *genetic ancestry* actually is—that is, genetic identity by descent. This latter concept has routinely gotten conflated with all manner of forms by which people might see themselves as "being from" a people, a place, or linked to cultural practices (like religions), with varying degrees of actual shared genetic ancestry with others from the same specific range of possible "origins." The recommended term they offer is simply "genetic similarity."[21] In justifying such a phrase, they go to great lengths to explain that *genetic ancestry* and *genealogical ancestry* are not the same thing—although they may overlap.

In essence, because of the way that our genetic material recombines when any two people make a child, we do not inherit all of the DNA that both of our parents possess. Therefore, we surely have not inherited all of the DNA carried, chromosomally crossed over, passed on (and *not* passed on), as concerns our many direct genealogical ancestors. In this way traditional notions of a person's genealogical family tree do not necessarily equate to what one carries in their inherited genome. This can be confusing, especially when people think of their recent ancestors as being from X or Y place. To compound the issue, commercial genetic ancestral "admixture" products, scientific modeling of "admixture" for population-based medical studies and everyday popular ancestry talk in this day and age often conflate all of the above.

·　　·　　·　　·　　·

In 2020, after months of the pandemic and shelter-in-place orders that left many people glued to their screens daily to ingest more news than ever about social inequities that the fallout from COVID revealed, on Memorial Day in the United States a white police officer callously murdered George Floyd as he begged for his life. The shock of witnessing the nonchalance of such white-on-Black dehumanization, coupled with the increased knowledge and viral video publicity that scores of Black Americans have been killed in such encounters because of their race, sparked worldwide protests. In the days and weeks following Floyd's murder, in an outpouring of long overdue acknowledgment and sympathy, fifty of the highest-valued companies (including Bank of America, ExxonMobil, Google, Nike, Pfizer, and Walmart) pledged $49.5 billion to address racial inequity.[22]

In that moment smaller entities, such as genetics departments at various US universities, posted statements on their homepages denouncing the past use of genetics for racist ends. For their part, genetic ancestry testing companies, among them 23andMe and Ancestry.com, held what they called "black outs" to signal their solidarity with those protesting Floyd's death. The obvious role of the field of genetics in promoting racial harm that culminated in multiple periods of eugenics has, at the time of this writing, only recently resulted in an acknowledgment and apology from the American Society of Human Genetics in the year 2023.[23]

Perhaps more consequential, however, in 2021 the National Institutes of Health sponsored and commissioned an ad hoc committee, under the auspices of the National Academies of Science (NAS), to assess and provide a final report detailing "best practices on the use of race, ethnicity, and genetic ancestry and other population descriptors in genetics and genomics research."[24] When the process completed and the document was made available in 2023, several of the NAS committee members, as well as others in the field, interviewed in the press commented on how the suggested guidelines were just that. It will now be up to geneticists themselves to "comply" or not with moving away from racializing study designs. Only time will tell.

Thus this book comes at a moment when the field of genetics and its public funding agencies are taking the idea of racialized ancestry head-on after two decades of scientists working with such concepts as a matter of routine. I entered into the worlds of key scientists in this domain in 2003 and carried out lab-based fieldwork until 2011 to better understand the organizational structures, rewards, and socially compelling reasons that inspired people to design their studies in these ways. I continued to follow scientists' debates, choices, and ideas at conferences and through assessments of their later publications, while I documented trends in their larger field until 2023.

CHAPTER OVERVIEW

In chapters 1 and 2, I explore scientists' narrative-based approach to particular forms of genomic world building that rely on specific storylines of Iberian imperial history to make sense of how people and their genes mixed to create the contemporary world they want to now act upon through specific forms of genetic medicine. Although *how* American Latino genetic similarity at the DNA frequency level might have occurred was often left unexplored, the lab scientists often assumed it to be the result of direct genealogical ties in the form of gene flow in some instances, racial belonging to broad source populations in others, or a combination of the two more generally.

Similarly, in the realm of trying to use genetic variants for precision medicine, or pharmacogenetics, racial groupings have been a principal way that researchers have organized rare allelic changes that affect some aspect of different individuals' (and assumed groups') varied drug responses and adverse reactions. It is in this context that chapter 3 explores how scientists who may have been born and raised in other global contexts, or whose parents hailed from other places, found themselves faced with the tricky task of classifying their study subjects by race when they personally expressed confusion about how they might fit into the very same racial categories deployed in their lab. Different personal frictions created sticking points for Black American scientists who tried to create a scientific home in a lab focused on Black health disparities, detailed in chapter 4. Furthermore, sometimes living and working in a scientific space focused on African ancestry and genetic risk also meant trying to play by certain rules of mapping genomic territory through scientific claims-making and subsequently risking getting caught in the racializing habits of differentiation that might open conference doors to include them but that also ended up making them feel ghettoized along with their findings, examined in chapter 5.

The focus on health disparities between US racial groups can overlook key intersections with sex and gender if these are not made explicit. Chapter 6 chronicles how larger social dynamics between Black men and women in the cultural sphere could also be reflected in how scientists understood the potential tensions around their preferences for genetic studies to pursue, as well as the ways they grappled with the politics of sexism in the social space of the lab itself. In each of these instances, practitioners often accepted that it was necessary to classify the DNA they worked with via a racial nomenclature couched in terms of ancestry. This tendency was sometimes tied to their own professional if not personal subjectivities vis à vis a cultural schema that effaced some of their deeper desires for social recognition regarding broader aspects of who they are. This happened through habits of power that they themselves came to embody as they instantiated racialized DNA work as a potential factor that might reduce health disparities and that also helped them to express their care for communities in need of more equitable medical interventions.

Their personal stakes, the political priorities of their science, and the human data they collected were framed within the language of inclusion. The simultaneous erasure of social and historical inequity that created (and maintain) the dismal numbers of scientists of color, and that shape the embodiment of inequality we know as common health disparities, was coupled with a more naturalizing racial branding *imposed upon* the slate of consequential histories of racism that were too often left obscured from view. That fusion-as-tension, of an invisibilized broader complexity absorbed by racial descriptors and labels, characterizes the tabula raza more broadly.

Displacing a more complete vision of people, DNA, subjects, and objects via racial frames were not always permanent or unquestioned, however. In each chapter there are insightful instances of when scientists as the utterly human beings that they are pivoted to a stance of conscious abstraction—when they saw problems with racialized recruitment, race-tagged data, and reductive notions of DNA. It is important to understand the need for abstracting at some level that is required for most science but to also recognize that it can be done with a mindful eye. It was always instructive to witness people's reflective turns, when they thought deeper about the back and forth of their own positions as racialized subjects in various ways, while working on objects of race that implicated some aspect of their own subjectivities.

The final two chapters chronicle other sites that illustrate the persistence of race in genomics more broadly, even when scientists have tried to resist participating in racial schemas for a variety of reasons. Chapter 7 follows the inception and launch of the Personal Genome Project (PGP), where visible leaders in the field hoped to focus on full genome sequences and all of the inter-individual variation that such scans would unveil. This, they thought, would obviate the need for an emphasis on racialized groups. In the end, they realized that the highly revelatory degree of identifying information that can be gleaned from such scans made it hard to convince people from underrepresented groups (or anyone who might fear genetic surveillance, discrimination, or economic repercussions) to join. Those who were attracted to participating were mostly white, male, and relatively privileged. And so, despite the PGP's initial stance, those in charge of recruitment ended up creating calls for racial "diversity."

Chapter 8 takes on many of the book's themes regarding racialized assumptions about populations by focusing on the ontological fact of missing data and what it reveals about how racialized populations in the here and now are functionally utilized to extrapolate about the origins of ancient DNA. At stake is a reconstructed genetic hominid presence so old that it has been called a phantom, or "ghost DNA," in some African populations. Taking the specter of the ghost even further, I query the more common idea of ghost DNA that is meant to signal a missing population as a means to delve deeper into the ways that missing data entails a feature of most population genomic studies.

More crucially, however, I connect these ethics of reconstruction to the facts of missing data in other political contexts. I focus the second half of chapter 8 on the many Indigenous tribes and Native geneticists who may choose to "go missing" from studies since their actual ancestors were violently excluded through genocidal campaigns, racist assimilationist policies, and cultural erasures that created gaps in their actual lineages and ancestral lines that cannot simply be artificially re-constructed. I revisit the protestations of Indian tribes, activists, scientists, and everyday people who refused participation in earlier population genetic studies, like the Human Genome Diversity project. Several of the points of negotiation that geneticists from various tribes are demanding with regard to DNA sovereignty—that is, to control how their data is collected, stored, used, and interpreted—has forced the US government to go about its large-scale *All of US* initiative with more care than earlier state-led genetic projects. I believe that the actions of scientists who have created collectives like Nativebio raise issues that are making people more aware of the need for increased education on what exactly can be done with DNA data for the larger public going forward.

Despite my attempt at a brief roadmap for the reader, the detailed narratives are what give the chapters their depth. Similarly, any "theory" of the tabula raza only goes so far. It is one attempt at an ethnographic device that tries to explain the ongoing cultural legacy that inheres in scientific practices that make the genome legible in terms of land, race, and origins. I therefore write these accounts to allow the lives of scientists to breathe beyond my own anthropological schemas. Where possible, I have tried not to succumb to insular academic jargon and refuse to write in the dense form of most ethnographic texts (including some aspects of my own prior work). In this, I hope to share with nonspecialist readers the many ways that scientists have struggled to address key problems of unfairness and Euro-white supremacy in genetics, while also at times struggling to confront gender, class, historical narrative, and other biases in their worlds of work. In this, I follow them in trying to broaden insular academic pursuits to be of more social relevance.

SCI NON-FI

There are times when relaying the work here that I experiment or draw from other genres, other languages, and other fields to perceive links and allow forms to emerge that may help navigate the ineffable or inscrutable dynamics of cultural life. Science journalists' techniques for profiling new discoveries can be very useful to the ethnographer. As an anthropologist of science, however, I am not merely practicing the craft called "science writing" that many reporters do so well. The distinction I see between the anthropologist and the journalist centers largely on the ways that ethnographers attend to details with a particular intention to limn the cross-fertilization between scientific pursuits and features of modern power.

   Although ethnographic texts like this one are technically works of nonfiction about science, in keeping with the larger goal of this book to trouble the neat dividing "cuts" that humans make to delineate, order, and produce knowledge about their fuller worlds, I want to include in that the anthropological academic tendency to make our own separations (and we make many) regarding acceptable forms of ethnographic writing in light of other, "different," genres.[25] In this spirit I join the myriad others in the field who experiment and depart from the notion that there are clear conventions for ethnography, especially within medical anthropology.[26] In this, I want to explore a layering that is, more importantly, a signifying truncation to invoke a writing modality that I call "sci non-fi." The expressive potential of this "science nonfiction" remains in constant conversation with its corollary "sci-fi"—or futuristic science fiction—as a genre by relational necessity. The difference is that what I describe here is not only real but also happening in the contemporary world in real time, even as the present gets conjugated in the near future-tense and very much relies on specific visions of the past.[27]

OF PASTS, PROLOGUES, AND PIPE DREAMS

On April 4, 2022, a public meeting was hosted by the aforementioned National Academies of Science committee as part of their effort to bring together experts to figure out whether or not (and how) "population descriptors" should be used in genetic studies. Geneticists Aravinda Chakravarti and Charmaine Royal cochaired the afternoon session.

"I've been around in genetics probably longer than most of you," Chakravarti announced as he drew upon his forty-plus years of insider status. Following up on presentations where the presenters were mostly emphasizing the importance of keeping what had become traditional racialized population labels, he braved a basic question: "In the genetic clusters [of groups], deriving these clusters depends also on other reference data. How do we get around that circularity? [...] How do we get around it?" He wanted to make clear that he was asking the question, as a cultural anthropologist might, because certain behavioral tendencies in his professional tribe worried him. "We infer somebody's proximity to some population or region based on existing data, which itself is biased. . . ."

Citing Shakespeare off the cuff, Chakravarti prefaced his observations with "I know many people say 'the past is prologue,' but I actually want to move through this prologue . . . If we can."[28]

Several geneticists responded thoughtfully by arguing for combining methods that looked at how genetic markers are inherited within families and to robustly track such markers in a more granular way by looking at clear lines of identity by descent, rather than just broad continental catchments. Bioethicist and geneticist Malia Fullerton ended her presentation to the committee by informing the public audience that the Human Genome arm of the National Institutes of Health (called NHGRI) listed as one of its "Bold Predictions" for the year 2030 that research in genomics will have moved beyond using race.[29] But, as Fullerton told the committee, she was "pessimistic." Her own "boots on the ground experience" led her to believe that, in her words, "this may well be a pipe dream."[30]

I wondered about the term "pipe dream," so I looked it up. It refers to the murky haze of hallucinations brought on by smoking opium, for which pipes were used. Such trips were also called "opium dreams." The haze as a metaphor for the elusive efforts to pin race down to DNA, a haze that might materialize through the swirl of layers of inconsistent labels that the committee eventually found, did seem to fit. It might help, however, to track back to an earlier scene in the dream sequence.

In the period shortly after the first genome draft, in 2002, the US government funded another major effort called the International Haplotype Map Project (HapMap for short) to discover human genetic diversity in people from around the world. Although the aims of the HapMap were "to determine the common patterns of DNA sequence variation in the human genome and to make this information freely available in the public domain," the populations sampled were a highly reductive representation of the said "human." This was apparent in the language used in one of the research consortium's first publications on their results, where the map is referred to as "patterns across the genome" consisting of "the genotypes of one million or more sequence variants, their frequencies and the degree of association between them, in DNA samples from populations with ancestry from parts *of Africa, Asia and Europe.*"[31] The focus on populations from select regions in Africa, Asia, and Europe was not simply a "natural" choice. Yes, the geography of these different lands, the fact of their separation, have been politically cemented as real. Yes, people's biological and bodily physical differences in physiognomy, most visible as surface traits, are often experienced as real. Yet neither of these are fixed and unchanging. This pertains both to their classifications and to their forms.

Pinpointing these particular geographic regions where people exhibit present-day phenotypes that conform to certain contemporary ideas of how humans sort into distinct types would, as we will see in the chapters that follow, do the work of enlisting physical DNA base pair variants (single nucleotide polymorphisms, or SNPs, chosen to differentiate populations) into a science focused on teasing out the small frequency differences mentioned earlier. These would then be analyzed as diagnostic of population belonging, rather than being seen for how they are differentially distributed among populations in terms of their relative occurrence, or frequency. For instance, one group could have a genetic change at 2 percent (leaving 98 percent with the common type), and another group could have the change at 30 percent (leaving 70 percent with the common type).

When aggregating such markers, the goal was to construct a picture of homogeneity among Africans, Europeans, and other groups that could be used for different types of modeling for genetic studies. Many of these "markers" came from the resource panel made possible by the HapMap that had already demarcated human genetic variation into categories that followed the exact same conceptual lines as the centuries-old idea of three large races. It was in such contexts of reductive thinking where particular bits of human biomatter came to constitute a distended lens through which many scientists saw the human genome. In reality, what was amplified conceptually, with such a narrow spotlight on difference, was a miniscule part of the genome that some geneticists zeroed in on to answer questions about disease risk, population relatedness, and how people's mating behaviors over time have created patterns of genetic variation (called "population structure").

At the same time, scientists' laser focus on the extremes in the genome where these large continental groups differ in observed patterns of genetic variation reflect their own cultural and political choices to highlight differences between people as "populations" that most professionals in these field must realize reaffirm flawed everyday notions of race. They must realize this even if they hope that the social consequences of their work do not fuel racists' ideologies that racial minorities and white-descended "Europeans" should live in separate countries. They must realize this even if they hope that they are not arming racists with data to support their arguments that minorities are genetically less intelligent and more prone to disease. They must realize this even if they hope their work does not inspire even one racist to commit mass murder of innocent Black people on a Saturday morning in Buffalo, New York, after posting an online manifesto where he cited genetic population studies about human differences.[32] *They hope, they hope, they hope.*

Perhaps ironically, the HapMap organizers crafted statements about the care that researchers should take not to reify the groups that they selected in terms of racial categories, nor to generalize findings found in the "Yoruba," "the Han," or "the Utah" samples.[33] Yet the fact that the US-led international effort chose these particular representatives for human diversity—in several HapMap publications the consortium authors themselves called their populations generally "African," "Asian," and "European"—was puzzling. It was not surprising, then, that when scientists and others offered up the language of major continental groups that are often associated with Western notions of race (as essentially "different" peoples from key, yet separate, continents) readers interpreted the HapMap data on genetic diversity to indeed be about racial difference.

Almost two decades later, in May of 2023, a brand-new consortium interested in global genetic diversity published a draft of what head researchers have named the human "pangenome." In its first iteration the draft consists of forty-seven individuals' sequences, now deciphered with a dizzying array of methods that provide more insight into the gaps and heretofore undetected large genomic regions of diversity that were completely missed in the human genome reference sample (70 percent of the original human genome was based on one sole individual's DNA). The pangenome consortium members have long insisted on the need for an "update," which includes a focus on variation well beyond SNPs but also emphasizes including individuals from places that were not represented in the HapMap or the subsequent 1000 Genomes Project (yet it must be noted that the pangenome draft heavily relies on cell lines and data from the latter).

The goal for the new initiative has been to compile a composite reference made from multiple genomes, 350 of them from people around the planet. So far, however, in the consortium's 2023 *Nature* publication on the pangenome draft, the initial graphic still emphasizes the continental divisions of "Africa," the "Americas," "Asia," and "Europe," despite the authors' stated intention to explore the vastness of human diversity in a true global sense.[34] Nonetheless, in multiple *Nature* articles accompanying the announcement, as well as the press coverage, a sense of change prevails. A new era seems, once again, to be emerging. Yet with continental dividing lines still used as one powerful sorting mechanism, this future may look a lot like the past.

It might be said that there is power in these various future-trending phantom solutions, those forward promises that someday group-based racialized approaches will be amended, that there will be a consensus to abandon racializing terms when examining human DNA. It may well be a pipe dream, but it is one where the future tense allows researchers to straddle a threshold between reverie and reality to revive imagery and concepts from the past to construct a twenty-first-century genomic New World for various ends. In these spheres a contemporary fugue of race, ancestry, genes, health disparities, social justice, racial science, minority uplift, white nationalism, roots tracing, and forensic sleuthing all merge messily—and yet peculiarly with little effort at all.

1 Genomic World Building

Imagine a world where only certain geographies mattered
and only half of the continents existed, a world where history
started in 1492, a world where human bodies were seen for
their colors and textures, but only some physical fragments
of a person—skin, hair, freckles, and eyes—mattered most.

These traits (details of captivation) would be signals to
white-coated seekers who made it their life's work to survey
your shared species, signals that told them that your DNA
was valuable for their country, for their sick, for their sense
of self and for their newly sown data trades.

All of this prompted them to look even further . . . for
genetic marks that aren't visible to the eye, but, in their
hands, would reinforce the racialized aspects of you that are
. . . so much so that you would be consistently arranged, via
a bit of willful statistics, into races, caged by lines and bars.

These racial cells would include you and exclude you
based on some shared aspect of
   your coloring,
   your stripped history,
   and a desire to matter in a different kind of world.

From 2006 to 2010, I agreed to be part of a series of conferences on
genetics and ancestry on the East Coast of the United States in a town
called Cambridge on a campus called Harvard. Many such gatherings
were taking place at the National Institutes of Health, at universities and
research centers across the country and around the world as the concept

of genetic ancestry seemed to be steering breezily onto the terrain of race since the mapping of the human genome a few years earlier.[1]

These particular gatherings brought together a lot of highly accomplished people from a range of fields. There were social scientists, human geneticists with varied specializations, working both in academia and private ventures, and scholars in the humanities. As more than a few people in the room were stars in their fields, recognition of who was who weighted the atmosphere. Status within academic discipline (with some geneticists being the most revered) functioned like a glorified name tag. At first, the alpha ambience didn't necessarily foster warm hellos or genuine conversation. The invitee roster changed with each gathering, but there were many consistent returnees.

During the first year, after the ice-breaking welcome from the host, who by comparison with the rest of the group was a jovial bolt of high spirits, people started to loosen up and the sense of competition, if not rivalry, among a few of the younger geneticists began to fade. They seemed reassured once they realized that the host genuinely appeared to be validating all of them with kind words. One year, early in the proceedings, a veteran geneticist who made a big name for himself in the 1970s stood out. His very presence seemed to physically irk yet awe several younger scientists. The older man was now an icon of sorts. When his name was announced, some of the seated, settled, bodies could be seen leaning toward those next to them to whisper opinions. The little commotion was silent enough to still count as polite.

It had been a few decades since 1972, when the famous geneticist had made his worldwide citation-worthy statement that would be repeated by many for years to come: "It is clear that our perception of relatively large differences between human races and subgroups, as compared to the variation within these groups, is indeed a biased perception and that, based on randomly chosen genetic differences, human races and populations are remarkably similar to each other, with the largest part by far of human variation being accounted for by the differences between individuals."[2] With that startling idea, which came on the heels of the late 1960s cultural revolutions, the edifice of skin color–based beliefs in a biological concept of race slowly started to lose some ideological sway. Many teachers and researchers across varied disciplines, from universities to preschools, incorporated this harmonious notion. It served as a corrective to the biological misnomer that race might be thought of as rooted in one's genes. The geneticist's pronouncement became the concrete for the idea that race was a social construct.

While the older gentleman was still at the mic, one younger scientist in particular could be heard sighing, seen shaking his head, and looking around the room to get a pulse on other potential dissenters to the famous man's live words. It was still early, and so we braced ourselves for a long day. The renowned geneticist was of course the late Richard Lewontin. Because his ideas about the relationship between race and genetics are publicly known and had not changed, I include them here as a staging mechanism to begin to tell the story of a different kind of New World, one of the twenty-first century where race was being granted a new lease on biological life in some genetic circles.[3] This was a *mundus* in which many scientists sorted themselves into distinct clusters—where some geneticists embraced the idea that DNA markers might reveal racial groupings, while others opposed such flat reductivism. Their professional clustering reflected the behavior they hoped to optimize in genetic marker panels, the objects of the day, that some of them had culled from specific world populations—Africans, Asians, Europeans, and Native Americans—that increased the likelihood that racial delineations might appear in "the data."

I share the familiar ideas of Lewontin in part because they long preceded the initial conference itself, which was declared by the host to be "confidential." The host made everyone who attended agree not to write about the presentations. Some of the speakers already had much of their thinking and their science in print, but now there was a mutual crafting of science and history—double helical in expression. In writing this book, many years later, I find a confidential pact for a conference as curious now as I did then, back on the Island of Harvard with its clear rulers and rules. I have nevertheless written up, and promptly blacked out, a symbolic key moment of the gathering below for my own recollection and catharsis. I do this to honor the confidentiality agreement and also to find a way to illustrate that in this twenty-first-century world of genomics and race there are some elements that people did not want aired.

Once I blacked out the text below, the first person I showed it to suggested that I reveal a few random words—one here, two there—to show the reader that written text actually lies beneath the black lines, yet now unseen. This visual technique reminded her of an aesthetic choice that some poets deploy. I reflected on how poetry can free us from having to grammatically make sense of an idea and to feel through everyday absurdities instead. In letting these fragments stand as they are for the reader, the invitation is to piece together the bits of language contained in the redacted prose not as a secret code—but as the partial bones of an unearthed poem. This form of seeing the unseen helped me to craft creative sense out of a kind of secrecy that served the powerful as they discussed the many obstacles plaguing their genetic enterprises. These peeled back openings are also an invitation to engage, by other means, an obscured set of events about which the author cannot share all that she knows.

based in "biogenetics"

modern humans

people on the continent as

genetic story

oversell[s] the idea

oversimplifying human

borrowed metaphors

.   .   .   .   .

One of the principal cultural barriers that divided social scientists and geneticists at the conference above, and in many others I attended between 2003 and 2022, had to do with the increased ease with which scientists seemed to pick and choose specific continents as bins of sorts in which to place people when doing genetic studies. Their divisions made sense in a way. Surely Africans live in a place we know as Africa, Europeans in lands now called Europe, and so on. The problem of course was, and remains, that the broad brush of naming people by continental demarcation too easily glossed over the vast heterogeneity of each place, while creating modern-day notions of *humankinds*. In labs where I conducted ethnography for this book, this happened to varying degrees.

One of the technologies that allowed scientists to do this, and that convinced more and more people as the technology became routinized in medicine, law enforcement, and ancestry testing, is the aforementioned model of Ancestry Informative Markers (AIMs). For the system to work, each marker is carefully selected and tested in reference populations to make sure that enough people who are labeled as, say, West African have the variant in question to call it a "West African" marker. In addition, this construct requires that people from other labeled groups do not *also* possess such markers at relatively elevated frequencies. The vast majority of the markers (mostly SNPs) are simply "filtered" out because one perceived group—again, let's take West Africans—have these base pair changes at different frequencies among themselves. Genomic variation among the same group is called genetic heterogeneity. In essence, to construct the AIMs technology, all of the so-called "ancestral" continental groups cannot enter into the world of the model with their naturally occurring human variation intact. Each "ancestry" by continent is thus artificially made to be "homogeneous."

Conceptually, Ancestry Informative Marker technology reorders human interrelation along axes of differentiation. Ancestry genetic marker panels are set up in a way that requires that the user hold most of the genome at bay while imagining AIMs themselves as having a life of their own. What gets featured in this theater of genomic animation are genetic variants stripped from their biological function, environmental history, evolution, and potential cultural relevance regarding what social arrangements allowed for them to be passed on and shared among people or not.[4] Most geneticists I spoke with would agree that it makes little sense to think of an evolving genome rooted in global ecologies and vast human and animal living systems as static. Given what AIMs are—many are adaptive traits—to freeze them in a racial frame conceptually oversimplifies their stories. What we are dealing with here are often bits of DNA that code for proteins that do things like protect against *vivax* malaria, or that modulate eye, skin, and hair color along with freckling, or that underpin the functions of numerous drug (xenobiotic) metabolizing enzymes, and the list goes on.

Regardless of whether researchers used AIMs or different techniques to answer their questions about human genetic diversity, they tended to focus in on, *to separate out*, some aspect of human genetic difference. In various ways, in a range of labs, I witnessed well-meaning professionals attempt to theoretically isolate part of our species' shared biology to focus on what was *not* in common. Their theoretical excision of humans' genetic biology from its larger context, while still embracing the reality that the human race is singular, made its way into their hypotheses and studies. They did this in order to emphasize a point, albeit expressed differently for each lab, that there might possibly exist fundamental genetic differences along ethnic-racial lines between people. The problem, however, was that the "part" that many geneticists have happily been willing to theoretically blot from view is the vast majority of the human genome itself, otherwise known as the 99.9 percent. It is the 0.1 percent of rare variants that currently excites many within the field. Anthropological engagement on how this emphasis solidifies, and how to better understand the social and health consequences of such a focus, is sorely needed.[5]

On a more general register, I also witnessed in certain labs a willing suspension of disbelief, and an invitation for me to do the same, regarding the prominence and importance of some continents over others. Here a big four and sometimes big three came to represent—as a scientific account—who the humans are who matter for our genetic understanding of future health. As a first order of genomic world building, the planetary setting in question consisted of Africa, Asia, Europe, and pre-Columbian Native America as elements of a more or less static "Old World" that were necessary to understand the mobility and mix regarding people in the present.

The obviousness of this cosmological stance, where it is these four continents that matter, reflect the fact that the professionals I studied were mostly American scientists situated in the United States. What I show in this chapter and the next is that in certain instances how they drew the lines that divided historical time, geographical space, and DNA-based group belonging relied on reductive modeling that even one of its inventors acknowledged was a kind of scientific fiction. He described it as a "limited New World technology." In essence, faced with the anthropologist, he consciously relayed its abstraction: that it was conjured in labs—a man-made mockup. Yet it was built via SNPs, statistics, and a deep social sense that there are different kinds of humans. The amalgam of biomatter, math, and cultural-political understandings of racialized ancestry brought this technology's results to life—made them feel all-too-real. This is one instance of what I call genomic "sci non-fi," or the lived unfolding of everyday science *non-fiction*.

Because scientists' deployment of AIMs nonetheless carry real substantive social effects, I mark my own engagement with their work as a specific kind of genre that seeks to understand the otherworldly—as well as the inner-worldly—nature of their projects. My use of "sci non-fi" here is to underscore how fictive world-building elements (based on models of imagined realities) can yield simultaneously utopian and dystopian consequences of racial science. These can be forward-looking yet very much rooted in visions about the past that can be both empowering and stultifying. As we will see, a newer generation of geneticists want to change how science is produced, while their bid to be included in what they call "the genetic revolution" can come with a tax.

What some of my interlocutors took to be their right to be "represented" very often meant that their presence within their larger field was racialized. As this newer generation of geneticists hoped to bring to the fore social conditions of exclusion that were habitually left unseen, ignored, or effaced, they created more visibility for themselves and the populations they wanted to help by putting race front and center—through the tabula raza—as a clear tactic for "revolutionary" change. Yet revolutionary action can have multiple facets. While attempting to topple old systems, one could also find themselves spinning the revolving door of old patterns of thought.

THE DOOR TO INCLUSION

When I carried out fieldwork in the Bay Area–based Genes for Health Laboratory in 2003, its new equipment and enthusiastic team reflected the moment.[6] The human genome had just been mapped, and it seemed to many people in the field that a different kind of "New World" awaited explorers and pioneers to detail and claim it in scientific terms. The lab itself was housed, almost anachronistically, in a historical brick tower of a local university teaching hospital. The lab director, whom I will call Dr. Gabriel Morales, was a no-nonsense, sometimes combative, yet affable physician-scientist who easily shared a laugh or fishing story with the janitors in the corridor. In lab meetings, between serious moments of discussing sequencing issues, Morales could precipitously code-switch to street slang (to refer to an enzymatic solution the lab had prepared as "home-brew," or to talk about exploitative power dynamics in science as "pimping"). Most people in the lab got the references—those who did not might simply let them pass, while others asked for clarification, much as they would with an interesting p-value. Morales refused to talk and act like a researcher who was isolated from "the real world." Others in the lab appreciated his informality and told me that they too also wanted to "keep it real."

Morales was known on his campus, and later in the larger world of medical genetics and precision medicine, for his work on the genetics of "minority populations," "the reality of race and genetics," as he often put it when on a dais, as well as his talent for growing DNA banks of Black and

Brown populations. One such bank stored his own research samples of DNA taken from over two thousand Mexicans and Puerto Ricans to research health disparities. In comparing these two populations of "Hispanics," his lab wanted to show that these populations could not be lumped together due to their underlying ancestries.

Morales was born and raised in what was at the time a largely immigrant neighborhood. "We were in the Hispanic ghetto," he told me. He added, "and were right on the edge of the Black ghetto. So that's where I grew up. [. . .] I've always been keenly aware of race, ever since I was a child." When asked about his research and academic life, he often talked of being a "product of affirmative action," and having a deep awareness of racial inequality. He emphasized that he himself was "mixed" of "white and Mexican" descent. Affirmative action, he was proud to say, opened the doors to him of several of the most prestigious universities in the country, where he did his medical training and residencies.

Morales's main patient recruiter (also an MD), the manager of his first DNA bank, one of his bioinformaticians, and the project's data manager were also "Latino," though of different origins spanning South and Central America from Argentina to Mexico. The other lab workers were all "hand-picked," as Morales recounted, for their "diverse backgrounds," which for most also happened to be humble. In addition to his regular team, a few more collaborators, some of whom provided access to state-of-the-art genotyping platforms and others of whom shared more advanced statistical expertise, visited once a week at lab meeting. The lab also took in two minority high school students every summer through a mentorship program that aimed to increase the dismally low numbers of minorities in US science.

Seventeen-year-olds, who had only heard about polymerase chain reaction (PCR) and other laboratory technologies, were given the chance not only to learn techniques but also to conduct research. During my stay Morales hosted two young scientists. The first student in the program, whom the team described as "scientifically brilliant but socially awkward," was an African American high school junior. The second, who all agreed was "perfectly social but less driven," was a more mature Latina in the same year of school. Like many of the researchers and technicians in the lab, the students had their DNA extracted fairly early during their stay. As a sort of modern-day initiation into a family that would now share this DNA for tests and training purposes (so as not to waste precious DNA from patient samples), the students, like the others, began to watch their genetic signals appear fluorescently on gels for particular genetic ancestry markers of interest to the lab.

Other examples of what I came to understand as processes of "genome-kin," where bonding, sharing, and positioning themselves as subjects who were also implicated in their own research goals, revealed the degree to which these scientists imagined themselves fused with each other as well as with the patients and donors who comprised their databases. In short, their commitment to resolving health disparities—through methods of zeroing in on how genetic variants might determine whether any given racial or ethnic group exhibits a higher probability of contracting a disease or responding negatively to a drug—was often phrased as not only helping but also recognizing these communities at long last. It furthermore became clear that this penchant to help, ultimately framed as "saving lives," expressed their deep commitment to people to whom they felt themselves to belong. But that is not the full story.

Many of the scientists that I engaged for this book, especially those from US underrepresented groups in the Morales lab and the lab of Rick Kittles, might also be said to be on a mission to *save themselves* as members of Black and Latino communities more broadly. They, like their imagined participants, had individual genomic variation within larger group-based genetic patterns. They, like their imagined participants, also had nonwhite skin color and Latino and African heritage. In part because of these aspects of who they are, they, like their imagined participants, lived lives faced with bias when navigating US institutions and everyday life in America. The neglect that they wanted to redress, therefore, also included remedying the marginalization of scientists like them in their fields. An unstated portion of their ticket price toward inclusion was tied to their ability to enroll people in genetic studies for their universities.

Being scientists of color themselves, and being perceived through that color to share experiences with people targeted for genetic study enrollment, made many who worked in these labs reason that potential recruits would trust them in ways they might not trust white scientists. In practice, this translated to a point of leverage within a larger system of medical genetic labor. One day while shadowing Morales at the hospital, I observed younger residents whom he was mentoring ask him about his successful career trajectory and his ability to land a job at a top medical school. He homed in on the genetic databases that he was building for his own research as part of why he was so valuable to his institution. In his light-hearted yet trenchant way of keeping it real, he joked slightly about being used but then commented on how it was a win-win situation. Indeed, Morales has become an impressive leader in his field.

I understood both Morales's willingness to do this work and his belief that amassing large amounts of family DNA on common ailments could one day help people. But I couldn't help but wonder about this recruitment by "same-race work" as a kind of tax—an extra burden on him and other researchers of color who are hired (however willingly) as liaisons of sorts to convince everyday people from their same racial background, and who may be hesitant to give a part of their person as a sample for a database, to indeed give . . . and to do so assuredly.[7]

RACIAL LABOR, TRIBUTE, AND LEGITIMACY

In her book *Taxing Blackness*, historian Norah Gharala details how start-ing in the late sixteenth century the Spanish Crown demanded different rates of tribute, or tax, for Indigenous Indians and Black subjects who were granted their freedom in what was then called New Spain. Viewed as an act of "largesse," the Spanish viceroyalty saw its gesture of granting certain non-European subjects the possibility of living in freedom as an act that should be repaid as a sign of loyalty and obligation. This was a complex system where many free subjects rebelled and refused to pay, while others saw the tribute system and their ability to work within it as a marker of their political legitimacy within the *casta* system.[8] Through vast bureaucracies and visitations from tax collectors, Black and Brown people in the territory that eventually encompassed what was then Alta California and reached as far as the treaty line of Oregon Territory in the early nine-teenth century, could be grouped and exploited economically now not as "slaves" but as "tributary units."[9]

The system worked differently for Indians whose humanity was asserted by the papal bull *Sublimis Deus* of 1537, which began their emancipation when enslavement had occurred, than it did for Africans who might be able to buy their freedom, be born to free parents, or who, in the course of generations, could prove that they had enough nobility or status via successful ancestors to acquire freedom. When free, many African-descended subjects were considered *mulatos* with African ancestry but who could also have European and or Indian heritage and therefore might be seen as only partially African, which attenuated the tax they were required to pay. From the late sixteenth century to the late eighteenth century, heritage was parsed meticulously through genealogies. This was especially useful for Afromexicans during the Bourbon rule from 1700 through 1808. If they could link their genealogies to a "Spanish ancestor," "a conquistador," or even a "noble" Black or Brown soldier who may have fought for or served Spain, they might change their tribute status altogether.

The Crown recognized that mixing was inevitable in the colonies, and therefore the arrangement was largely an economic one that generated a sizable tax base very much reliant on keeping a robust system of race-infused "qualities" (*calidades*) of ancestry in place, even as its tax-obliged vassals could potentially rise on the social ladder. Successful and savvy mixed-heritage subjects paid less or nothing at all through all kinds of negotiations of their genealogies, which took on new racial aspects depending on their social entourages or if others could vouch for their reputations. Remarkably, these mixed Afromexicans could petition tax attenuation by being associated with the most powerful aspects of the system itself—European or Spanish power through descent or strategic alliance—all the while acknowledging, reporting, and still embracing aspects of their non-European ancestry that were partially configured in a fluid opposition to it.[10]

I saw parallels centuries later in the New World that Morales and other scientists were drawn into through enticements for inclusion, representation, and a kind of scientific legitimacy within a system dedicated to racialized, now scientific, *calidades*. Aspects of their ancestry were not explicitly taxed as tribute, but their ascendance to power entailed working within a highly racialized system without disturbing its contours, while recruiting others into that system and therefore expanding a pervasive structure of human genetic science that seemed to allow them room to maneuver while keeping aspects of their Black and Brown ancestry intact as a means to enter into the strata of the powerful. In this world, as in the historical one described above, the overall schema might be characterized by its overwhelming normative whiteness, which currently includes political and ethical obligations to make room for minorities to belong via similarly compromised notions of "freedom." The racial terms through which this happens hinge on the hope of fitting Black and Brown scientists of exemplary status into a system that was once characterized by their exclusion. Now the door to inclusion sits slightly ajar.

THE MESSY GOVERNANCE OF RACE

All of the databases that I witnessed scientists constructing in the course of research for this book were made by human hands that sorted fellow humans' DNA by continental racial categories, or alternately, by US census categories of race and ethnicity. The idea of "white," "Black," "Hispanic," "Asian," or "Native American" deoxyribonucleic acid is a concept that the world's largest funder of genetic research (the National Institutes of Health, NIH) has both simultaneously rejected and encouraged. This has resulted in a strange kind of schizophrenia where the split institutional mind seems to function quite productively. It works to address issues of inclusion in science. It works to address health disparities. And it works to bring racialized others into a system that further racializes them, now via their genes, and does so in ways that feel progressive, all the while keeping a social and medical system of race-based human difference firmly in place.

One key driver of this productive contradiction started with the passage of the well-intentioned 1993 Revitalization Act that pertains to almost every aspect of human health research funded by the US government. It reads: "Members of minority groups and their subpopulations must be included in all NIH-supported biomedical and behavioral research projects involving human subjects. . . . NIH funding components will not award any grant, cooperative agreement, or contract or support any intramural project that does not comply with this policy."[11] As sociologist Steven Epstein has shown, the inclusion of race and other points of difference (such as gender, sexuality, and age) were not natural givens with regard to the type of body that has historically been included in human medical and scientific research.[12] Through various forms of activism, and explicit intervention on the part of key women scientists at the NIH and African American members of Congress, women and minorities lobbied for, and succeeded in obtaining, this congressional legislation mandating that publicly funded research include them as study populations.

Part of such action was to contest that white men, traditional research subjects for much of the twentieth century in the United States, could stand in for all of humanity. Yet as Epstein pointed out: "The new emphasis on inclusion and the distrust of extrapolation across social categories were not without opponents. . . . Concerns were . . . raised about the problematic business of defining medically meaningful racial and ethnic categories. In its implementation of the NIH Revitalization Act's directive concerning 'minorities,' the NIH follow[ed] the path of other government agencies by adopting 'Statistical Directive No. 15' of the Office of Management and Budget (OMB). . . . [And] Directive 15 specifies the racial and ethnic categories used in the census."[13] During discussions among the scientists I followed, it was clear that they saw themselves fulfilling a much-needed service by recruiting "diverse" human subjects. The ethos of the Revitalization Act was sometimes explicitly referred to, as was the reality that some portion of scientists' projects were funded by the NIH, which, as part of the US federal government, would require the use of OMB categories.[14]

Shortly after the human genome map was published, I interviewed Dr. Rochelle Long, the NIH-granting program director who oversaw a large pharmacogenomics research network that funded some of the scientists in this book. In our conversation it became clear, however, that she and others were concerned about using race/ethnicity as a biologically based sorting technology for pharmacogenetic discoveries (i.e., whether or not certain groups have genetic variants that affect how they respond to drugs) on the part of network scientists. Yet in discussions her office had with specialists in the fields of genetics and pharmacy who were invited to "advise" on the issue, they decided that it made sense to use racial ascriptions with their imperfections, given that NIH grants and reporting were *already* structured around their importance since 1993.

The decision to use race/ethnicity went forward despite the fact that, according to the Census Bureau, "the Office of Management and Budget's standards for maintaining, collecting, and presenting data on race . . . generally reflect a social definition of race recognized in this country. They do not conform to any biological, anthropological or genetic criteria." Similarly, in the NIH's own words: "The categories in this [the OMB] classification are social-political constructs and should not be interpreted as being anthropological in nature."[15]

Morales and his collaborators were therefore caught in the thrall of what Karl Marx in *The German Ideology* would call the "material result" of this particular point in history. This swirl of seemingly contradictory elements, where race in their labs is ascribed to DNA by people who have traditionally been left out of leading scientific enterprises, and where science has historically been used against them, culminates in "a sum of *productive* forces." Specifically, the dynamisms in play are "a historically created relation of individuals to nature and to one another, which is handed down to each generation from its predecessor; a mass of . . . capital funds, and conditions, which on the one hand, is indeed modified by the new generation, but also on the other prescribes for it *its conditions of life* and gives it a definite development, *a special character*."[16] Morales and his collaborators' conditions for life indeed possessed a special character—they were the ones who would be exploiting racialized DNA as a progressively inflected kind of capital embedded within a larger ethos of social justice genetics.

In the beginning of my fieldwork, I found this incredibly curious. How was it that despite some scientists' own strong senses of their anti-racist consciousness and work, this genome-kin bonding of seeing themselves in their science emerged in part from their involvement to decipher methods to isolate DNA markers said to reveal genetic ancestry as African, Asian, and European—what have been called the "Big Three" races?[17] The methods that several of my key informants were developing were the fruit of collaborations within a network of geneticists where, it must be said, the scientists training them were not from underrepresented groups. Within these collaborations they used and continue to use population descriptors tied to geographic divisions that no matter their granularity—ethnicity, city locale, country, or continent—always fit within a larger notion of racial types. For Morales and his team, this tension, contradiction even, was tempered by the fact that—again—they saw themselves doing the essential social justice work of including minorities in science. As a call to arms of sorts, Morales would often repeat: "We can't be left out of the genetic revolution!"

Representation within genetic studies has indeed been dismal. This fact has mobilized many in the field to try to address the gap in the numbers. We can review the stark divides, but in so doing must still ask the question: What does inclusion in genomics as it is currently structured entail? How does an obsession with inclusion at all costs in a system structured by categories of human difference captivate talent that might be better applied to broader problems that drive health disparities that do not racialize people's biology but that might begin to reexamine systems of inequality societally? And how is inclusion captive in terms of giving scientists of color an "opportunity" to enroll more Black and Brown bodies into genetic databases whose visualizations on computer screens reveal a tabula raza passively and aesthetically structured by rows and dividing lines that contain cells of information that seem to speak race into biological reality?[18] And why does it matter that such data may not stay put for strictly biomedical uses, and might not be as private and anonymous as people are told given that re-identification of DNA in such contexts is more than feasible?[19] We are currently witnessing a push ahead motivated in part by researchers' focus on a data-centered sense of quantitative "equity."

INTOLERABLE NUMBERS, BEGGING FOR "REVOLUTION"

When researchers in the UK tallied large-scale genome-wide association studies (referred to by field insiders simply as GWAS) from the years 2005–2018, they found that the proportion of samples from "diverse" people was less than 5 percent for African, African American, Caribbean, and Latino populations combined. On the flip side, more than 80 percent of the large-scale genomic studies in the same period were conducted in white populations. Studies on Asian populations came in at a far second, hovering just above 9 percent.[20] Throughout this key period in genomic science ascendance, just three countries—Iceland, the United Kingdom, and the United States—accounted for nearly 72 percent of all human genetic studies on the planet. According to a real-time data in motion tool called "GWAS diversity tracker," these stark imbalances persist as I reread my draft copy of this book in 2022.[21]

Revolution might first require that those wanting change must reveal the ways that people are structurally estranged from their own labor that generates the capital that they produce for others. Again, going back to Marx, such "estrangement" might be abolished when it culminates in an "intolerable power"—that is, when the contradictions of a system are no longer bearable.[22] At base, how are any of us ensnared within processes that alienate us from what might be ours to keep? The goods to consider for safekeeping may be our own bodies and biologies, as well as labor and information extracted from them that do not always foster health and well-being, or that might reduce us to narrow versions of who and what we are. It is worth asking: when are we enticed, or forced, to hand over in objectified form some part of our body, group identity, or potential social belonging in ways that keep us running in circles of subjugation? On the one hand, Morales explicitly called out the larger habits of power within medicine and science that result in exclusions—which often fade into the background of social injustices amid everyday racism more broadly. On the other hand, scientists doing such work strictly within the frame of ancestry genetics are more often than not *mediating* the contradictions that might truly bring about a revolution from within.

Morales's focus has been on working with and within a system that racializes biology with the mantra of inclusion that has been written into funding guidelines at the level of the NIH with very little emphasis on exploring the social determinants of health in concert with genomic data. Other questions that might be asked include how racist bias affects people's health but also extend to how genomic data is sorted, organized, and eventually allowed to stand in for, *or to absorb,* larger societal patterns that it can render invisible when genomic explanations take center stage.[23] These layered issues may actually contribute to the persistence of health disparities and further drive scientists, who at times have expressed to me their own exasperation and impatience with the stubborn professional structures they inhabit, to concentrate on what they consider the easier and more fundable work of locating genetic variants rather than focusing on community health and "environmental" issues. In their attempts to be successful as they work within these structures as scientists of color—if we put the onus for change on scientific policies and structures and not the individuals—people in these fields are not wrong to focus on mechanisms of genetics.

A 2018 critical review carried out by researchers within the National Institutes of Health analyzed what they called a "funding gap" along racial lines. They found that white researchers won grants to research health problems *at a rate of approximately 70 percent more often* than their Black counterparts.[24] One of the biggest factors distinguishing proposals that made it through the process successfully concerned what the authors called "topic choice." Research applications focused on more than 150 varied topics, ranging from "molecular biology" to "community health" to "vaccines" to "disease prevention." Within the overall disparity, topic choice accounted for over 20 percent of the funding gap after those doing the analysis controlled for other variables, including the applicant's record of achievements.[25]

In the press coverage of the study, one of the researchers involved commented that the most successful applications "were really about molecular mechanisms—cells, or parts of cells. Words like cilium, DNA polymerase, chimeral chemistry, ribosome." He followed up with: "It's not absolute, but it's really quite a striking distinction."[26] The study authors also flagged the insularity and circularity of prestige accumulation, what is known as "the Matthew effect." Taken from the Parable of the Talents from the biblical book of Matthew, this economic theory is otherwise summed up in the saying "the rich get richer and the poorer get poorer," due to what people start out with and their ability to invest and accumulate capital monetarily as well as socially, which are often linked. The study authors ventured that the Matthew principle often overlaps with racial bias against scientists from disadvantaged backgrounds (scientists of color and also women) who do not start out with the same degree of resources and connections.

It is not really surprising, then, that people will do all that they can to get a seat at the table, a spot on the conference roster, an opportunity to publish a "hit" finding (arguably much easier to do with genetic sequence data, several informants told me, than with environmental factors) and obtain high enough review panel scores to win grants. All accumulate and translate into even more prospects for advancement in the field. I witnessed many scientists of color pursuing these aspects of their careers with a determined vision to make inclusion a reality for themselves, which often tied back to the communities they represented as well as their unique position to recruit from them as their own obligation but also as an obligation to do so with funds from the government agencies now dedicated to health inclusion. As Morales told me: "I think it's our responsibility as physician-scientists, as members of this population, and as tax-payers who are funding the NIH, that we require the NIH to study minority populations."

I am deeply sympathetic to the motivations for their work and the reasons for their emphases on genetics to better understand health disparities in a funding world where preventive public health and community health, without molecular or "mechanistic" variables, are simply less compelling to reviewers within the NIH funding machinery. Yet this does not detract from a wider ethic of care and responsibility that I also have. That is to clearly voice that framing health and other disparities in genetic racialized terms, that promote reductive thinking about types of people who are biologically distinct in absolute generalized terms feeds dangerous racist ideas. At its tragic extreme this thinking results in racists hoping to annihilate those they see as different from them, as in the Buffalo New York shooting mentioned in the introduction.[27] But the subtler rebuffs, everyday dismissals and recurrent violence that people act out on racialized others are also at issue.

Concerning biomedical research, a strict focus on DNA and human difference may actually overlook the root causes of many health problems leading to premature death—aside from white supremacist gun violence— as social epidemiologist Nancy Krieger has shown in her work on the fact that Black women who simply lived through Jim Crow are more susceptible to aggressive deadly breast cancer. In a similar vein, epidemiologist David Williams's work reveals that when people of color live in highly segregated neighborhoods they suffer a range of life-shortening common diseases, including higher levels of type 2 diabetes, heart disease, breast cancer, and obesity.[28] The focus on increasing genomic representation in quantitative terms leaves intact deeper problems that plague the very premise of separating groups by simplified continentally marked genetic difference—a premise that reinforces an idea that race (now in the more palatable semantic guise of "ancestry") has a meaningful genetic basis.

This is not to say that genes are useless. It is to say that it is a lucrative yet grievous reach to rely on them to explain racial health disparities and to catalog genetic variation at all costs by the social lines of race. Given the above, what kind of revolution is under way exactly?

POLITICAL AND GENETIC CONVERGENCES

Morales was very much aware of the fact that a notion that white peoples' genetic biology might differ in some fundamental way from that of non-white people gives ammunition to white supremacists—individuals and collectives who sometimes call themselves separatists or white national-ists. He told me that when some of his first publications appeared, people with white supremist beliefs contacted him about the implications of his work. It was early morning when he first shared with me this jarring story. Morales had stopped to pick me up in Oakland on his commute into San Francisco from farther out in the East Bay. As I opened the car door, he cleared the passenger seat and tried to wrestle his sandy fishing poles away from my side. He squeezed them into place between the gap in the seats over the console. The poles' ends poked out freely through the moon roof as he drove off.

Morales multitasked our interview with traffic and the beating wind that whipped through the car as we headed for the Bay Bridge. "We got praise and criticism from, I would say, four different groups," he began. "On the praise side, we got praise from people who were really interested in studying minority populations and who really believed that we were on the right path. That was group A of praise. And group B of praise was the white supremacists [who said], 'Right on! You're confirming what we believed all along.' And the argument there is, you know, 'Yes, only a single base pair is required to cause cystic fibrosis, sickle cell anemia, Tay-Sachs, well, it's probably only a single base pair that's required to cause violent behavior in Blacks, criminal behavior in Blacks, cheapness in Jews.'"

"Are these white supremacist scientists?" I asked him. "Do you mean to say that white supremacist groups are sitting around reading *Genome Biology*? Do you know who these people were?

"One of them—well they probably wouldn't identify themselves as white supremacists, but one of them is trying to link intelligence—well looks at head size and intelligence. And he has published papers—"

"So he's a scientist?" I interjected.

"Yeah. He's published a paper showing that Blacks have the smallest heads, and Asians have the biggest, and that was correlated with—I don't know if you know about that?"

"Nineteenth-century science."

"No, it was *recent*."

". . . I mean ideas carried over from that era."

"Oh, yeah. And then, on the David Duke site, their research, or what they talk about—I'm not sure that they do research—is very similar to what we said in the *Genome Biology* paper. And that's where we found out that we were referenced on their website."

Four years after Morales's first controversial papers were published in 2003, a San Francisco–based newsweekly ran a story on race and medical genetics titled "First: Do No Harm?" The story included an interview with Morales that repeated some of the themes he had covered in his conversations with me a few years earlier. In the same no-nonsense tone, Morales publicly told the weekly that he got unexpected praise and criticism from both racists and anti-racists, stating: "The sociologists are afraid that one group will use this sort of information to try to subjugate another group. . . . That's the fear. I mean, David Duke [former Grand Wizard of the Ku Klux Klan] probably loves the kind of research we do because it seems to play right into his supremacist views."29

Duke had long shed his white sheets by that point and upgraded to a three-piece suit. With small forays into US politics in the state of Louisiana, two unsuccessful US presidential bids, and extended stays in Austria, Russia, and Ukraine, Duke spent years rearticulating and refining some of the explicit hate speech of his Klan phase to appear more polished, compelling, and perhaps rational. In 1980 after he parted ways with the Klan, he started a foundation, strangely mimicking civil rights pillar the National Association for the Advancement of Colored People (NAACP), not so ingeniously called the National Association for the Advancement of White People (NAAWP). Duke's new surface makeover would center on white separatist cultural and "national" spaces. Still, the Anti-Defamation League has described him as "perhaps America's most well-known racist and anti-Semite."[30]

The day after the article appeared, Duke posted an open letter to Morales on his website, davidduke.com. The former Klan leader first corrected Morales's choice of identifiers for him before commending him for his work on the "reality of race." Duke wrote:

> I am not a supremacist ... [but] I do believe that different population groups have different characteristics in important areas that reflect everything from disease rates to tolerance for medications—even psychological characteristics and tendencies.
>
> I do believe that your work and others who show real biological differences between races is important. You show that race is real, not a societal construct or some sort of conspiracy theory.[31]

The question was (and remains), how do scientists in the cultural scientific spaces that Morales and others occupy show that race is real and not a societal construct? Do they succeed in bracketing the societal as strictly separate from the biological, or is the biologistical use of frequency differences on a statistical level informed by, and infused with, societal constructions of race?[32]

When the contradictions were so glaring, Morales told me that he didn't respond to the white supremacist scientists who wrote to him. He shook his head, softly sighed, and explained that he couldn't control how others might use, or misuse, his research. He more or less wrote off the likes of Duke as extremist and just put his head down, got back to work, and refocused on trying to uplift those who mattered most to him. In the following years, Morales worked diligently to gain expertise in epidemiology, undertook long training stints in various statistical methods, obtained a master's in public health, and engaged more explicitly in trying to understand the epigenetic and environmental factors that could also be driving the health disparities. In a 2021 article that researchers interested in these questions penned called "Race and Genetic Ancestry in Medicine: A Time for Reckoning with Racism," the authors started out by explaining the political and historical contingencies of racial categories and the dangers of an overreliance on a biological notion of race. Throughout the paper they go on to select an array of examples that demonstrate "ancestral" genetic risks (in continental terms, again, African Ancestry, Asian Ancestry, European, etc.) to discuss what are now routinely used ancestry markers, and conditions associated with ancestry, that researchers use to make the case for a more general acceptance of racialized "population specific" genetic risk.[33]

Curiously, their 2021 attention to how race is constructed socially, with which they open the paper, did not reflexively apply to a now long habit of organizing and interpreting population genetic data via racialized categories, or bins. In the first part of the piece, the authors laid out a careful step-by-step discussion of how race is a census category, a political entity, an important social demographic to track epidemiological incidence, a means to document racism in medical care and treatment bias, and even an embodied phenomenon that may influence how racism manifests biologically. They cited census data over time to point out that Latinos are the fastest growing group in the United States, which has been the case for several decades. From there the authors emphasized that in the world as they live it, and as contemporary taxpayers, Latinos should be represented in the genomic revolution. Yet toward the end of the paper there is a strange turn: the authors wed themselves committedly to genetic studies even when there is no obvious health reason to insist upon genetics for disparate disease burdens. They write: "Genetic studies of non-European populations are important even if genetic variants are not responsible for overall differences in disease incidence or outcomes."[34] One might understand this aspect of "the genomic revolution" through a different set of revolutionary terms, perhaps metaphorical. These are the terms where habits of power are mechanically reproduced and literally re-volve, as in a revolving door, on the same old hinges, within the same old frames.

DEEP TRAITS AND OTHER STORIES

Back in the lab, the high school students were chaperoned by a Canadian researcher, whom I will call Katie, for a few hours a day. She talked them through writing the abstracts for their research projects and walked them through how to carefully run gels. I was at the bench with Katie as she trained the aforementioned African American "socially awkward" student, whom I will call Jerold. He was instructed on which primers to use in order to analyze his own DNA for the *FY* gene referred to in this lab and others where I worked by its more commonplace name, Duffy. Named after a hemophiliac patient where it was first found in the 1950s, the Duffy blood antigen system has several genotypes and phenotypes of interest to scientists who attempt to infer ancestry. Among other functions, Duffy antigens also provide entry points for the plasmodium vivax malaria parasite to penetrate red blood cells. Yet one particular Duffy genotype yields a phenotype resulting in what is called "Duffy-null" or "Duffy-negative" (*FY*-null), which, it turns out, provides a dramatically helpful trait for humans since the parasite has no means to enter the red blood cells and infect them. Today there is very little vivax malaria infection in West Africa, where large swathes of the population carry Duffy-null.

The Duffy-null marker is said to be one of the most "informative" of the DNA markers that researchers have collected to infer ancestry and also to determine the race of a sample left at crime scenes in forensics. This is because West Africans have this particular genotype at high frequencies. I never heard Duffy's environmental selection story told in labs where I observed researchers routinize the use of AIMS. Instead, what was referred to simply as "Duffy," and rather flatly, was often brandished as proof of a population "specific" allele of "African ancestry" even though populations such the San in South Africa and others in the Horn of Africa do not have the null variant in large numbers.

During then Senator Barack Obama's early 2007 campaign, as he was trying to appeal to Black people based on shared racial heritage, I heard researchers in Chicago toying with the idea of offering Obama the AIMs test that might break down his ancestral composition into percentages of "African" and "European," as they routinely did with African American data. They decided against it, saying that, as an East African, Obama might not have the right Duffy variant, and the whole thing could backfire if the idea was to make the political newbie more familiar to Black folks who were leaning toward Clinton. This was despite Obama's clear *actual* (not inferred) African ancestry.

I always wondered how markers like Duffy might be recounted differently and if that would complicate how people viewed and worked with these now widespread DNA base pair changes used to infer ancestry. What if Duffy itself could be rendered as a protagonist with a biography, a history, and most important, an environment? How would the Duffy story be chronicled? What might be featured in a more complete world—a world where Duffy alleles were not simply seen as "ancestral" but, more to the point, selected and passed on when humans reproduced because one version of the Duffy gene protects their progeny from a severe ecological disease threat?

Such a larger explanation would bring into view the dynamic nature of this DNA sequence change—one that is not actually fixed in some people's deep nature that distinguishes them "continentally" from others. Instead, it would reveal a biocultural interplay of life, within specific geographies and the ways that scientists use contemporary signatures in the ever evolving human genome to infer racialized ancestry by continent. Why some people living in West Africa have the Duffy-null genotype and those living in South Africa, East Africa, or Europe may not stems from a story where a protective trait (with stark frequency differences between specifically selected groups of people) is collected for ancestry tools. Duffy is one of the moving parts that makes this ancestry technology work. It is one of the alleles that makes up ancestry panels and gives us a window onto how these markers are too often absorbed into pat surface readings. The Duffy-null allele could just as easily be thought of as an "Environmental Informative Marker" found in people who live throughout West Africa. That is, if the environment rather than ancestry was front of mind.

Jerold seemed a bit nonplussed by his result, for which Katie had primed him, saying, "this one is found in Africa, so most African Americans probably have it." The *FY*-null signal from Jerold's DNA made for a dull picture when mixed with the gelatinous agar gel derived from red seaweed on the electrophoresis PCR plate. He looked around and gave a small shrug as he examined the small tray filled with the strange Jell-O that displayed something about his body, his race, and presumably his ancestry. The training ritual did not immediately translate to a vision of high-tech genetic science promised by the famed Human Genome Project that everyone was talking about in biology class at school.

Katie approvingly congratulated Jerold on his work with an encouraging smile. She set him up with journal articles to read, instructing him to "pay attention to how they write the abstracts." The papers featured studies that provided background to the lab's ongoing projects. The student visitors' pace was slow. To their delight, it was often necessary for their mentors to leave them and pursue their normal rhythm of efficiency in the lab: a frantic hustle replete with coachlike chants from Morales who wanted "three papers out for review before the end of the summer!"

As an example of how their professional efforts were not separate from their larger lives, scientists in these labs regularly discussed their own DNA traits, with admittedly more interest than the newly arrived Jerold showed. The teens were excited to be there on the first day, but they, like most teens, had their own preoccupations. They were not immediately inducted into the cultural discourses that often defined the world of the lab where scientists frequently discussed the dire need for their specific involvement in studies to overturn what they saw as an unfair focus on white populations. One of the recruiters explained to me that their lab did not perform the same diligence when recruiting "Caucasians" as they did with people of color in their studies, saying, "I don't think it was a big issue for [the team] because ultimately what we want to see is more of the ethnic groups." The students listened but mostly stayed quiet.

The older scientists effortlessly moved from conversations on recruitment numbers in the community to talk about their own features that might be ascribed to their "Indigenous," "white," "European," "Black," or "African" DNA. They would sometimes include details about the physiognomy of their family members. "My mother was really dark-skinned, and looked more Indigenous," one shared with the group at certain points when discussing admixture. Another told me: "My mom's background is more of the European. My grandfather was blue-eyed, blond hair." They contemplated their own upwardly mobile social positions and ever-expanding cultural backgrounds as their work world required new kinds of social passports to traverse sometimes stuffy university meetings, or to attend fancy philanthropic donor events. But they never lost sight of the importance and their own obligation to use their social entrée to do outreach in working-class and poorer communities throughout the Bay Area from Fruitvale in Oakland to Redwood City.

Sometimes their own progressive politics of health would likewise enter into their conversations about the motivations that drove their work and their desire to recruit study subjects. These were deeper charges than simply gathering the valuable human commodity that DNA samples (in the aggregate) have become. There was a sense in several of the labs that they were uniquely positioned to do what became a broad notion of "outreach." It was often framed as a way to extend social bonds of care to "the community" where their people had been marginalized within the history of US medicine and its mistreatment of people of color. Several labs offered some form of preventive health testing. The Kittles lab also sometimes gave nutritional and medical education to DNA donors and potential volunteers. And at different points, all of the labs translated consent protocols into lay terms as well as into other languages when English was not people's mother tongue.

These efforts were done in order to attempt to build trust. They were also attempts to engage with people in a more sustained way so that they might see a tangible benefit to "giving" their blood and genes without necessarily "receiving" any immediate improved health outcome from the exchange. Extra efforts to offer a modicum of care were to combat community mistrust of the researchers as scientists tied to universities, which all of them were, or simply to offer assurance when uncertainty about the complexity of genetics and what one can do with such data arose. Still, recruitment, which is always hard work, posed difficulties. I detail some of these encounters in chapter 3. How the samples found a home among others, how they were organized, how they live on, and relate to the past first requires exploring still other aspects of this world.

2 From *Mundus* to Model to *Mundus* Again

THE ART OF ANCESTRY BETWEEN WORLDS

As the Morales team rushed around the lab office space discussing their latest results, a reproduction of Diego Rivera's *The Flower Carrier* hung on the wall (figure 2). One afternoon when Morales slowed down after taking in hopeful news that his postdoc excitedly shared about patterns she saw in their data, I asked about two paintings that gave the space both sense and color. The paintings integrated a past world into the present. They rooted the geneticists and researchers among a long line of people, who in all of their labor, greatness, and historical weight, constituted a larger sense of *la raza*.

"I can't help but notice these. Can you tell me about them?" I asked Morales.

"I have this one to remind me of all of the people who have come before me, including my mother, who was a migrant worker [. . . .] It's a picture of a man on his knees. His wife is loading a very large basket of flowers on his back, much like you would load a mule. To me, this symbolizes the struggle that we have and continue to face."

A few feet away, another reproduction of one of Rivera's murals that contained more busy details of daily life than were legible at first glance displayed the caption: "La base de un gran futuro està en nuestro pasado"

Figure 2. The Flower Carrier, Diego Rivera, 1935. Courtesy of the San Francisco Museum of Modern Art. © 2023 Banco de México Diego Rivera Frida Kahlo Museums Trust, Mexico, D.F. / Artists Rights Society (ARS), New York.

(The base of a great future lies in our past). The original mural is part of a series reproducing two thousand years of Mexican history that Rivera was commissioned to paint for the Plutarco Elías Calles regime in 1929 (figure 3). Although Morales didn't have a ready explanation for it, and simply said that he "got it from the California Lottery," the significance of the mural's placement as the first thing one saw when entering the lab revealed the importance of this history for their understanding of the DNA patterns that emerged in its presence. The reproduction of the mural

Figure 3. The Great City of Tenochtitlán, Diego Rivera, 1945, at the National Palace in Mexico City. Photo courtesy of El Comandante, Wikimedia Commons.

evoked both the long past and a faith in a future where the scientists and those they were studying would be represented anew. The mural depicts Tenochtitlán, the ancient Aztec capital, now Mexico City. The Aztecs were sophisticated mathematicians. For the lab scientists, these artworks signified a distant source of modern Mexican origins that were now being rolled into the broader term "pre-Columbian Native American" for the DNA marker panels that they and their colleagues at other sites used during this early period as the human genome was first mapped.

In a conversation with Rick Kittles, one of Morales's collaborators at the time, I asked why, as "minority scientists," they framed their analyses of ancestry and risk with racialized genetic categories, rather than attempting to isolate social conditions that lead to asthma, cancer, and the many other ailments that continue to constitute racial health disparities. Before assuring me that he does incorporate social aspects, Kittles first corrected me by saying: "I don't use the word 'minority.' I use African American. I don't want to give up my identity to simply be seen as a 'minority.'" I asked what he made of the fact that his colleague used the term. Kittles replied: "Gabriel's nation building." This phrase was meant to underscore the idea that, as Morales always made sure to say in his many public talks, Latinos were the fastest growing group in the United States by census category and that research on health disparities that disproportionately affect them was grossly underdeveloped. Morales's genomic research "empire" was arming to address both realities. The murals, the explicit "peopling" of Morales's lab, and the reference to science as "nation building" were some of the first indications to me of what these researchers held dear. How these signs translated into genetic studies would prove to be a more complicated matter.

Morales and his team were excited to use the tool of Ancestry Informative Markers when I observed their lab in 2003. The actual technology was borrowed from a white American collaborator on the East Coast who shared with them his set of markers. In order to explain how this conceptually powerful tool works, I ask the reader to accompany me on a technical journey for the next few pages to see inside the AIMs machinery. Despite its scientific detail, this section of the chapter should in no way be taken for a mere data dump. My hope is that in flagging this turn in the text for the everyday nonspecialist, readers will grant themselves the time to review parts of it or to go slower when needed. I argue that AIMs are an essential pillar of the built world of scientists who pursue questions that link race, ancestry, and genetics. By laying out geneticists' narrative logics, their thought traditions, their cultural mores and norms of data making (and shaping), we can begin to see how racial thinking pervades the life world of the laboratory and influences the products that are created and exchanged between scientists at different sites within the US and abroad.

At the beginning of the AIMs construction process, researchers made sure that each selected marker had a key characteristic: it needed to contain a single nucleotide (DNA base pair) change that appeared at a higher level in at least one—but not all—of the three groups under study (Africans, Europeans, and pre-Columbian Native Americans).[1] This focus on differential frequencies relied on a calculus of what the researchers called a notable "delta value." "Delta" in this context simply means a subtracted difference, or the change in value that occurs between two numbers (e.g., the delta of 65 minus 30 is 35). In this case its value is the frequency of an allele that appears above a certain threshold in one designated group compared to the frequency of that same allele, below a certain threshold, in at least one of the other two groups. When compiling markers that might be meaningful for their models to infer ancestry, scientists in this field work around the fact that people across the globe—that is, at least some members of a population group so named—will possess even these carefully culled markers to some degree. The goal scientists have in making ancestry panels is to compile and document genetic differences at a population level, where the populations in question—from Africa, Europe, and pre-Columbian Native America—are demarcated by continental borders (see table 2).

Only when this selective world map was pieced together did the scientists I studied use that frame to calculate what the "delta" or "difference" in allelic frequencies for each group (compared against the others) might be. These values were always rendered in precise terms, usually expressed as numbers down to a decimal point to portray exceedingly exact percentages of ancestry. Integers—those lesser expressions of excessive precision—were rarely seen.

What is clear is that ancestry as a statistically built fact, tied to the logic and naturalization of the map, might also be understood as an art: a wittingly whittled talisman whose parts gesture to a fetishized construct of human genetic difference in a highly artificial world. In reality, rare is the variant that is completely population specific. That is because rare is the population that has been in complete isolation for millennia—a phenomenon that might allow people to form wholly new polymorphisms only shared among themselves.

A brief word is in order on the issue of rare variants. A common conceptual conflation that affects how many researchers, journalists, and people of the lay public think about racialized difference has to do with talk around "population specific" markers. These are sometimes called risk variants if disease is involved. I should state at the outset that genetic mutations that appear in some groups and not others certainly do exist. If they didn't, humans would be extinct. The issue turns on how these genetic variants are interpreted, which is that oftentimes they are generalized to a *whole group*, or population. If one glides through these nuances, the concept of population genetic *frequency differences* for specific variants can become conflated with *race*.

I would like to unhitch these mutations, many of which have illuminating stories, from the racializing hook. It might be helpful to think of them as having their own biographies where they are protagonists, of course

Table 2 "Parental Population" Samples

Continent	Sample Size
Africa	
Nigeria (Benin City)	100
Nigeria (Sokato City)	46
Nigeria (Yoruba)	100
Nigeria (Hausa)	120
Nigeria (Kanuri)	100
Nigeria (Bini)	100
Liberia (Kru)	80
Ghana (Akan)	100
Central African Republic, Bantu	49
Sierra Leone (Temne)	98
Sierra Leone (Mende)	181
Camaroon [*sic*]	150
Europe	
Spain (Valencia region)	90
Spain (Basque region)	100
Irish (eastern coast)	90
England (London)	48
Germany	80
Hungary	50
Lithuania	50
North and South America	
Dogrib (Northwest Terr.)	70
Navaho (Southwest US)	37
Pima (Southwest US)	35
Keres (Southwest US)	24
Tiwa (Southwest US)	28
Cheyenne (Southwest US)	33
Central America (7 groups)	300
Bolivia (Aymara)	70
Peru (Ketchua)	75
Surui (Brazil)	23
Pehuenche (Chilean plateau)	120

Asia	
Southern Chinese (Taiwan)	300
Northern Chinese	200
Southern Chinese (Han and Minority Populations)	300
Japanese	600
Insular Southeast Asia	600

SOURCE: This table is reproduced from a photocopy of a document given to the author by technicians in the Morales laboratory. I have kept the wording of the original document. These "parental population" samples were used for the AIM technology that Morales's early collaborator initially shared with him.

joined to our stories of who we are and have been over the long durée. The now all-too-common slippage occurs, however, when specific DNA changes experienced by some people in a given "population," which can usually be traced to a fraction of that populace living in a designated geographic zone, get extrapolated to a ballooned notion of any representative hailing from said region or their descendants. This is especially true as concerns variants that may contribute to health disparities in the United States.

One important case concerns variants in the gene for apolipoprotein L1 (*APOL1*). These code for a lipid in the LDL family that can cause severe cellular kidney damage leading to blood and protein leakages that impair kidney function. The risk variants, called *G1* and *G2*, are only found in people with recent African ancestry—and in 13 percent of African Americans. These genotypes, in combination with other environmental factors, contribute to one of the many health disparities seen in the United States where African Americans suffer from end-stage renal disease more than any other group.[2]

APOL1 has a recessive pattern, which means that *G1* and *G2* need to be inherited together in some configuration together to confer risk (i.e., *G1/G1*, *G2/G2*, or *G1/G2*). A third nonrisk type has been named *G0*, which has a dominant effect so that any combination with *G0* (*G0/G1*, *G0/G2*, or *G0/G0*) does not make people sick. A critical detail here is that 13 percent of African Americans carry a risk combination—but, it has been shown, only a subset of those will go onto develop kidney disease.[3] This means that 87 percent of African Americans do not carry the risk genotypes. Moreover, a portion of people even with the high-risk muta-tions can also be grouped on some level with the earlier 87 percent in that they will not develop *APOL1*-linked kidney disease.

This is where we need to slow walk, rather than mentally run, to avoid the slips and slippages of tongue and mind that enter into talk about these highly specific risk variants in generalized racial terms. Even "population specific" might be too broad of a term, since not all of the Black or African-descended "population" carries these risks. The other piece of the story is that these variants are not solely disease risks. Like other genetic changes that human genes, cells, and blood have made in the face of parasitic and ecological threats (think sickle cell anemia and beta-thalassemia hetero-zygous states that confer protection against malaria), the *APOL1* "risk" variants are also advantageous immunologically in different environmen-tal and cultural contexts beyond the United States.

In their 2010 *Science* article on the *G1* and *G2* genotypes' role in kid-ney disease, a team led by Giulio Genovese, David Friedman, and Martin Pollak at Beth Israel Deaconess Medical Center and Harvard Medical School determined that the variants, especially *G2*, were also highly effec-tive in preventing illness and death from the parasite *Trypanosoma brucei rhodesiense*. Transmitted by the tsetse fly throughout Africa, *T. brucei rhodesiense* causes trypanosomiasis—more colloquially known as sleeping sickness. Even when plasmas were highly diluted, those containing the *G2* mutation cut through the cellular membrane of *T. brucei rhodesiense* and destroyed the trypanosome organelles.[4] Interestingly, *APOL1* does not protect against the much more common *T. brucei gambiense* form found in West Africa, which is a bit of a mystery perhaps indicating that the West African parasite population numbers exploded when humans acquired such an efficient genetic protection against the East African species.[5]

Although the Harvard team and others who have worked on *APOL1* for over a decade now advocate using their genotyping discoveries to better assess the clinical manifestations of renal disease in Black American patients, they emphasize that the genotypes in question, *in and of them-selves*, are not sufficient to cause chronic kidney disease (CKD).[6] Again, some people with the high-risk variants *G1* and *G2* do not get the spectrum of kidney disorders brought on by *APOL1*.[7] It turns out that viruses, like HIV and parvovirus B19, seem to be necessary ingredients for many severe forms of CKD in African-descended people.[8] Other risk factors that play into health disparities in kidney disease include diabetes, hypertension, heart disease, and obesity. There are patterns noted in what is called *APOL1* aggravated hypertension-attributed end-stage kidney disease (H-ESKD) in that people affected also have hypertension, thus the name. Another serious form of renal disease, known as focal segmental glomerulosclerosis (FSGS), may be linked to one of the leading hypotheses about how *APOL1* *G1/G2* combos result in a phenotype that does severe damage.

According to a 2021 review by Friedman and Pollak, although the exact mechanism is not known, "a leading hypothesis is that APOL1 risk variants may create pores in kidney cell membranes in much the same way as APOL1 punches holes in trypanosomal organelles."[9] Still, HIV infection in those who carry two risk alleles remains the most significant trigger.[10] Today medications are being developed to block *APOL1* from doing its damage to kidney cells since nonfunctional variants are still compatible with life.[11] On the flip side of the focus on disease in the United States, one of the concluding insights from the Pollak team's original paper was the suggestion that the *APOL1* genetic variants' protective mechanisms might actually inform therapeutic designs for sleeping sickness, which still affects people in many regions in Africa.[12]

Another intriguing story of human adaptation and highly specific allelic change concerns Indigenous people living in Chile's Atacama Desert who, over seven thousand years, have gained the ability to ingest exceedingly high levels of arsenic in water—levels that are one hundred times that deemed safe by the World Health Organization. Arsenic is a metallic poison spread through volcanic activity that contaminated the river waters in this most arid area of the world. In the mountain pass and river basin within the Atacama called the Quebrada Camarones, mummified human remains from the Chinchorro culture (ca. 7000 BP) as well as from an Inca population that settled Quebrada Camarones millennia later (600 BP) showed significant signs of arsenic poisoning with extremely elevated measures of arsenic in bone and hair tissue.[13] Overtime, these levels decreased in archeological finds, which cued a team of geneticists and physical anthropologists led by Chilean geneticist Mario Apata, writing in 2017, to propose: "Such differences in the levels of arsenic could be explained by the Camarones population's adaptation to this toxic environment with an increase in the efficiency of their metabolic detoxification processing of arsenic."[14]

The *AS3MT*, or arsenic (+3 oxidation state) methyltransferase, gene contains a series of mutations (grouped into haplotypes) that are associated with a lower risk of developing cancers when people are exposed to organic arsenic. Through epigenetic and genetically induced methylation, the tolerance to this toxin that poisons most humans allows certain groups in this region to better metabolize it. The Chilean team found that two groups of people, those from Quebrada Camarones and those from the Azapa valley, had haplotypes that contained protective alleles at frequencies of 68 percent and 48 percent, respectively, allowing them to efficiently break down arsenic and excrete it without incurring cancers. Another group from further south, in Huilliche, possessed the same variants at only 8 percent.[15] When researchers sampled people who lived at greater distances from the Camarones, the less frequent the protective alleles for arsenic in their DNA became. In this Andean case there were several haplotype combinations associated with adaptive tolerance to arsenic, which again gradually declined by region the further one lived from the Atacama Desert. In this example we see the plasticity of the human body and the specificity of biological change that can in no way be extrapolated to all Indigenous people of the Atacama region, to all Chileans, Andeans, or to the Latin American population as a whole.

These are a just a few examples that illustrate the importance of interrogating labels of human difference, and the necessity of mapping the pedigree of genetic variants' social, cultural, and geographical lives—that is, to treat them as protagonists in the story of human genetic diversity in order to avoid reductionist thinking.

If the scientists who were constructing the genetic ancestry panels for AIMs had entertained sampling techniques that follow what we know to be clinal global gradations of allelic frequency (which simply means that one can observe that genetic traits gradually change as one moves across the terrain of the earth), then the current post-genome Mundus Novus would look very different. A more gradual, globally comprehensive, and meticulous sampling approach, known as "grid sampling," never gained traction due to the difficulty, as many scientists pointed out, that such a global engagement might pose logistically.[16] Instead, a few specific sites of "reference," for geographies and populations of interest, were chosen for many projects that would feed into each other conceptually and at times materially, reproducing a certain reductivism across datasets. These included the Human Genome Diversity Project, the International Haplotype Map Initiative, and the Human Genome Project (HGP) itself since both the NIH as well as the private venture of Celera Genomics wanted "diverse" populations.[17]

Even though humans share most of their genetic make-up, as mentioned in the introduction, those at the helm of both the publicly funded HGP, as well as the private venture, launched by J. Craig Venter, were said to sequence the DNA of a few select people from "diverse" US racial groups. Venter spoke openly about his company's emphasis on diversity for the genome map, offering details on the individuals sampled, which the NIH was not willing to do. As Venter told journalist James Shreeve in his book *Genome War*: "It would be fundamentally wrong to end up with five white men."[18] Therefore, Venter's team were supposed to examine the genomes of three women and two men who self-identified as "African American," "Asian," "Caucasian," and "Hispanic." (Later it came to light that Venter mostly sequenced himself.) Although this imperfect scan of human difference performs a gesture of nodding to the notion of the globe, and correlates to the logic of geographic origins that the AIMs researchers in this book took for granted as key points of planetary human diversity, it is in fact a highly specific American vision. This local notion of what human genetic difference might mean was another register on which the public was invited to see a relationship between genes and race, even if both the private and public arms of the Human Genome Project announced at their joint unveiling of the first draft map that such a relationship did not exist.

GENESCAPES WITHIN THE LABSCAPE

In panels where racialized continental separations structured the categories of *who* possessed *which* alleles, and when samples that were meant to represent general continental groups shared a common allele for any given ancestry informative marker (but the degree to which they expressed it varied in frequency when the groups were compared and also compared crosswise), then those were instances when a DNA marker was in business. That any evidence of sharing markers was read as what is called "gene flow" (the inheritance of a trait through relatedness) rather than the many other explanations of why humans might display the prevalence of a specific genetic variant (genetic drift, genetic convergence) was often not fully entertained. In other words, allelic frequency differences for AIMs are teased out as much as possible for this science through a layered process that practitioners in the field call "selection and validation," which some have described in detail.[19] In this process they do their best, as they say, to "exclude" markers that people across their chosen populations all share at high levels. In essence, this is an effort to construct "homogeneous" groups. As part of this process, their methods exclude markers that show too much "heterogeneity" within any one of their designated populations.

In a key paper where researchers from the United States in collaboration with scientists in Mexico, Spain, and various South American countries produced an AIMs panel for public use, they offered a graphic of what they called their "algorithm for selecting AIMs."[20] At the top, or at the starting point of the visual algorithmic path, the viewer is presented with three distinct boxes. The first reads "European populations," the second reads "African Populations," and the third, "Native American populations." Next the algorithm directs the reader through the research team's actual marker building process. This happens via black indication arrows that jut downward from each population box, pointing to the next step in the process. Here the team instructs the imagined user to "calculate allele frequencies." From there the formula proceeds to "calculate pairwise statistics," which is the difference, or delta value, in allelic frequencies when two populations are compared.

At the bottom of the viewer's algorithmic journey, the final box reads: "Check for Exclusion Criteria." The ultimate item on the list warranting exclusion for an ancestry informative marker is the ever-present notion of "heterogeneity." In other words, the algorithm asks researchers to sift out that key population characteristic that Richard Lewontin established for the public more than fifty years ago—that there is more genetic variation within groups than between them. How, the algorithm asks, could geneticists working with AIMs override this pesky problem?

Table 3   Genetic Markers and Frequency Differentials by Population

SNP rsID	chr	Position	A1	A2	NAM_AF	EUR_AF	AFR_AF	Population
rs12085319	1	10952065	G	T	0.347	0.817	0.015	EUR
rs41009	2	8012609	A	G	0.011	0.154	0.879	AFR
rs4109078	3	3215356	T	C	0.111	0.858	0.903	NAM

SOURCE: Joshua M. Galanter, Juan Carlos Fernandez-Lopez, Christopher R. Gignoux, et al., "Development of a Panel of Genome-Wide Ancestry Informative Markers to Study Admixture throughout the Americas," *PLOS Genetics* 8, no. 3 (2012): supplement, https://doi.org/10.1371/journal.pgen.1002554.

NOTE: This table presents a few lines of data reconstructed from this group's AIM panel. On view are comparisons of continental populations of interest to the researchers, along with the frequency differentials of SNPs for this instance of the tabula raza.

This conceptual and literal mapping of scientists' algorithmic New World expedition is not hidden away, or black boxed, in the scientific literature. It is colorfully diagrammed in plain sight in some publications and meticulously described in others. The point is that geneticists consciously and bio-logistically construct these continentally-based markers to explicitly work with a model that maximizes—focuses in on and renders larger than their lived realities—the genomic points of difference among so-called ancestral groups.

Table 3 presents excerpted lines of data from the AIMs panel for the same paper. The first column contains three different markers indicated by their "rs" numbers (reference SNP cluster IDs). These are followed by the chromosome where they are located, their position number, then their allelic change. The delta values for these are listed under the columns that contrast NAM (Native American) to AFR (African); EUR (European) to AFR (African); and strangely—due to a typographical error in the chart—AFR (African) to African. The last column should convey a comparison between European and Native American, since according to the text of the article, the "markers were calculated between each pair of ancestral populations (African/European, European/Native American, and African/Native American) based on reference allele frequencies."[21] In going through the published marker panel, I excerpted only the first SNPs said to indicate ancestry in each of the continental groups for chromosomes 1, 2, and 3 from a dense grid of 446 markers in this particular tabula.

This information comes from the supplemental table offered on the part of the international collaboration for their intended public resource. As far as dividing up the world into "Old world" and "New" goes, a key table in the paper lists the "ancestral populations" used for the study. Those meant to represent Old World Europe are an interesting amalgam. Further in this chapter I take the reader through some of the pre-1491 history of the Iberian Peninsula. There it may become clearer why I was left scratching my head when I saw that this paper listed the European contribution to Latino "New world" ancestors as 619 "Spaniards from Spain," 44 "Toscani in Italy," as well as "56 Utah residents with ancestry from northern and western Europe," a reference sample that I often heard scientists refer to as "Utah whites."[22]

In this built world of genomic points of difference, people/samples were read through these DNA base pair changes in select parts of the genome that were optimized to display variance by race—recast as a highly abstract, almost spectral, form of statistical evidence that projects possible predecessors. This phenomenon of recreating progenitors as a statistically collective concept is now known as "population genetic ancestry." It is helped along through an artful performance of visualizing differences in humans. This scientific effort was not about burrowing down far enough to unearth a natural fount of a biological reality that geneticists might excavate. Instead, what we are invited to see in this process concerns how the "real" DNA—that is, the physical world referent of genetic material—is culturally and carefully chosen to enact a relationship of racial differentiation. The end point is one and the same as the beginning. It is a circuitous path where a cultural vision of who should count as different from the other influences the products (from ancestry tests to forensic analyses) that give us algorithmic results of continental human difference in stark racialized terms. This meticulous selection of traits creates the sense that people with these ancestries contain quantitative genetic differences that are racially consistent in "nature" instead of differences that are assembled through human scientific design. In this way the social construction of race has gotten more scientifically complicated, even if its a priori bases have been remarkably similar for centuries.

MUNDUS MODELS, COLORS, AND PAINT

Morales hoped to use an early version of this scaffolding of genomic structure to decipher the genetic drivers of asthma health disparities. At issue was his observation that asthma affects Mexicans and Puerto Ricans in different ways. His lab worked from the premise that Puerto Ricans and Mexicans were both "Hispanic" groups, but that their base ancestry contributions, on average, might have something to do with the fact that Puerto Ricans had more severe asthma, more scarified bronchial tubes, and responded less to steroids in bronchodilators that are found in pharmaceutical drugs like albuterol. The working hypothesis of the lab was that Puerto Ricans had more severe asthma because of higher percentages of their underlying African genetic ancestry, whereas Mexicans, it was thought, had very little African ancestry and instead had sizable "pre-Columbian Native American ancestry." Both groups of course possessed "European" ancestry.

Morales was not shy about his theory. He told it to me, to his public audiences, to his mentees, and to his broader team and collaborators. The theory could also be understood as a simple story: that African Americans had more severe asthma than any other group in the United States, which might lead one to believe that African genetic ancestry was driving severe asthma in both African Americans and Puerto Ricans at the genetic level.[23] The complexity in this thinking was that the racial/ethnic block identity of "Hispanic" would have to be deconstructed into its component parts. The simplicity of it was that such component parts were decipherable in black, white, and red DNA.

When I sought out Morales's collaborator who offered the group his panel of AIMs for their studies, it became clear that there was both a commitment to thinking about ancestry in hard factual terms and a simultaneous noncommittal stance vis à vis the historical realism and genetic naturalness of the AIMs technology. This was when the collaborator readily admitted that he optimized the AIMs markers as a model to detect ancestry in people in the "New World." And that this choice was contingent upon specific assumptions about population composition that worked as a closed, "restricted," system. In other words, many postulations and beliefs had to be held constant to imagine the world in this way. One of the most obvious elements of this reasoning had to do with the way he responded to my queries about purity and racialized percentages of ancestry that relied on holding people in the Old World static—that is, somewhat fossilized, and pure. Conversely, those in the New World were seen to be mobile and busily mixing.

It was during this conversation that he explained that the AIMs model was built as a conceptual replica of the world as of a specific point in time: that being one pivotal historical marker of European conquest, or the year 1492: "So, we know that people who were separated before 1492 started to come together after then. So, if we want to just take a very distinct time and space—and limit on that—we can now measure in many instances where your ancestors came from before that point. So, how many of your ancestors were from West Africa, how many were from Europe, East Asia? [. . .] These are distinct geographic regions that had been separated. Now this process [of separation] was reversed by technological advancements, people learning about the New World in this case, people learning how to travel across the oceans. . . ."

Back in San Francisco, Morales explained to me the other side of that limit, which was important to him. He homed in on 1492 as a key point, but he sped up to a sharp focus on the present. Impatience seemed to be part of his work mode and in this discussion I found the source of his hurried nature: that researchers like him had to act "before it was too late." Here, tempura pigments came to stand in symbolically as distinct DNA patterns in ancestral groups that he feared were fast disappearing due to New World mixing. "It's like freshly mixed paint," he told me. "You can see the different colors for a while, until they are too mixed up to see any more. We think that the lines will be clear for ten generations. Ten generations, that's recently admixed. After that it's too mixed up to see anything." The "paint" was meant to refer to the chromosomal chunks that this lab called European, pre-Columbian Native American, and African. Those were the increasingly vanishing lines. Ten generations translate to roughly five hundred years by Morales and other scientists who use AIMs ancestry inference estimates. When he referred to this block of time, he too was evoking 1492.

I, like so many social scientists, wondered if the people using AIMs took them as seriously as some of their publications made it seem they did, and also as serious as some of the media coverage of their analyses. These ranged from BBC documentaries to PBS series to *Frontline* specials to the *New York Times* science pages, which parlayed genetic ancestry precision in racial terms to wider publics—statements that mixed math and race in ways that readers will be familiar with by now—where someone might be 24 percent African, 36 percent European, and 40 percent Native

American. In a conscious portrayal of the way Americans often volley between genes and race, a 2009 PBS show called *Faces of America* featured comedian Stephen Colbert who submitted his DNA for an AIMs ancestry analysis. His result was presented in jocular fashion that he was "100 percent White Man."[24] Ironically, years later in 2017 and 2022, Colbert would have several segments on his own show that criticized the same ancestry genetic tests and the ways that people habitually conflate genes with race. His focus was on white men, white supremacists to be more precise, some of whom believed genetic ancestry algorithms could validate their sense of racial superiority based on false notions of purity.[25]

Years before Colbert's public vulgarization, I asked Morales's collaborator about some of the specific claims, such that a person could actually be "100 percent" genetically a social category, like "European," and the self-referential nature of the results that can only reflect predetermined ancestries—that is, what is in someone's database. How open, or closed, was this facsimile world that they were portraying? In a conversation about the model and its assumptions, he talked in full paragraphs. I remember slowly raising a single finger as a signal of my confusion and the need to ask the following burning question: "Okay, if someone is '100 percent European,' it means that they have tested in a certain way for the alleles that you have that indicate European ancestry. So, it's based on *those* markers. So, if there were other markers, that maybe people shared between populations, then, it would be harder to discern [European] ancestry?"

The collaborator didn't miss a beat in his response. He was quick to make it clear that this new world—in his hands—was just a model. And a highly cultural one at that. "Oh yeah. Yeah," he said. "These are markers that were—I mean, these markers don't tell us anything outside of the model that we're assuming. . . . Right now, it's pretty much an *American* technology—in a way." He later followed up with: "I guess a *New World* technology might be the easiest way to think about it, and on a restricted level too."

At that time, I hadn't yet realized the great lengths that these scientists had to go to exclude markers in the aforementioned "selection" and "validation" processes. I did witness conversations between Morales and other collaborators about exporting this American model to other parts of the "Hispanic" world. There too researchers would work to "optimize," "tease out," "highlight," and "structure" studies with the most extreme markers to avoid as little overlap in the predefined population groups as possible. In short, these methods are geared to excavate the tenth of a percent of the human genome that is not shared among the human race. When tabulated in comparative rows and cells, such quantities, in the parlance of population genetics, are the frequency differences of allelic changes. These are what drive most thinking, discourse, and practice, that constitute what can be called "racialized genetics" today.

If the reader takes nothing else from this book, my hope is that it is clear by now that the phenomenon of mere frequency differences in genetic variation that most groups share is how autosomal genetic ancestry tools separate any one group from another. When we realize that we are dealing with frequency differences, it is easier to cut through what most scientists themselves know to be true: that genetic absolutes rarely apply to continentally defined human groups. That is, unless artificially derived absolutes are explicitly made to stand out by design through selection, validation, and exclusion criteria that create a perception of genetic homogeneity—despite the heterogeneity—in human populations who are cast as similar.

Albeit rare, there were times within the first decade of the genome map when geneticists weighed in on the methods and algorithms used in these kinds of studies. For instance, genetic anthropologists Kenneth Weiss and Jeffrey Long took a meditative overview of estimating ancestors from these aspects of genetic variants and wrote: "This kind of analysis seems to make intuitive sense. But what is it? Whether the parental populations are externally user-defined or internally statistically defined, we cannot distinguish the analysis from a search for ideal types. Such analysis may use modern genetic data, but the statistical output is not conceptually different from classical racial analysis based on morphology."[26] Herein lies the frame and cultural portrait of racialized genome biology as it has been practiced in the United States for the past twenty years, irrespective of how sophisticated some methods have become.

A WRINKLE IN AL-ANDALUSIAN TIME

The complexity of how both geography and time have borne out human variation has been drastically simplified in what sociologist of science

Karin Knorr Cetina has termed the "epistemic machinery" that makes the AIMs technology increasingly appealing to the wide range of lay, scientific, and law enforcement clients who are now using it.[27] While the girth of the globe has been flattened to a small area of West Africa, sporadic points in North and South America, and even sparser points of Europe, researchers within this field have incorporated "time" into the model as that pivotal year in world history when Columbus arrived on the Caribbean island that the native Lucayan people called Guanahani (which Columbus renamed San Salvador).[28] Yet one key publication showed that scientists working with AIMs acknowledged both the complexity of mixing and ethnic pluralism in fifteenth-century Spain. They wrote that the population consisted of "Celts, Greeks, Romans, Sephardic Jews, Arabs, Gypsies, and other groups," yet they nonetheless still strangely amalgamated such a range of diverse backgrounds under a unified "European" label.[29]

The acknowledgment of Old World ethnic, religious, and genetic diversity in the paper is laudable. But it's still a bit bewildering that such a description of Spain, an area with a well-documented complex and dynamic history, especially leading up to 1492, would be blithely brushed with the veneer of a mono-European label. Many readers will have heard of "Moorish Spain," that long period starting with the invasion of a largely Berber army under Arab-Muslim leadership in 711, the various successions of Muslim rulers centered in the key medieval cities of Córdoba, Grenada, Seville, and Toledo, the eventual fall of the last Islamic city-state of Grenada in 1492, followed by forced conversions and finally the expulsion of all identifiable Muslims between 1609 and 1614.[30]

Known in Arabic as *al-Andalus* (*Andalusia* in Spanish), the territory was home to a population that was itself diverse both ethnically and religiously, with only some being termed "Arabs" from the Middle East, of the Umayyad lineage, while many were Indigenous Berbers from Morocco and Tunisia who at times resisted Arab subjugation and who were not always Islamized to the same degree as other converts between Damascus and North Africa. Yet in the early period of al-Andalus, Arab generals negotiated with various Berber tribes, and in essence hired them for their skill as warriors while offering them a share in the conquest of the Iberian Peninsula in the mid-eighth century. A quantitative look at genealogical records shows those already living in the region began to adopt Muslim names in a decades-long massive conversion of the local population from Christianity to Islam—while the people of different origins and beliefs also mixed and reproduced. In the year 800 a mere 10 percent of those in al-Andalus professed Islam but by the new millennium 80 percent claimed the faith.[31] In the beginning of the twelfth century, after a period of Muslim rule and intercultural flourishing that made both Christians and Muslims refer to al-Andalus as the "the ornament of the world," highly conservative Christian leaders formed armies to slowly overtake the Islamic *taifas* (medieval kingdoms).[32] The remaining Muslim rulers appealed to the deeply conservative dynastic Berber Almoravids from Mauritania and Morocco to help them stave off Christian troops from seizing more of the south. This set off a series of Almoravid incursions from 1086 through 1147 and led to another three-and-a-half centuries of official Muslim rule in Grenada.

This long durée where a plurality of Muslims, from different geographical regions, maintained power, and where groups often battled violently for territory, was one where the ruling emirs often permitted Christians (known in the early period as Visigoths) and Jews as "people of the book" to continue with their religious faiths if they did not resist Muslim authority (and paid a larger share of taxes). Still there were feuding factions among Muslims themselves (the fall of the Córdoba caliphate came at the hands of more conservative Muslim Berbers in 1009) as well as between the northern Christian kingdoms. Then there were other invaders, like the Vikings, who also captured, enslaved, and traded people who were "Irish and Anglo-Saxon" as well as "Frankish and Slavonic" throughout different periods within this near eight-hundred-year Muslim-Arab presence in the Iberian Peninsula.[33]

In addition, by the end of the 1400s there was a significant Black African presence in the peninsula that would continue to grow over the next two hundred years.[34] The enslaved, largely made up of women but also children, were brought from the sub-Saharan regions of Senegambia and the Gulf of Guinea. They were human commodities that were bought and sold but often rented and sometimes freed along with those from the Black Sea region, southern Europe, and the larger Mediterranean.[35] Following the first sale of African slaves to the Portuguese, Seville was

Europe's most important trade center for captives after Lisbon. By 1492, some thirty-five thousand people from West Africa had been sold into slavery in the Iberian Peninsula.[36] As historian Leo Garofalo writes, "as many one in four or one in five people in the southern Iberian port cities such as Cádiz were of African descent by the end of the 1600s."[37]

All of this is to say, what scientists call "admixture" was surely already very much alive and well in Spain before the highly conservative Reconquista effort to force a Christian hegemony in the late fifteenth century. Any notion of a purely "European" Hispania should make us recall processes of homogeneity in-the-making—yet this instance of such a tendency of course took place centuries before geneticists constructed AIMs algorithms to sieve out heterogeneity on a molecular scale in more seemingly banal ways. In al-Andalus peoples and cultures (religions and ethnicities) comingled throughout an extensive and vibrant period of what was known as "la Convivencia" where Christians, Muslims, and Jews lived among each other, exchanged, shared, and created poetry, music, philosophy, and sciences in what was a long-lasting syncretic Andalusian culture.

Recorded examples of mixed lineage that occurred through the dynamics of wars, hostage taking, and concubinage highlight the physical traits of prominent elites—those whose lives were documented. One such emir was 'Abd al-Rahman III, who ruled during a critical fifty-year period (912–961) when "the Spanish-Islamic State reached the peak of its power and renown." 'Abd al-Rahman III's eminence followed a distinctive lineage that was in no way consistently (purely) of one people or another. Historian Richard Fletcher writes:

> 'Abd al-Rahman III's father Muhammad was born of the union between the amir 'Abd Allah and the Christian princess Onneca or Iñiga, the daughter of a king of Navarre [Basque peoples] who had been sent to Córdoba as a hostage in the 860s. 'Abd al-Rahman himself was the child of a union between his father Muhammad and a slave-concubine, a Christian captive possibly from the same Pyrenean region, named Muzna. . . . In his immediate ancestry, therefore, the new ruler was three-quarters Spanish, or perhaps, more accurately Hispano-Basque, and only one quarter Arab. He had blue eyes, a light skin, and a reddish hair. We are told that he used to dye his hair black to make himself look more like an Arab.[38]

Other scholars have pointed to many such instances of people of different ethnicities and lineages mixing that made certain emirs stand out because of their physical traits—again, usually focused on blue eyes, blond or red hair, throughout this period.[39]

That said, in medieval Spain religious tolerance, rather than pluralism, often proved the norm, while many instances of warring violence and internal strife characterized everyday life.[40] With the conquest of Grenada, the last Islamic *taifa*, in 1492 Queen Isabella of Castille and Ferdinand II of Aragon intensified the long effort to make the whole territory Catholic with the expulsion, torture, and killing of Muslims, Jews, and also Black Africans who were prosecuted for practicing "witchcraft" and "magic" as part the horrors of the Inquisition, which started a decade and a half earlier.[41] It was during these same years that Christopher Columbus began soliciting the royals to subsidize his voyage to what would be called "The New World"—a request that was at first rebuffed because the monarchy was financially strained by fighting to bring down Grenada.

Narratives of history as having a single arc where a people and their ancestries are portrayed as relatively consistent, despite their "mixing," which can be seen as messy or even contaminating, have definitely marked the telling of medieval Andalusia. Historians make a powerful case that in the late twelfth and early thirteenth century the influential archbishop of Toledo, Rodrigo Ximénez de Rada, who was a chronicler and orchestrator of Spain's formal history, constructed a centuries-old Spanish Christian identity to make the Reconquista appear as old as al-Andalus itself.[42] De Rada's telling is a classic revindication-creation narrative that championed a single small enclave of Christians in the northern mountains of Asturias led by a king named Pelayo who resisted Muslim domination in the eighth century. De Rada took their existence and reimagined them as a resilient force that spearheaded the survival of the "true Spain." In so doing, he reframed the Reconquista as a persistent effort whereby Pelayo served as a point of origination that many Spanish Catholics have incorporated into their national narrative to this day.[43] Still, national creation narratives like this are instructive and give us some understanding of the impulses of those who were more zealous in their view that their religion should dominate (De Rada was also a Crusader). This impulse was shared among more puritanical Christians, as well as powerful and often despotic Muslims, who were frustrated at losing influential footing.

The anxieties and aggression that led to the "reconquest" are too detailed to recount here. One of the more striking elements, however, lay in some people's religious conviction that the earth was approaching its "messianic age," and that the seven-thousand-year timeline for humanity's stint on earth, from the date of creation, was predicted to end with the apocalyptic second coming, which various people—from Isadore of Seville to the Jewish scholar Isaac Abrabanel to Christopher Columbus—calculated as sometime between 1503 to the mid-1600s.[44] In addition, the scourge of the bubonic plague—where a third to two-thirds of Europe was decimated—also presaged details in the book of Revelations. These signs—the mass forced conversions of Jews, Muslims, and "pagans" to Catholicism, the Inquisition, as well as Columbus's landing in the Americas, where even more people could be converted—were all seen to be part of a great unfolding drama. Here Columbus saw himself an active participant in catalyzing the end of human history prophesied in the second coming.[45]

Even in death, the Spanish royals and Christopher Columbus retained their stance as Crusaders for Spain. As historian Christian Lowney writes: "Despite the Catholic Monarchs' many accomplishments over their long reign, the epitaph chiseled onto their sepulcher pointedly memorializes just two: 'Destroyers of the Mohammedan sect and the annihilators of the heretical obstinacy [i.e., of the Jews]."[46] Similarly, Columbus made sure to express his conviction for the eradication of Muslims into many of his surviving journals and letters whereby he requested that after his death a portion of his wealth be used for Jerusalem's liberation from Islamic rule.[47] Four decades after Columbus's passing, when more conquistadores were sailing to South America, the Christian Spaniards' Old World battle cry "Santiago y cierra España!" (a call to Saint James to close ranks for Spain!) was deployed by Francisco Pizarro's men against the Incas in Peru.[48] This marked an apocalypse, an ending, for many Indigenous people that was indeed put into motion by Spain in the New World. From the Westernized retelling of that history from today's vantage, however, it is framed as all-important beginning.

Columbus, his financiers, and the many others who would follow him emerged from a world steeped in a religious philosophy and tendency to exterminate en masse despite its recent intercultural past. The Inquisition ushered in an obsession with what was termed *limpieza de sangre*, an exacting surveillance not only of inner spirituality and a commitment to the Catholic faith but also an essentialized fixation on Christian identity and even "Christian ancestry."[49] In the New World this religious preoccupation concerning purity of belief melded with purity of blood in terms of the type of people one belonged to: Indigenous, African, Spanish, and eventually the various mixes that marked peoples' statuses. In this overlay, *limpieza de sangre* became the psychological, social, legal, and bureaucratic fiber of the highly racialized *casta* system.[50] Thus, starting in Spain with the razing of Muslim sites and the expulsion and denial of Muslim and Jewish populations, we unearth yet another example of the tabula raza. And with this element of it we see that there is no beginning to the "New World's" history in the year 1492. The racialized erasure of lands, peoples, and cultures were merely extensions of prior obsessions within an anxiously created hegemonic "Europe," now fueled by new bloodshed and Christianizing vigor.

So now we must ask, how do "Spain" and the "New World" as geographic references and conceptual entities for ancestry testing work today? In what ways does the past as it is selectively known make its way into the built world of ancestry genetics in the case of Latinos in the twenty-first century?

EXILE AND INCLUSION AT THE LEVEL OF THE GENOME

When I saw the key paper that was meant to be published as a resource for scientists more globally, I expected to see regional sampling for Spain, since the original AIMs panel used by the Morales lab at least had "Valencia region" and "Basque region" listed (recall table 2). It should be clear by now that population labels used in genomics are strange things. This is often due to a mix of factors that range from convenience sampling to cultural ideas of human difference and diversity (or lack thereof) to the use and reuse of proxy populations taken to be static and somewhat pure as good enough stand-ins for people in similar regions. (See table 4.)

Table 4 Ancestral Populations

Population	Designation	Sample Size	Platform(s)
Utah residents with ancestry from Northern and Western Europe (HapMap Phase III)	CEU	56	Affymetrix 6.0/ Illumina 1M
Toscani in Italy (HapMap Phase III)	TSI	44	Affymetrix 6.0/ Illumina 1M
Spaniards from Spain	SPAIN	619	Affymetrix 6.0
Yoruba in Ibadan, Nigeria (HapMap Phase III)	YRI	53	Affymetrix 6.0/ Illumina 1M
Luhya in Webuye, Kenya (HapMap Phase III)	LWK	50	Affymetrix 6.0
Aymara from La Paz, Bolivia	AYMARA	25	Affymetrix 6.0
Quechua from Cerro de Pasco, Peru	QUECHUA	24	Affymetrix 6.0
Nahua from Central Mexico	NAHUA	14	Affymetrix 6.0
Maya from Campeche, Mexico	MAYAS	25	Affymetrix 500K/Illumina 550K
Tepehuano from Durango, Mexico	TEPHUANOS	22	Affymetrix 500K/Illumina 550K
Zapoteca from Oaxaca, Mexico	ZAPOTECAS	21	Affymetrix 500K/Illumina 550K

SOURCE: From Galanter et al., "Development of a Panel of Genome-Wide Ancestry," 3.

NOTE: This table describes which populations they used for a large public resource intended to determine AIMs "in the Americas" concerning people in Latin America. The author summary includes the following: "Individuals from Latin America are descendants of multiple ancestral populations, primarily Native American, European, and African ancestors. The relative proportions of these ancestries can be estimated using genetic markers, known as ancestry informative markers (AIMs), whose allele frequency varies between the ancestral groups. . . . We have made the panel of AIMs available to any researcher interested in estimating ancestral proportions for populations from the Americas."

Here certain groups are taken to share similar socially recognizable phenotypes, illustrated by the recurrent use of white people from Utah in study after study when researchers are interested in some aspect of European ancestry. But I had not yet seen the label "Spaniards from Spain" as an ancestral category. It leaves a lot to be scientifically desired, but so does the use of "Utah residents." Then, perhaps of course, most strange of all, there are no Middle Eastern or North Africans to speak of in the panel. The long Arab presence on the Peninsula is once again exiled, this time at the level of the genome.

This "old" versus "new" world terminology holds for sub-Saharan "African ancestry" as well. As a clear example of what social epidemiologist Nancy Krieger has termed the "politics of time," in the first AIMs panel used by the Morales lab it was assumed that, for instance, present-day Yoruba and Mende people were older than "African-Americans," "Puerto Ricans," and the more than a few "whites" with "African ancestry."[51] We see the Yoruba samples show up again in the later panel as well, now complemented with East African samples. In both instances one is led to believe that those in the New World with "African" or "European" ancestry, detected by the test, have actually *inherited* genetic material from ancestral Yoruba, Spaniard, or Utah populations. Yet these reference people are contemporary humans not "ancestors." To clarify matters, I prefer to call them "the Today People" to cut through the veneer of distant time in such world-building models.[52]

With the first AIMs panel that circulated to the Morales lab there were other problematic assumptions as well—for instance, that Mexican Americans' and Puerto Ricans' ancestral populations currently still exist. Using putative Native American "parental source populations" can overlook an actual history of genocide of peoples who no doubt contributed to present-day populations in the Americas, but who were targeted and killed in all manner of ways.[53] To complicate matters more, Mexicans often have more Indigenous heritage than the presumed referent groups who are posited as their Native American ancestors. For political and historical reasons many Native Americans need only possess one-eighth (12.5 percent) demonstrable Native ancestry, whereas Mexicans may have considerably more.[54]

For the later panel in the "Admixture in the Americas" paper, the authors included four Indigenous populations from what is now Mexico, one from Bolivia, and one from Peru. Totaling 132 Indigenous samples to represent the Native aspect of present-day Latino ancestry, they are 25 Aymara "from La Paz, Bolivia"; 24 "Quechua from cerro de Pasco, Peru"; 14 "Nahua from Central Mexico"; 25 "Maya from Campeche, Mexico"; 22 "Tepehuano from "Durango, Mexico"; and 21 "Zapoteca from Oaxaca, Mexico." There exists much overlap with linguistic groups and some aspects of human divisions, such as ethnic affiliation. To get a general picture of the overall diversity in the region of "the Americas," it was helpful to consult the Ethnologue database that monitors the world's languages by country and also breaks down those that are living, extinct or in trouble.[55] Currently there are 31 living Indigenous languages in Bolivia, 91 in Peru, and 291 in Mexico. As for Italy, Spain, Kenya, and Nigeria, the numbers are as follows: 20 established languages for Spain, with 15 of those being Indigenous; for Kenya the numbers are 69 established, with 61 being Indigenous; and for Nigeria the figures are 530 and 520, respectively.

Perhaps unsurprisingly, there is no data for the US state of Utah.

3 Making Race

PHARMACOGENETICS AND ITS NECESSARY PEOPLE

With time I came to realize that American scientists' ideas of how people might be grouped by genetics, and race, were circulating the globe among international researchers in ways that sometimes surprised me. This was happening in places like Brazil and Mexico but also in places like China and India, where researchers conducted studies on health and collected DNA for what has become the fungible commodity of health data.[1] For ambitious scientists in these latter nations, with general populations large enough to greatly "statistically power" studies (as of 2023, there are 1.426 billion people in China and 1.425 billion in India), some scientific professionals have attempted to organize their data by US racial categories even though it makes little obvious sense, given who lives in their countries, to do so.

This aligning, or intention to approximate commensurability with US racial groups, demonstrates how racialized genetics is seen to add a value, however distorted, to human bio-samples that scientists even outside of America's borders see as a means to promote their work. For instance, the anthropologist Kaushik Sunder Rajan reports on stark scenes in his book *Biocapital* to drive home the geographical and market variegation of how such terms come to be articulated and weighted alongside longer histories of class and labor. Sunder Rajan chronicles the state's investment in start-ups, from granting land for research parks to giving biotech researchers rare freshwater resources in regions where drought-struck farmers were regularly committing suicide due to the stress of crushing debt. He situates this despair within the recent history of the industrial textile industry where, in the central area of Mumbai, over one hundred thousand textile jobs disappeared starting in the 1970s. This dramatically increased in the following decades. Eventually, with the rise of the biotech industry, "employment" as a clinical trial subject gained traction. This was especially true as Indian research teams with links to Silicon Valley began to see their populous compatriots as a bulk "commodity form," or large data-set, for genomic studies.

Sunder Rajan quotes one researcher's description of envisioning the subcontinent's diverse peoples as modifiable to American study designs, with the hope that Indian scientists might be able to construct genetically targeted pharmaceutical products that would be commensurable in bodies categorized by the most simplistic of American racial ascriptions. Sunder Rajan's business-minded interlocutor—who was also the director of India's flagship public-sector genome lab and a board member of a key genetic start-up with ties to the Nicholas Piramal India Limited pharmaceutical company—offered Sunder Rajan a cool supply-and-demand acquiescence to the "Western" market for racialized biospecimens. "If they want Caucasians, we'll give them Caucasians; if they want Negroids, we'll give them Negroids; if they want Mongoloids, we'll give them Mongoloids."[2] The pharmacogenomics emphasis, Sunder Rajan tells us, was "explicitly conceived of as research that can be of interest to Western biotech and pharmaceutical companies that might wish to contract clinical trials" out to Indian research hospitals and start-ups. "But the resource in question that would make this attractive," he added, was "not just the emergent pharmacogenomic capabilities in India as a result of state investment in biotechnology, but the *population*."[3]

This scene suggests the ease with which certain connected elites working in genomics in India can lump globally diverse individuals and groups who could just as easily be divided by certain DNA sequence markers and in fact usually are. What was also striking here was that an idea of the big three races was the focus of Sunder Rajan's interlocutor. He peddled in a continentally based pseudo-scientific schema in a setting where surely local human diversity flies in the face of such reductive outmoded divisions—notions that are even more glaring when colored by such archaic language.

Similarly, medical anthropologist Katherine Mason has written about how Chinese public health officials restructured local epidemiological studies after SARS I to compete with US health research teams. During this time a renewed culture of innovation and scientific creativity was communicated to researchers at every level and career stage. One of Mason's young scientific interlocutors lamented her lack of creative, competitive edge in a study on diabetes because she had difficulty fitting China's ethnic diversity into an American racial framework. Mason relates the researcher's sentiment: "Yet, although she also tried to follow the American model by studying racial disparities within China—taking as her racial minority groups the Hui (Chinese Muslim), Meng (Mongolian), and Zang (Tibetan) groups because they were 'the most different from the Han' majority—she had not yet managed to obtain a statistically significant sample from those groups [because they did not meet the enrollment criteria]. . . . Even if they had, Hu sighed, the study would still not be as good as those conducted in the United States because China's racial diversity was simply not 'fengfu' [rich, fruitful] enough."

The researcher elaborated on her dilemma to Mason: "Of course it would be better if we could do it as in the United States," she said. "[. . .] We don't have many white or Black people here!" Speaking directly to her reader, Mason writes: "It is worth asking why a society that was at least 90 percent racially homogenous according to their own definitions (and nearly 100 percent homogenous according to the American category of the "Asian" race) was using race as a category for the study of diabetes." She refers back to the young researcher in the story. "Hu had a simple answer: Racial differences were a hot topic in diabetes research internationally, and she wanted to publish her research in a respected English-language journal."[4]

These difficulties and social pressures that Chinese scientists face as they strive to compete on the global stage, where the United States often sets the terms regarding what counts as important scientific findings on human health, includes specific modes of strategizing to create value in terms of "creative" hacks to transform their population into a subject pool with racial disparities in biology. The effort to shape and curate their data to fit Americans' racial understandings is tied to potential recognition by professionals in the United States. Such recognition could translate into the scientific goods of publications in reputable journals, contracts for pharmaceutical development, and/or opportunities for postdocs and professorships at prestigious universities like Stanford in wealthy, sunny Silicon Valley or MIT in the intellectually venerated region of Cambridge, Massachusetts, whose research park joins it to Harvard. All three colleges identify themselves via their logos with a bold vermillion red—the color symbolizing good fortune and joy in China. Even if these cases offered by Sunder Rajan and Mason are few among the unstudied many, they are insightful in that they provide a glimpse into some scientists' desires to make people's biology in various global settings conform to the potentially racializing aspect of genetics in the United States.

One might imagine that moments of pause, interruption, assessment, and reassessment figured heavily into the ways these scientists processed how to interpret difference, race, and substitutability, as well as how to re-interpret these within the conceptual confines of a potential specific study question. For many of the scientists I worked with in researching this book, interpreting genetic sequence data through a US racial lens for American DNA donors could indeed present certain moments of pause, interruption, assessment, and reassessment, but usually only when I asked them about it. In many instances the goal-driven pace of the labs, their race to get publications, and the funding incentives on the part of the National Institutes of Health to "include" minorities in research, and that called them on many levels, fueled a "full steam ahead!" ethos.

This pace became a kind of habit energy that allowed people to focus on what seemed to interest them most—the "differences" in the genome that might explain some aspect of human health. Although some did pause and liked to think about what they called "higher order mechanisms," "cellular functions," "gene environment interactions," and even "epigenetics," I found that in the everyday doings of lab work, the forward flow was about not letting the deeper questioning slow down the research team, or as one lab head told me, "to get them off track." The main goal at hand was to detail genetic differences that might be correlated with a phenotype of interest.

The focus for the scientists in this chapter was to potentially develop "ethnic specific" drugs, also known as the aforementioned pharmacogenetics, or "tailored medicine."[5]

When I observed researchers who came to the United States from China and South Asia, and in one case a researcher who was a "mixed" person with one parent "from India" and one "from Czechoslovakia," however, they were less than zealous about making *themselves* fit the American racial understandings by which they had dutifully organized their large pharmacogenetic project. Was this a contradiction? Cognitive dissonance? Or simply their attempt to deal with an evolving social fact in the making—of the American brand of race for the genome?

Several of the researchers who were foreign nationals, or in some cases first- or second-generation immigrants, were actually in an ongoing process of learning *how* to classify humans in the US by race. They seemed to feel much more comfortable maintaining some distance from the whole issue—that is, it was often easier for them to place anonymous mail-order DNA samples from "African-Americans," "Caucasian-Americans," "Asian-Americans," "Mexican-Americans," and "Pacific-Islanders" into categories of racialized continental origin than it was to broach the quizzical task of classifying their own biology, and person, in such ways.[6] I asked the postdoc with Indian and Czech parents about this: "Do you feel that the racial categories that we use, for example in the United States, Black, White, Asian, and Hispanic, etcetera are too simplistic? Would you expand these if you could?"

"It always—as a child it always bothered me," she told me. "I hated it on tests—you know, when you took your SAT test, or when you apply for college, they ask you to check off this box: are you Asian, Caucasian, African American, or Other? You know, and I didn't feel like putting Other. You know, because Other just seems like, 'Gosh, I don't belong to anything.' It kind of makes you feel—" she interrupted herself to think for a moment. She continued, but began talking a little faster. "I just thought it was weird. I never liked the fact—no. I didn't like the fact that you had to define yourself as one race. . . . So, I guess it depends on the situation. I guess in social type things, I don't even want to have to think about it, if that makes any sense? . . . So, I have two extremes: when I'm doing my genetic type of research, I want things very well defined, and in a social setting I don't even want to think about it." She laughed, and added: "I don't know if that makes any sense?"[7]

That is not to say that the postdoc and others did not have their own notions of how people in their prior home countries were different, ethnically and culturally, or that the status of socially marked groups could also be associated with certain biological or seemingly inherent traits (as we will see later in this chapter). When they would eventually have to collect and work with DNA from real people in the San Francisco Bay Area, things got even more complicated. First there were the issues of how they had been trained to enroll people in studies based on parentage and the participant's own family's consistency of identifying themselves with racial or ethnic labels for the past three generations. Then there was the question of how to create a working definition of purity (read "not mixed") as an eligibility criterion for the study at hand.

Who could be considered "African American," "Chinese American," "Caucasian," or "Latino?" Would it make a difference that many African Americans were by historical circumstance "mixed," or that they were recruited from homeless shelters, or from areas outside of the county hospital where the city's most indigent sought care? What did it say about the local political economy of research subject-making that the team's fliers with information to join the study offered a twenty-dollar payment for the "blood draw" and were hung on the Plexiglas covers of bus stops not far from the ER where many with no health-care coverage were seen? Or that many of the "Asian" recruits were science and medical students at the local prestigious medical school? In what ways would the US research industry also create research subjects based on longer histories of race, value, and human material need?[8]

FIELDWORK IN THE WEEDS

In the spring and summer of 2003, I studied a second California-based genetics lab that focused on how people metabolize what are called xenobiotics (drugs, chemicals, plants, and food elements). That collective of collaborative scientists started what in 2023 is now a decades-long project on the pharmacogenetics of membrane transporters (PMT). During my fieldwork with PMT, I shadowed many members of the team, some of whom rarely interacted. After one meeting, someone took me aside and said, "You're probably the only one who has a bird's-eye view of all the moving parts, besides Kathy." Kathleen Giacomini was the graceful leader of their many projects.

My approach to writing about laboratory life is not to focus solely on the big people at the top, however. This is not just because I was often a "little person," as a non-geneticist, floating somewhere near the bottom to the middle in what is quite simply a hierarchical space based on smarts, degrees, publications, and status recognition in the field. As a non-geneticist, but anthropologist conversant in the acquired language of genetics for my ethnographic work, I had some freedom from the norms that solidify these vertical social lines that influence how many scientists interact with each other. For better or worse, and at times a bit of both, there were a range of additional ways in which I was interpellated—from outside observer, social scientist other, ethicist, and racial minority critical of race use in genetics to sensitive chronicler, potential auditor, and/or upwardly mobile academic (with National Science Foundation and Robert Wood Johnson funding to study these issues and also positions at New York University, the Princeton Institute for Advanced Study, and Harvard University over the course of my field stays).

I found it much more insightful to see how graduate students, postdocs, summer interns, recent undergraduates, and in some cases, people with technical training who were not on an academic track learned and lived these sciences. Another dynamic, which Harvard geneticist George Church put to me bluntly when I was in his lab, was that he as a project director, principal investigator, and entrepreneur, like many senior scientists, had to sell ideas to the outside in the same way that the start-up companies he endorses, invests in, or serves on boards for, do. He termed this unfortunate but necessary entrepreneurial "hype." After some time I was able to get most senior scientists off of their optimistic sales-pitched spiels and to spend time with me in the weeds of scientific uncertainty. But there was another obstacle to shadowing them as closely as I followed the people under them: they were incredibly busy. I learned very soon that it was the more junior lab members who were actually doing most of the hands-on science in real time anyway. They were the ones dealing with the everyday dilemmas that emerged from conducting genomic studies on racialized populations.

PHARMACEUTICAL SUSCEPTIBILITY, MARKETS, AND THE ALLURE OF RACE

One person who stood out early on in my PMT stay was a PhD candidate (whom I will call Tian) who was from a part of China that he described as "rural and mountainous," as opposed to where he went to school for his MD—an urban center in the central south region of the country. As a young researcher, Tian was interested in the idea that various xenobiotics would affect people of different ethnic backgrounds who lived in diverse environmental settings in part because he himself had lived as an ethnic minority in China until he moved to the city for medical school. When we were discussing one of his early papers where he and his colleagues looked at drug metabolizing enzymes in the Chinese ethnic Han majority compared to those who identify as Zhuang, Tian proudly said that he was able to find differences in how each group metabolized a common drug called Omeprazole based on single base pair changes in a common drug metabolizing enzyme called CYP2C19.

Omeprazole is a widely prescribed proton pump inhibitor in China, the United States, and elsewhere that is used to treat gastric issues ranging from *H. pylori* infection to other stomach and esophagus issues, such as gastroesophageal reflux disease (GERD). One overview of Omeprazole's market report for China published in 2018 reads: "Although there are many anti-ulcer drugs on the market, because of good effects, Omeprazole stands out from its many competitors and secures its leading role. As the lifestyle changes, the number of patients with gastric ulcer continues to rise in China."[9] Global sales of gastric ulcer drugs currently exceed $8 billion. Tian explained how his research trajectory in medicine led him to his PhD dissertation research. He broached the topic of how Han Chinese people are represented in genomics as *the* Chinese. At the time, people who identified as Han were the only group with a Chinese appellation in datasets at the Coriell Institute for Medical Research in Camden, New Jersey, the premier repository in the United States for researchers to store and share anonymized DNA for a fee.

Tian's PI, Giacomini, and select members of her team were currently collaborating with the Morales lab to collect samples in order to build their own local, San Francisco Bay Area, database with the goal of re-contacting their DNA donor participants to come back for in vivo (i.e., in their bodies) drug studies. For this dataset, called SOPHIE (the Study of Pharmacogenetics in Ethnically Diverse Populations), people of Han ethnicity might have still been assumed to be the majority, but Tian did not know this for sure. The now published criteria for SOPHIE states that volunteers "must self-identify as Asian with parents and grandparents of the same ethnicity. Must be healthy taking no over the counter medications or prescription medications." There were more criteria for certain prospective donors, including those who identified as Chinese. For now, let it suffice to say that Tian was not sure how well questions of "Asian" ethnicities, beyond those that Americans lumped into the category, would be documented. Currently there is no detail about different Asian ethnic groups in the now public dataset. They are simply classed as "Asian."[10]

Shape-Shifting Identities in Life and in the Lab

When Tian was pursuing his MD thesis back home, he had started out with a more expansive idea of intra-Chinese diversity where ethnic minorities were important to study because environmental factors might come into play regarding how people adapted to xenobiotics or how they expressed genes differently. But during his time in the United States, as well as in the Han majority medical school he attended, he began to see his own ethnic minority status as "changed." He told me his story as he poked around on his computer to pull up some of the English-language papers he had published in the Chinese scientific literature.

"So, for my MD thesis, we looked at the Han and two minority populations, and there were differences in the drug metabolism enzyme activity for omeprazole."

"And who were the groups?" I asked.

"They were Zhuang and Miao. Actually, I chose them because their living environment is very different from the Han."

"Okay."

"And you can see some physical differences too. The Zhuang are generally shorter than Han. The Miao live in a remote area not too far from my hometown, and their diet is very different from the Han. They eat more preserved vegetables and more smoked meat."

"What about the other group?"

"They also eat more preserved vegetables, fermented, but their diet is more sweet. The acids in the preserving can cause esophagus issues, maybe make them more prone to cancer. I don't know. But the Zhuang had a difference in drug response compared to the Han."

"Okay," I said, "so back in China you were looking at pharmacogenetic differences between different Chinese groups, and here the Coriell data you are using are 'Han,' the label is '200 Han Chinese.'"

"Yeah. You know, this '200 Han Chinese' you mention is interesting because I wasn't always Han. My parents are Tujia, but my grandmother was seen as Han because of how she dressed. I became Han when I left my area and went to the city to go to school."

"You *became* Han?"

"There really are no differences between me and the other Chinese here, Han Chinese." Tian glanced up from his computer where he was looking at his paper he wanted to show me. "Can you *tell*? Do I look different?" He was serious, but also didn't appear to put too much stock in waiting for an answer—perhaps knowing that I had no idea how to assess such an ask. I shrugged awkwardly and smiled. I hoped he would be fine with my non-answer as I leaned over a bit to engage him about what was on his screen.

"There are fifty-six minorities in China," he began. "So, many of them live together with Han. So, we are all 'mixed up.' Sometimes you can tell. One group in particular looks similar to Turkish people. They have a more apparent nose. But most people look similar." This advanced student effortlessly showed how the category of Han is a social construct, without saying as much, but he was also studying how people of different ethnic belonging metabolize a common drug in China based on changes in the genetics of a drug-metabolizing enzyme, which had already been reported in the United States.

On the one hand, Tian's study is an inquiry into how the science of Omeprazole needs to account for ethnic specificity and broader geographies beyond the United States and Europe. On the other hand, in doing so, he is going beyond just lumping all Chinese as "Asian" or even Han. In his discussion about where he registers ethnic difference, however, Tian cited one example of reading traits off of bodies but mostly referred to lifestyle as well as regional differences—both of which are changeable, as his own case showed. He started our conversation off keenly aware of the strangeness that the Han are taken at face, *and race,* value. After his initial probing of the category to show its porousness, Tian simultaneously nevertheless joined an Americanized version of "the Han." He did something similar in San Francisco to what he did at university in China. He demonstrated the ease with which he could learn to fit into this important, major, representative group. His mobility and perhaps even emergent cosmopolitanism had something to do with this.

MODERN COSMOPOLITAN CENTERS AND PAST RACIAL PERIPHERIES

cos·mo·pol·i·tan
/ˌkäzməˈpälətn/
adjective

1. including or containing people from many different countries.
2. (of a plant or animal) found all over the world.

I invite the reader to look back to table 1 in the introduction, where some of the first examples of the world-building blocks of the tabula raza are introduced. The table denotes a series of allelic changes in a cell membrane transporter gene as a kind of tabula. It reproduces a screenshot of an intranet interface in the PMT lab where researchers are interested in genetic variants that bear on how drugs interact with cell membrane transporter proteins, those little bits of the human body that allow substances to be pulled into or pumped out of cells for all kinds of functions that keep us alive. A few examples are organic cation transporters (OCT 1, OCT 2) or organic anion transporters (OAT 1, OAT 3), but there are dozens and dozens more.

That key table 1 features the genetic location of multiple transporter variants and the types of changes they've undergone in the first half of the graphic. The second half is ordered by race. The ways that the PMT researchers highlighted frequency differences concerning mutations in people's DNA—that is, by their racial census category identifiers—visually and technically expresses a literal genetic tabula raza, or racial table. This kind of classification and conceptual organization conveys an American obsession with naturalizing racial difference, as a potential genetic fact, to which Tian and his US colleagues' work has contributed.

Scientists' curation of this data reveals several aspects that demonstrate the social relationships that inhere in the DNA they've purchased from the Coriell Institute and have also collected for their own SOPHIE database. The snapshot of their intranet screen shows how they preserve in bold-font base-pair changes that they originally highlighted in red for emphasis. Each of these bold numbers indicated a frequency of note for the team. Many of the differences, or changes, happened in all groups, but the occurrence as a frequency value was varyingly different for each. For those that occurred in only one group, the bold lettering furthermore flagged that change as a potential study target. Such instances were always referred to as "ethnic specific" variants by the team. When variants appeared in more than one group and exceeded 1 percent, the team would brainstorm about how to test drugs or drug substrates in the different "populations," as they might be called.

For instance, they would try to design studies in "Caucasians" versus "African Americans" who at different phases of one key postdoc's work seemed to possess notable OCT 2 genetic variation that appeared to track by race when she looked solely at the Coriell samples. To her and the PI's surprise, the earlier pattern that the postdoc observed "switched" on her when she later included their local California samples. The significant patterns that were initially seen in the first dataset of African Americans were later more prevalent in the local "Caucasians." Still exasperated and somewhat confused a few months later, the postdoc complained to me that "some of the haplotypes completely changed to the opposite race."[11] As they themselves knew, from their own lived family histories, individuals and groups do not stay put on the globe, and genetic variation that might appear in a small subset of people from one self-defined group may not sit still either.

Significance, in both the statistical sense and in the everyday sense, would have to be re-thought as more people were brought into the count.

As scientists who often hail from international backgrounds, bilingual families, with parents and grandparents who were born in countries on continents other than the United States and North America, some of the scientists doing this work were uneasy using American racial categories. A different postdoc from Delhi racked her brain and paused for nearly thirty seconds of our recorded interview for a definition to explain race when I asked her how she understood the concept. She somewhat bashfully offered that she didn't have the cultural background to attempt an answer before following up with what to her was an obvious segue: "In fact, in India it is forbidden to work on genetics and caste." Once the interview was over, she went right back to writing a code "script," or algorithm, to assess racialized genetic markers that her PI wanted information on ASAP!

The postdoc with the haplotype dilemma, who identified as "mixed with Czech and Indian," was equally uncomfortable "using race in social types of things," which perhaps led to the strange moment when she said that the haplotypes switched "to the opposite race." What might it mean for any one race to have an opposite? Although the answer could take many directions, the implication here was that Black people and white people occupied opposing poles of human biology. This makes little sense scientifically, but it has much conceptual purchase worldwide.

While I found these moments fascinating, what interested me even more was how these very cosmopolitan scientists christened some of the genetic variation that they witnessed in the data as *itself* "cosmopolitan." This naming happened almost as a reflex. Their "cosmopolitan" framing was meant to describe rare DNA changes that transcended race-based or "ethnic specific" isolationism. It pertained to variants that were seen in all or nearly all the groups.[12] So, if they were uncomfortable classifying themselves by the racial categories into which they organized their racialized study subjects' samples, it was by contrast an effortless exercise to call DNA variants that had no unique racial specificity the same word—"cosmopolitan"—that might be used to describe themselves. This was the case even if no one actually went around calling themselves cosmopolites. Instead, they talked about their personal global trajectories that tied them to family, educational opportunities, research, and travel. For several, their multilingualism and their mixed ethnic and cultural heritage came up when discussing how they themselves "would not be able to participate" in their own current genetic study. This was because their research required more "homogenous" humans.

For Giacomini, who told me that she was of mixed Filipino and Italian heritage, even one single base pair change, in one solitary person, was of interest to her if the change produced a serious result—like a stop codon. In such cases the common DNA message for a protein to be made would be absent so the body just wouldn't make that essential albeit miniscule part. "I want to know what is happening in that person," she said, "especially if there is a loss of function." Such extreme one-offs were termed "singletons." Samples that were labeled by each of the five census categories, where nearly all possessed some degree of an uncommon genetic variant, were also of interest when a small minority (fewer than 1 percent) of any group possessed the particular change, but at different rates.

By way of background, it must be said that people of all groups usually possess what geneticists not unproblematically call the "wild type"—that is, the most common allele at a specific locus in the genome. That parlance refers to the known variant of an allele that has been taken to be the "norm" in the "wild," in samples mapped to date, before, during, or after the Human Genome Project.

The base-pair changes in the snapshot image, and their frequency differences, were localized in samples from the Coriell repository. PMT's initial database and exploratory establishment of which genetic markers occurred in whom were also assessed using these Coriell specimens. In the early days before SOPHIE was up and running, when the team wanted to do in vivo studies in a human model, they would use HeLa cells or create part-human chimeras using frog oocytes to see how stop codons, or other decreased or accelerated changes in function, worked in the race-tagged Coriell DNA.

Giacomini told her mentees that she didn't want their work to just consist of uploading sequence data to the National Institutes of Health (NIH) communal database for pharmacogenetic information, called PharmGKB. She didn't want the lab's precious potential work "to just be another Human Genome Project," or a catalog of As, Ts, Cs, Gs, and their sequence positions for known genes. For her it was the in vivo studies—how variants translated to typical or atypical aspects of an organism's biology and if this biology would require different kinds of attention, and ultimately pharmaceutical care—that held value. What followed was a sometimes stern focus on getting cells to stay alive, grow, and to possess the transporter proteins that the lab scientists primed the cells to assemble—and to have these respond to drugs in the cell media. Each step amounted to an individual feat, and each took its own time.

THE CELLULAR LIFE OF RACE AND FROGS

On any given day, graduate students and postdocs in the lab went between their computer screens to look up technical details, consulted their own databases to examine sequence data, and went back to their lab tables, or "benches," to work with living cells. They used Madin–Darby canine kidney cells, human embryonic kidney cells known as the HEK293 cell line, derived from an aborted female fetus in the Netherlands, and for much of the lab's life, they used the *Xenopus laevis* oocytes, or eggs from the African clawed frog.

Working with frog eggs was more obvious and tangible for a graduate student whom I will call Sara. The cell lines were abstractions in a way, and the young researchers often had to do searches to get updated about their features, new studies, and what people were able to do with them.

"Are you interested in seeing how we retrieve the oocytes?" Sara asked. "Sure," I responded before I fully absorbed what this might entail. She went to fetch the female amphibian and methodologically anesthetized her. I watched the small body go limp in Sara's gloved hands. Immobile, the frog still gleamed with the shiny freshness of life.

I wanted to spend time with all of the scientists, as effortlessly as possible, and not draw attention to my own emotions or reactions. It was often the case that students and postdocs were anxious or even slightly depressed. Many, if not most, experiments do not pan out. Paper deadlines are always looming, and the threat of being scooped (someone publishing a finding that they were working on before they did) were all constant sources of angst. For this reason I instinctively adapted an attitude to listen when they vented, and to keep my own reactions to myself when possible. Through my curiosity about their work as well as my own excitement about what I was learning from them daily, I usually tried to offer a positive, open, or neutral tone to our interactions. This was not an unnatural way of relating, but rather a layer of conscious intention to be as easy of a presence as I could. At base, I did not want for my eye, my questions, my difference as an anthropologist studying them, to be an emotional burden on top of what they were already dealing with.

On this particular afternoon, it must be said, I was uneasy about the frog. Some part of me dreaded watching what I imagined would evoke the gore of dissection. Feelings of avoidance came rushing up. So I focused on Sara's face, her reddened cheeks, her neat glasses, her young gray hair. I distracted myself with my human interlocutor to avert the splayed little body on the table that refused to be ignored. Sara talked as she worked. "Oocytes are definitely my preferred cell system."

Despite Giacomini's hope that the lab members get accustomed to working in as many cell systems as they could, Sara was allowed to continue her experiments mostly with frog germ cells since her attempts in the other systems were proving unsuccessful. As I listened to her frustrations with the pressures to expand her skills, she calibrated where to insert the cuts to the female's gray underside. Despite my avoidance, I did begin to glance down. I got a sense of the blade on the gleam of the animal's epidermis. These little bursts of light on such a bleak body were a sensory force that eventually connected me back to the full scene. Sara delicately made the slits to harvest the eggs.

"There are a lot of ethics about using animals in research and we always want to be as respectful as we can. It may seem strange, but it makes a huge difference to realize the sacrifice these frogs make." She continued. "They're more than frogs. They're doing something for us," She paused. "Maybe. Eventually." For Sara, the emotional toil of the violence of her scalpel was lessened by imagining the greenish-brown creature doing spiritual work for their experiments. As for me, I felt queasy until she put the animal away and we were just focused on the cells, now somewhat abstracted.

Next Sara showed me how she artisanally fabricates the needle that she would use to inject the cells with RNA, which would transcribe different versions of a membrane transporter that vary between some of the lab's race-coded samples. She sat down and put all of her tools under a microscope in the yellowy light of the room where the cell cultures grew. As she drew out a heated filament of pliant soft liquid glass for the needle, she eyeballed her desired length and cut the strand, leaving it to cool briefly. Each detail was rendered truly visible only by a microscope. Sara determined the miniature size of the needle's liquid uptake hole by blowing air in it via a calibrating machine and gauging the hole size by the size of the bubbles produced on a sample medium.

She gave me a peek into the microscope at various points along the way. This part of the process now felt special. Her dexterity and grace were exquisite. She began injecting the Human DNA instructions of the CNT1 transporter protein (from cDNA) into a poxvirus vector. As she worked, she told me that some on the team were nervous about handling viral vectors, such as the poxvirus that she used and also a related vaccinia virus, because of deeper fears of contagion. Too often, this aspect of cell science goes overlooked. Researchers may have poisoning or contamination worries that they speak about among themselves, or to the anthropologist, but rarely does the psychological hurdle that they are conditioned to overcome in using dangerous forms of life that can meld with their own appear in their published papers about their scientific and all-too-human processes.[13]

Next, Sara injected the mixture of viral plasmid and the genetic messengers for human CNT1 variants into the oocytes. She anticipated the cells growing over the next few days, and expressing the proteins in question within the week, "two max." She said coolly: "We are in the process of making chimeras—part human, part frog now."

The team could see from their sequence data that genetic variation in CNT1 might indicate that a *very small* subset of people in different racial groups would make altered versions of this protein that could affect drug responses. Their functional studies in the frog eggs would bear that out in a few individuals. The oocyte as a blank slate, as far as its *potential* human element was concerned, became literally imbued—injected—with racialized genetic variation in CNT1 through Sara's glass needle, the pox vector, and her steady human hand. This cellularly inscribed form of the tabula raza, the team hoped, could result in a finding that Sara's chimeric creations would lead to better medicine for racialized groups.

Even more exciting, however, was that with the prospect of SOPHIE recruits, the local volunteers who had consented to be called back for truly in vivo human studies would show how CNT1 variants were actually working in the body with all of its cells and higher-order mechanisms. Now a different kind of exquisite finesse and patience would be needed.

TO BE DISTINGUISHED BY JUST TEN SNPS

The volunteers that Sara imagined joining the human in vivo trial would still remain anonymous, or so she thought. They would be listed by a code and by no more than ten of their "interesting" genetic variants, or SNPs. Morales's lab was responsible for keeping the records of who was who. He and only one other person had participants' traditional identifying information—their actual names and contact details to be able to reach them. So they "of course" would not be identifiable by name . . . but would that be enough? It became clear that—well, no. With genomics, identity by the usual means—facial photo, social security number, name, address, telephone number, etcetera—all made for obvious but outmoded and incomplete elements to scrub from data to assure privacy. There was more to consider.

During one of the first multi-lab collaborative meetings that I attended, which brought together several teams that were working with PMT, Giacomini went down a list of items that she wanted to hear about from each collaborator. She was in the process of completing a report that was soon due to the NIH. She needed clarification about the construction of SOPHIE and asked Morales if they were going to list the genotyped SNPs found for each person and whether these would be visible to the larger world of users who were poised to consult PharmGKB, to which PMT would contribute SOPHIE participants' information. Morales looked slightly surprised and said, "Of course. Why not?"

Giacomini had clearly been harboring a concern that she was now ready to share with the group. Her voice got shaky. "To my knowledge, as far as I know, if you list ten SNPs that is like a genetic fingerprint. No two people will have the same 10 SNPs, right?" She turned to a respected expert on high-throughput sequencing.

Morales began to look frustrated. "I really don't think there's a problem," he offered.

Giacomini refocused on the high throughput lab guy again. "Help me out here on this. No two people would share the same ten rare SNPs right?"

The scientist agreed. "Yes, that's right." He tried to even out what he sensed might be brewing as a tension between his two colleagues. "*But,* you'd need to be looking for a match." He laughed slightly, as if to say that this kind of sleuthing was not the business they were in.

Giacomini came back with an even more serious tone. "So if the person committed a crime, then the police *could* go to PharmGKB and match up their SNPs?"

Another colleague offered: "They'd need a password. Access is not open to just anyone."

Morales intervened and assured the group that no one had access to SOPHIE participants' identifiers except for him and a recruiter who most people knew. Again, he stated: "This is not a problem."

Giacomini's voice got a little higher with concern. She turned to Morales then scanned the room to address everyone. "I've just heard that if you list a person, even anonymized, and then list their SNPs, that's the same thing as publishing their genetic fingerprint." Another senior scientist, also a woman, offered that adding their ethnicity to the mix would also narrow down the "suspect" pool. "Ethnicity" (as race) was a key feature of SOPHIE and would indeed be a primary aspect of each sample's public description and scientific value.

Earlier in the meeting, the group had engaged in a more detailed discussion about taking care not to include any clinical identifiers in the database because of HIPPA concerns. They agreed that they needed to safeguard against including any and all identifying information. Ethnicity came up, but it was agreed upon that this feature would stay in. That was the point of their research! And they were going to help the world understand how drugs interact differently in people from different groups. This was not about racism, profiling, or anything of the sort.

Now another high-level researcher, also a woman, asked Giacomini if she had secured "a certificate of confidentiality" for SOPHIE, which she said would hopefully protect anyone on the team from being obliged to give information to law enforcement or the courts. "We did that for an earlier depression study because of their conditions and other risks." Giacomini lowered her head slightly, before coming back up to face the team. "But no. We didn't do it for SOPHIE."

"Well—that's unfortunate," the colleague flatly offered. Morales tried to reassure the room. He confidently said that he had *not* done this for his other successfully recruited databases and that, again, he and only one other person had access to the participants' identifying information. Now others started to share Giacomini's concerns. Yet another colleague who had been mostly quiet up until that point decided that she needed to be the obvious translator (not that one was needed).

"What Kathy is asking is, *could* the information be used to identify someone—somehow—because no one shares ten SNPs. The *same* ten SNPs?!"

A response came from across the room from yet another of the male professors: "This is like what you'd read in a science fiction novel."

"I don't know." Giacomini responded. "I just go to all of these paranoid scientific meetings."

"Just stop going to those meetings," Morales abruptly injected. "I did."

Finally Giacomini's senior postdoc raised her hand and started speaking, realizing that she didn't need to be called on. "If every sample will have ten idiosyncratic SNPs, and then be identifiable—this just can't be. There has to be a way to list without identifying issues coming up."

"This is what the consent form says we will do," Morales offered in a legalistic tone. "We did what the IRB [Institutional Review Board] said we had to do," he said resolvedly. "We should be fine." Several people seemed to agree. They chimed in that "they," as a team, shouldn't have to think about these issues because the people in charge of PharmGKB, as one person said, "promised a method of masking things to prevent anyone from being identified."

Yet now no one was certain that anything could actually be promised or guaranteed. Morales wanted to get on with the research and felt that he and the team had crossed their t's and dotted their i's for the IRB. He was right. But so was Giacomini. It would be almost seven years before scientists would prove that anonymous samples in large databases could be re-identified—issues that I delve into more deeply in chapter 7. Giacomini's "paranoid" concerns anticipated this vulnerable new world. When it comes to genetic data, privacy is indeed difficult to guarantee and perilous to promise.

DEALING WITH REAL PEOPLE

With recruitment for SOPHIE, the researchers knew that "in vivo" studies also meant dealing with the larger human organism, or the person. Their vision of the human was still molecular, rather than deeply social-environmental, which would have invited a more epigenetic approach. Genes, drugs, and their substrates were still the focus. Because of this, participants would need to agree to alter behaviors and abstain from certain things: such as recreational drug use; avoid taking certain supplements, like the Saint John's Wort herbal remedy that many Americans take to boost mood; and even temporarily give up certain foods that interact with several pharmaceuticals, such as grapefruit, which can block the important CYP3A4 enzyme, resulting in too much drug in the body.[14] At the latter stage, the in vivo studies were done in only four groups. These were labeled "African-American (AA)," "Asian (AS)," "Caucasian (CA)," and "Mexican (ME)."

There was a sense that the team's data would be coming to life in exciting ways since SOPHIE required enrolling people who were out there walking around in the San Francisco Bay Area. The two researchers doing the recruitment had the idea that this particular database build would be relatively easy—in comparison to other ones they had built that required people to have a specific disease and two willing parents to enroll as well. The mix of people in San Francisco and its neighboring cities of Oakland to the east and Redwood City to the south made them feel that they would have ample Mexicans, Asians, African Americans, and "Caucasians"—their preferred word for people of European descent—to draw from. They explained that they were mostly focused on making sure that the study subjects of all groups had to be born of parents who both identified in the same way as the potential recruit did. However, ideally Mexicans not only had to have Mexican parents to qualify but all four grandparents had to be born in Mexico. The same latter parentage criteria initially held for Chinese recruits as well. Both parents had to also be Chinese, and the grandparents were required to have been born in China.

When I sat down to interview one of the principal recruiters, whom I will call Maria, she laughed slightly when she heard herself describing the uneven protocol to me when it came to "Caucasians" as opposed to the other groups. "There was no really tough eligibility criteria for the Caucasians, whereas for the other groups they had to be *Chinese*, from *China*."

"Right," I said.

"And Mexican, *from Mexico*, and so . . ."

I had heard the researchers talk about ideas of trying to approximate purity, which meant to avoid "mixing," by requiring what they called the "three generations" requirement described above. African Africans were also enrolled if they identified as such for three generations. I asked the recruiters about the difficulty of striving for consistency and what I saw as obvious pitfalls around the commensurability of their prospective sample groups. These issues could be raised for any of the populations they were working with, but it seemed glaring for Black American recruits, since the term "African American" is a relatively new one that took hold in the 1990s. We discussed how before that people were called "Black," "Afro-American," "colored," "Negro," while very few people understood themselves to be "pure" given the very flawed conceptual catch-all category in the United States based on "one drop" of "Black blood." The two recruiters agreed that it was not a very scientific way to go about things, but it was, as they saw it, "the best" they had. One spoke of often "eyeballing" Black people who they thought might be "mixed" if the potential recruit had blue eyes, naturally light or relatively straight hair. These were dead giveaways to them that the person in question was in fact not "purely" African American.

I asked Maria to explain the obstacles as she saw them, which of course went beyond race. Our conversation shows the extent to which she felt that she had to be on guard and actually hide the criteria for the study to prevent people from clamoring for the twenty dollars the team offered for "blood draws" (to take DNA).

"So with PMT, or the SOPHIE study, we thought this was gonna be really easy, "Maria explained. "We just really thought it was gonna be *so* easy." She laughed at herself, recalling the first days of their work. "So we sat down and thought, 'Okay, so, how are we gonna get these people? We're gonna to go to gyms. We're gonna to go to blood banks. We'll just post things.'"

"You were focused on *healthy* places," I offered.

"Healthy places! Exactly. The community, man, wouldn't you think everyone would be healthy? So, we kind of, like, looked at it and said, 'Okay, everyone should be really easy to get except the Chinese, perhaps.'"

This confused me, especially since by that time I had learned that they had recruited many of the Chinese people in the study from the local medical school. She continued: "We initially had in mind that we would recruit first-generation Chinese, versus Chinese Americans, perhaps. And you know, we did all of the literature in Chinese, and we did all sorts of things. We came across a couple people, physicians—who worked in that community—who were of Chinese origin. I worked with one and I told him about the study and asked him how he thought the recruitment was gonna go. He was like, '*Well* . . . ,'" Maria's pitch went high as she assumed both his voice and his doubt.

Continuing as her colleague: "I'm not too sure if people are gonna give you a stick [a blood draw], for twenty dollars. It's really uncommon in our culture to—give blood. You know, there's a lot of cultural sensitivity issues."

I nodded, wondering if this skepticism went beyond any one group. "You know." She carried on. "And so the other focus that we had was to get a team who was of Chinese origin, [another] who was of African American origin, so that we could be better able to serve these communities that we were trying to—*or attempting* to—obtain . . ." She continued. "And then came the obstacles."

I looked up from my notebook to give her my full attention. "People [were] coming in off the streets. And the idea was that we were going to be able to contact these people again. I mean that's kind of the whole premise of this all, because phase 2 of the study . . ."

"Was that you needed to be able to call them back for clinical studies?" I filled in.

"Yeah, it was like, 'How are we gonna do that?' It was difficult enough, already. People would call from all over the place. We had—our voicemail would be so full. We had hundreds of calls. People were really willing to do it. But people weren't checking out in terms of our eligibility criteria. People were just *not* checking out."

I asked Maria what issues were keeping people out of the study.

"With a lot of people," she explained, "a big thing was being on medication. Another big thing was mixed ethnicity. Then, IV drug use, current IV drug use, and then people didn't have an address." She paused to reflect back on when she realized the problems. "We were like, 'What do we do with this? *What do we do now?*' [. . .] Because we would get crowds of people. Say, we were like, we were seeing a patient—that patient would bring the cousin, the sister." She assumes the voices of the potential recruits: "'Twenty dollars! Free twenty dollars! Twenty dollars!' And this was—I would have to say—this was more of the African American community. Word got out in their community—whether they were living at a shelter or rehab center—the word was out that *this study was paying twenty dollars*—" She hit the table for emphasis on the last words—"*just* to have your *blood drawn.*"

My mind raced around its corners, where the voices in my head were clamoring to ask a different follow-up question. Amid my interior noise, I managed to stay silent since it was clear that Maria had more to say. I nodded for her to go on.

"And so, in trying to not let the criteria out, because if you *did*, then—" She paused for the right words. "Then that would *not* be good. Because people would *con*. And so it was almost like we had to be on guard."

I thought about the twenty dollars. I imagined the people who were being invited to give up their valuable DNA for what would be pennies on the dollar once that DNA made it into pharmaceutical drug trials. They, the "donors," were often poor and living between shelters, if they were Black in San Francisco. But for the recruiters, it was those targeted by the study who "would con." In the split second that my mind whipped around these ironies, Maria signaled with a shift in tone that she was now also going internal. She asked herself, with me nonetheless present: "'*Well, is this person telling me the truth?*' And there's really no way you can really make those judgments on people."

I wanted Maria and the team to just be honest with all of those individuals who called to join the study, which led me to somewhat naively ask her: "And why couldn't you say no drug use, no mixed descent . . . ?"

"Why couldn't we *initially* say that?"

"Yeah, in the ad," I offered, still torn by the many directions my questioning should go. I tried to let her continue to list the problems as she saw them.

"Because then people start *lying*," she said without missing a beat.

"Oh, okay."

"Because then they *know* the criteria."

"I see."

"So then they say, 'I don't have drug use' or 'I'm not half white, half Black." She paused with a sense of scrutiny as she narrowed her eyes, then continued: "Where you could *tell* that they are, but how do you sit there and argue with someone about their ethnicity?"

"Right . . ." I tried to suspend disbelief and enter into Maria's thinking. I pushed a part of myself into what I imagined to be her shoes on a hard day of recruiting as they met the pavement. I merged with their pitter-patter at ground level and Maria's inviting voice above as she hoped to enroll "pure" African Americans from homeless shelters, gyms, and the streets.

It was a confusing moment. One where I also paused. I really liked and connected with Maria, but I couldn't remember if I had told her that I myself was "half white, half Black." It didn't really matter that I wouldn't be able to join the study, even in theory, since they weren't interested in people like me. The point was that someone with my heritage, brought into existence by my parents' life circumstances that found them both waiting at a drinking fountain, on the grounds of their somewhat non-segregated high school in 1967, would by way of the generational domino effect of that circumstance, *not find a place* in the PMT SOPHIE database. It would be no consolation that scientists working on "admixture" and AIMs would later indeed ask me for my DNA when I worked in their labs—because, for their interests, I was "mixed." Those requests came with a different set of confusions and discomforts.

As the conversation went on, I realized that Maria and I actually had our mixed character in common, but it registered in different ways for her and the labs that employed her. There was one key difference: Maria would have been able to be in her own study because, as she told me, almost everyone from her country, Mexico, "is mixed." Therefore, the criteria of two parents and four grandparents identifying as the same thing was easier for these scientists to comprehend with regard to, as she said, the "race of 'Latinos,' if you want to call it like that." And she did indeed call it like that. The race of Latinos was one where admixture, or multiracial heritage, came to be featured as a kind of consistency, or purity, for them as a group. This understanding of mixing as the element that defines the perceived group was not extended to African Americans in the same way.[15]

A few years later I would witness scenes where African Americans were recruited into studies to assess genetic ancestry not for drug metabolism or asthma but for prostate and other cancers. Just over two thousand miles away, in the city of Chicago, the cultural dynamics of the lab researchers in question cohered not around nation building but rather around valorizing the many facets of Black life. Their cultural scientific world focused on how the powers of genetics might change when it was Black scientists who were searching for clues to complex diseases. At stake were specific genetic population variants for diseases that they saw as garnering less attention in the broader field because they affect people who are Black—Black like them.

4 For the Love of Blackness

WHEN SCIENCE CAN FEEL LIKE HOME

When I arrived at the lab one hot Chicago morning, Nefertiti, one of the first people there, was already deep in thought as she highlighted hard copies of articles, carefully brushing the text with a broad neon felt-tip marker. She was collecting bits and pieces for her facts and bibliography arsenal, whether for a future paper, a presentation, or a pitch to her boss about how they might extend or deepen their studies. Her desk was populated with the same genus of articles—scientific ones—printed on white paper with teeny black font and the occasional diagram or table. She was excited to learn more about genetics but, as she explained, her true interest was in the chemistry of foods, their health benefits, toxic effects, and how one substance that might be nutritious wasn't stably so. For Nefertiti, the precariousness of food went beyond just bad junk food. It was also about what people could afford to eat and how they had been socialized to like foods cooked in certain ways, even when they knew better than their tastes when it came to health.

Nefertiti's enthusiasm was palpable and contagious. With an openness, curious questioning, and warm laughter, she recounted much of her day, her plans, her weekend, her activities around cooking. Her sense of menu was broad. It consisted of what she made, who she ate with, or new spots she had discovered that weren't that expensive but were nonetheless high-quality "clean eats." It was even better if they had a vegan or diasporic flare, like the Trinidadian soul food place that was currently bringing her joy. It was clear that she would rather talk about food and health than simply read about it. I also had a deep nerdy interest in food, which she detected after a brief digression to collards, and how to cook healthy versions. Delighted with my take, Nefertiti swiveled back and forth in her chair as she leapt from chlorophyl-filled soul food to sources of selenium, lycopene, and heterocyclic amines.

At one point she showed me her packed lunch—a Tupperware container with stewed vegetables and lentils. As she recounted her weekend cooking stint, she remembered her family back home in Ohio. They were in the food, comforting her. She was thin and tall, with thick natural curly hair, full of energy and verve that could very well have been linked to her healthy diet (or to the fact that she was still in her twenties). As she seamlessly connected the food back to her family, she told me that she was raised in an "Afrocentric," "conscious" household with parents who were deeply aware of the systemic hurdles that so many Black children had to face on a societal level, starting with access to good food as the foundation for health.

"We were vegans until I was eight," she told me. "No dairy. No meat. Until it got too expensive."

"What'd you eat?" I asked.

"We ate a lot of tofu. We ate a lot of beans, a lot of rice. Wheat bread. The whole thing. . . . When I got older, we picked up milk and we picked up turkey and chicken. Just because there were five of us and it was really expensive to try to do the tofu and all that."

Nefertiti is a fast talker and a fast thinker. As she was putting her family's ethos on the table, I wanted to ask her about the expense of veganism, but she was a step ahead and reminded me that this was twenty years prior. *Before the corporatization of tofu*, I thought to myself. Back then, for lots of reasons, poultry could be cheaper than what was often small-batch artisanal curdled soy. So, she continued, as the family grew, they ate some animal products out of economic necessity.

I asked Nefertiti how she came to specialize in nutrition, her true love, in a genetics lab focused on prostate cancer. It wasn't a clear path, and she took me down its tortuous turns, which ended by her finding a home with Rick Kittles's team. He nurtured and supported her inclinations to dig deeper into how the foods people ate translated to potentially cancerous toxins. The social fabric of the lab, focused on health disparities, genetics, and common diseases like cancers as these differentially affect people with African ancestry, allowed Nefertiti to feel that her wider sensibilities about having meaningful work as a future career scientist would be possible. After graduating with a BS in the biological sciences as part of a pre-med track at a liberal arts college in Ohio, she worked in a hospital internal pharmacy, after several internships. There she regularly set up IVs and assisted during procedures.

"I liked research a lot more than I liked the idea of being a doctor. Blood has always been an issue." She laughed and squished up her face. "I like people, just not sick people."

One pivotal moment was when she witnessed someone getting a bone marrow biopsy. She had a deep emotional reaction. "I'm like sweating," she recounted. "I bust out of the room because I'm blacking out. The room is spinning, I began to see little sparkly things." She swayed side to side as she recalled the drop of her gut emoting an unusual disturbance of empathy. She racked her brain for the right words to describe the kinesthetic spin— words most of us do not realize we do not have until we are in situations of viscera. "I went home, and I sat still, and I was like, 'Is this really for me?'" She cut herself off to get to the truth of the matter. "I can't do this."

Later, Nefertiti was offered a provisional job in a virology lab at Battelle, a mind-bogglingly broad "scientific solutions" company. The position would have been as a tech in a molecular biology and virology lab that had biosafety and biohazards protocols dealing with what she called "experimental bacteria and viruses." During the job interview, however, she learned that she would need to get "experimental" inoculations. She speedily rolled through the story as if to get past it. "And I would be a temp. So they could always get rid of me, and now I have *who knows what*!" She laughed. "Scratch that!"

The pay would have been good, as well as the experience, but she couldn't get the idea of being infected with a deadly pathogen out of her mind. It was clear that these first forays into trying to find a research home were still alive, pulling on feelings not yet stored safely in the vault of another time. She reduced her hours at the hospital pharmacy as she looked for other positions. With the help of her boyfriend's father's family friend, a woman "who worked in research at Ohio State," Nefertiti started writing letters of interest and sending out her resumé and transcripts to PIs of labs where she might land a paid position, or even volunteer, within the university. Although she had not written to Dr. Rick Kittles, someone forwarded her query to him. Of the many times she pressed send that year, she only got one response.

"He wasn't on my list, so to this day I have no idea how he got my email. And he has no idea how he got it. [. . .] I sent them all my spiel, and *nobody* wrote back . . ." She paused. "Just Dr. Kittles. He sends me like a one-liner and says, 'Hi, we may have some room in my lab. Why don't you give me a call and we can set up an appointment.' I'm like, okay. That sounds great. I have no idea who you are, but if you have a lab, sounds good to me!"

After a few more exchanges, one mix-up about the timing of their meeting, trouble finding him on campus due to an outdated website, and a few more email conversations, Nefertiti started to sense that he was a Black man. This was before googling people became an acceptable aspect of human behavior.

"I hadn't seen any Black researchers really on OSU's campus, so I was like, hmm. This is interesting. . . . I don't know if it [the lack of Black scientists] is just an Ohio thing, or . . ." She paused. "But I don't think so. We meet. I tell him I'm looking for a job. We talk a little bit. And he's like, 'Well, I don't actually have a job to offer you right now because we are kind of in transition.' So I'm like, 'Would it be okay if I volunteer in the lab and try to learn as much as I can and get experience?' And he was like, 'Of course, 100 percent, not a problem at all, come in whenever you want as long as you want and let me introduce you to' . . . and says this is who you can work with." Nefertiti and her new mentor—a super relaxed, nice guy who knew the lab inside out—had an instant rapport. "And that was kinda cool."

She drew in a long breath. Her exhale on the next word she uttered signaled a transition. "Meanwhile . . . on television there was an advertisement for this PBS special where they were tracing ancestry." She told J, her boyfriend, "You have to tape this. I really, really want to see this. But I have to work. You *have* to tape this! He's like, 'Okay, okay.'"

"Did you want to see this independent of Kittles?" I asked. I was as drawn into her telling of the story as she seemed to be in its retelling.

"I didn't know he had anything to do with it." Her voice dropped down an octave to emphasize the bizarre coincidence. "I'm working, then I come home from work and J is like 'Dr. Kittles is on the PBS special.' I don't believe him. He's like, 'Seriously Nef, didn't you meet with a Dr. Kittles? I think he's in the—I'm sure it's him!'" She told J: "No, it couldn't be. Couldn't be."

"Okay," I offered.

". . . So I watched the tape and I'm completely shocked. I was like, 'Oh my gosh.' I'm thinking prostate cancer. He's in the Department of Molecular Biology, Virology, and Immunology. So, I'm like, 'Okaaay.' . . . I didn't have any idea about the genealogy side, or African Ancestry.com or any of this stuff. I just see this Black research scientist who has a lab—thankfully for me—and he's interested in studying Black people, with prostate cancer in particular. Then I'm like, this is cool!"

When Nefertiti went to the lab the next day, she was still in disbelief that her new boss was in such a celebrity-studded show, with Oprah Winfrey, Quincy Jones, and other famous Black Americans. Before she could bring it up with the new mentor that Kittles had paired her with, her new friend-to-be handed her a disc he had burned with the program on it. He beat her to the chase. "He thought it was a good documentary. He wanted me to have a copy. . . I think to give me an idea that Dr. Kittles does something else other than medical research."

Kittles started his career at Howard University, moved to Ohio State before moving to the University of Chicago, then to the University of Illinois–Chicago, onto Arizona State, then to City of Hope Cancer Center in the city in Los Angeles, and finally to the Morehouse School of Medicine in Atlanta, where he currently resides. When he got the opportunity to move his lab to the prestigious University of Chicago from Ohio, this was the move that allowed him to build his team in a more developed way. Chicago recruited Kittles as part of their community health arm and offered him more resources to hire several new people, with an inter-disciplinary range, including researchers interested in nutrition and epidemiology. When he saw Nefertiti's enthusiasm for the work, even as a volunteer, he offered her a fulltime job—if she would be willing to pick up her life and move to Chicago in a few months.

"I was really shocked," Nefertiti recalled. "I had to just take it all in. I didn't even say anything when I got home. I was just really quiet until I could decide what I was going to do." She would have to move her and her partner to a much more expensive city, three times more by her calculation, and follow a boss she had just started working for. The offer came with scant detail about the perks and what it would all entail. She decided to take the risk. "I would have to tell my family, which was going to be hard. They wanted me nearby. But something said, 'Do it!'"

Ever since seeing Kittles on PBS, as a potential "superstar," Nefertiti was endeared to him by the fact that he interacted with her and others in the lab as a "real person." His style, invitations for the lab to hang out after hours, to have drinks, do cookouts, and his plans to do many more community events in Black Chicago gave her a sense that the life of a scientist could be fuller and less compartmentalized than the pharmacists and research doctors she had been working with had led her to believe. "At the hospital pharmacy, they all just wanted to go home at the end of the day and clock out, just like I wanted to go home," she said, baffled. "I was like, but you're a doctor! You've made it." Contrasted with their burnout, she saw in Kittles an integration of life and work. Her first impressions of him were of someone with an effortless bravura of simply being oneself *and* a skilled geneticist that poked at the mystical biofilm that envelops so many scientists as entities who, in her words, are perceived as "superhuman."

"Being in Dr. Kittles's lab was definitely, you know, very different from any experience I've had. And I was like, wow, he's really goofy. You know, you see him on TV. . . . People are like, 'He's great. What is it like to work with him?' And I'm like"—she interrupted herself and laughed, dipping her head down. "Not like on TV! . . ." She rose, still laughing. "But he's fine. He's just a regular guy. . . . I'd find myself flipping through *Essence*, or *Ebony*, one of those magazines, and he's in there. And I'm like, eating and flipping through, and I'm like 'That's my boss.' It's weird."

BLACK SPACES, VISIONS, AND SACRED WORDS

> We have been captured,
> and we labor to make our getaway, into
> the ancient image; into a new
>
> Correspondence with ourselves
> and our Black family. We need magic
> now we need the spells, to raise up

return, destroy, and create. What will be
the sacred word?
Amiri Baraka, "Ka' Ba," 1972

As they moved to Chicago, everyone in the lab was in the process of making the South Side a new home and *coming into a new correspondence with themselves* and their Black family. The team would solidify more explicit language of kin in what they saw as the magic of the city of Chicago. They would forge projects wherein they traversed new terms and concepts—creative, scientific, and promising. Several ideas and hypotheses, or what Kittles called "the multiple stories" driving the lab, served to answer the question that Amiri Baraka asked in his poem "Ka' Ba": "What will be the sacred word?" These scientists had several. As their stories unfold over the next few chapters, it will become clear that their sacred words were not merely verbalizations of needed spells to rise up. Rather, they represented concepts deep in the Black body and its history— concepts mediated by skin color, genetics, and even cultural tendencies tied to Blackness, which they tried to chart along a knowledge pathway. This was a path that they hoped might be salvific for the community they were trying to heal, and also for themselves as Black scientists working within the field of genetics.

There was a palpable excitement about all that Chicago had to offer, the vibrancy of parades, day parties, festivals in various parks that all invited "beautiful Black people," as Kittles put it, out into the community. As they set up the lab, they were keeping tabs on what was going on each weekend, their own ability to familiarize themselves with their new neighborhoods, their need to establish personal ties and to map new networks for potential community outreach for their studies. The social life of the lab and its science were never clearly demarcated into discrete domains, nor was that even the slightest imperative. Quite the opposite.

After listening to Nefertiti's many interests, while I shadowed her and often accompanied her and the others for lunch outside during my first few weeks of fieldwork, I asked her how she decided to study the specific chemical elements in food that were currently driving her. She picked up one of the articles on her desk and put its dense script under my nose. I stumbled on the multi-syllabic key word "het-er-o-cyclic amines," trying

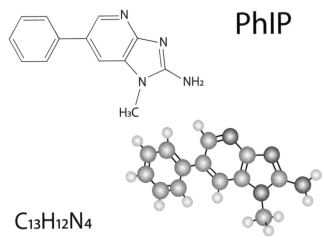

Figure 4. Heterocyclic amines (HCAs) are the carcinogenic and mutagenic chemicals formed from cooking muscle meats such as beef, pork, fowl, and fish. They form when amino acids and creatine react at high cooking temperatures, and are formed in greater quantities when meats are overcooked or blackened. "Heterocyclic Amine," Wikipedia, accessed April 24, 2021. A major HCA in meat is called PhIP, which is shorthand for its chemical structure: 2-amino-1-methyl-6-phenylimidazo pyridine. SOURCE: ARC Working Group on the Evaluation of Carcinogenic Risks to Humans, "Some Naturally Occurring Substances: Food Items and Constituents, Heterocyclic Aromatic Amines and Mycotoxins," *IARC Monographs on the Evaluation of Carcinogenic Risks to Humans* 56 (1993), International Agency for Research on Cancer, Lyon, France.

to decide where to place the accent (figure 4). She was already narrating its translation to their broader social life, connecting its implications to a barbecue at Kittles's new place the weekend before where the team was grilling and hanging out.

"I think where it started was when he gave me an article on heterocyclic amines," she said, the harsh-looking yet somehow special word rolling off of her tongue mellifluously. "I ended up doing a presentation on it this year when I first started in the lab. So as I'm reading it, the authors are talking about barbecuing meat and how bad it is for you. And I'm, like, *appalled.*"

As Nefertiti discussed the matter with herself in her mind, she seemed to go through the sequence of steps that led to her absorbing the article's message. And then: "But wait—" she interrupted her seamless internal download of the science story. "We just went to a BBQ at his house last weekend!" She shifted her position in her imagination and began to address Kittles. She proceeded to relive when she confronted him after the BBQ.

"How do you *know* this and you're *eating barbecued steaks*!?" She resumed herself for a second: "He was like . . . ," in a deep voice that was supposed to be Kittles, "I *like* steaks." She laughed. "I'm like, this is totally . . ." She was at a loss for words and then raised her pitch, as if speaking to him. "You have to be able to *change* in the face of knowledge!" She then picked up our direct conversation. "So we are going back and forth. He's making fun of me, saying I eat *grass*—or whatever." We both laughed.

Kittles was one to tease. His "playful" nature made him more approachable for some, at times, but several lab members would eventually feel that he could take things too far. At this point, Nefertiti, whom Kittles had begun to refer to as "the baby" in the lab family, was amused by it and accepted his banter by coming back at him with more nutritional knowledge that might help him, and the team, understand the results of the food frequency surveys they had participants fill out.

The heterocyclic amines that form in charred meat, a major one called "PhIPs," shorthand for its chemical structure, were being cited more often in conversations among the young lab members, with other scientists they encountered, and with people in the community whom they engaged. Despite his own penchant for barbecue, it was clear that Kittles had his eye on these chemicals that personally nagged him even if he led Nefertiti to believe that he would not allow the science to deny him his personal pleasures. He knew that the cultural challenge to get Black people to stop eating BBQ, starting with himself, would not be resolved by a public health messenger preaching "Don't eat BBQ. It might cause cancer." The problem was deeper because it involved depriving people of comfort, of pleasure, of sensibilities tied to sharing, and the chemistry that binds them to others—beyond the mere bonding of elemental compounds that results in mutagenic meat when cooked above 300 degrees F.

Another potential cause as to why this particular behavioral change might prove difficult to broach was that many Black men took pride in their sacred BBQ. The talent of the char, the pastime, the history of the food, its links to the South, the connection to family that it evokes, as well as to a mode of meal making that they could call their own, was too much to remove from a people who often struggled to keep what little power they had over their culture. The charred meat might be "bad," but Kittles also saw it as racially positive. It was a habit of taste that permitted a healthy and much needed sense of ease and gratification. I wondered about what parallels to this multi-sided "good" might be drawn regarding his scientific pursuits on African genetic admixture and the possibility of racializing Black people's genetics—which he also saw as a social good that nonetheless carried some potentially nefarious risks?

More often, however, Kittles had another compellingly forceful story he liked to tell. This one had less personal conflict, its plotlines were starker, and the chasm separating the benefit for Blackness versus the risk were vaster. This was the tale of vitamin D.

SUNLIGHT, BLACKNESS, AND POSSIBLY GENES

> We are beautiful people
> With African imaginations
> full of masks and dances and swelling chants
> with African eyes, and noses, and arms
> tho we sprawl in gray chains in a place
> full of winters, when what we want is sun.
> Amiri Baraka, "Ka'Ba," 1972

One of the ways that Kittles engaged race was through the obvious social marker of pigmented Black skin. As he would point out repeatedly—in lab meetings, in his lectures, or in everyday conversation with interested listeners—Blackness has direct biological effects with regard to the biochemistry of endogenous vitamin D production subcutaneously. The role of melanin in filtering UV rays and thus mediating vitamin D production in the body also affects hormonal precursors that influence prostate health. Kittles mostly focused on better understanding genetic variation in skin pigmentation as well as variation in a genomic region called 8q24 that is clearly involved in prostate cancer. But he was also constantly toying with the "vitamin D hypothesis" that low levels of this "vitamin," which is actually a hormone, might be implicated in the physiopathology of prostate cancer. On this count the food frequency surveys his team administered were an effort to see what sources of vitamin D people were ingesting. They asked study participants questions about sports or out-door activities they participated in, as well vacations they had taken, to get a rough idea about men's vitamin D exposure.

One key dynamic in the larger literature that compelled Kittles cen-tered on how dark skin hinders the amount of vitamin D that an individ-ual can make without supplementation. This, in addition to the fact that most minorities in the United States get inadequate sun, led him to ques-tion public health interventions like the application of sunscreen as a gen-eral protection for all people. This was an example, as Kittles told the team, "of the way public health is often geared towards white people." Individuals with higher concentrations of brown and black melanin (called eumelanin) as opposed to red melanin (known as pheomelanin) evolved to produce more of it, resulting in darker skin since this literal coat of cellular protection naturally filters out a wide spectrum of various ultraviolet rays. This phenomenon has been compellingly described in the physical anthropological literature—somewhat poetically—as the "hemi-spheric difference in human skin color."[1]

Interestingly, the idea that people who live close to the equator would have higher concentrations of black and brown melanin is true but not uniformly true for both hemispheres in the same way. As one measures the distance from the equator going north, the phenomenon of dark skin due to people's melanin concentration begins to wane faster than it does for those who live to the south of the planet's zero-degree latitudinal mid-line. The pattern of UV intensity follows a similar distribution—that is, there is a dangerously high UV index in sub-Saharan Africa for people without dark skin. This millennia-old and generations-long effect of dark skin that partly blocks out the sun's harmful (but also some of its *healthful* rays concerning vitamin D when one goes north) doesn't necessarily travel well—especially when dark-skinned people move to colder, overcast places situated far from the equator. A prime example would be the biting gray winters of Chicago—*wanting of sun.*

Kittles explained these scene-driven dynamics of how Black skin in cit-ies of dreary skies and bundled, coated bodies for protection from wind and snow might lead to more systemic disease. His retelling of the vitamin D story at one lab meeting took up such scenes: "Darker skin evolved in people living near the equator. Our melanin protected us from DNA damage caused by UV rays. When we were brought here, and then migrated north, being inside, being covered all the time—and then some Black folks didn't want to be outside or get darker . . . , it all affected our vitamin D."

Kittles and his team could now measure melanin index by seeing how much "tanning potential" people who were enrolled in their studies had by using a machine called a Derma Spectrometer. The machine measures light reflected off of the subject's forehead (one of the most exposed areas of the body to the elements, including the seasonal sun) compared to the subject's inner arm (one of the least exposed). Kittles wanted to look at general West African genetic ancestry to see what the relationships between the cases and controls in his studies would be on this count, and how melanin index, cancer, and genetic ancestry might actually correlate. This story had several elements of intrigue—deep history, migration out of Africa, a clear narrative of how skin color evolved in different geographical locales, due to environmental pressures, genetic and phenotypic adaptation for health, as well as a more recent possible explanation as to why Black Americans might be sicker with prostate cancer. Here race (solidified in part through social perceptions of skin color) and genetics (which partially code for pigmentation tied to type and degree of melanin content) intersected.

Kittles asked Nefertiti if she might incorporate vitamin D into her own research trajectory in his efforts to mentor her. "So we started to do research on vitamin D and other things," she said. "When we sat down to do my six-month review talk, he's like, 'All right, I want you to start thinking about what you want to do as far as your own research project, and in terms of graduate school and all that. I think you should try to put together what you really, really like. That's what you should look into.'" Nefertiti saw the opportunity to examine the social skin of Blackness alongside the genetic mediators of skin pigmentation genes on serum vitamin D levels that might lead to higher cancer susceptibility in dark-skinned people as a rare example of how to compellingly engage the genetic underpinnings of this particular racial health disparity—and they were off.

Even though she was often searching for the right words, psychologically pinching herself to see if her new job and responsibilities were real, she told me that she felt more hopeful and supported than she had thus far in academia. She was connected to a like-minded group and truly encouraged for the first time as a budding Black scientist at a top-tier university working with a nationally recognized African American geneticist—one of the very few in the whole United States at the time. This was in 2007.

SELF-SELECTION AND THE SEARCH TO BELONG

> For reasons not yet determined, Black people have a greater risk of prostate cancer than do people of other races. In Black people, prostate cancer is also more likely to be aggressive or advanced.
>
> The Mayo Clinic, 2022

Kittles has invested much of his recent career trying to explain what might be specific genetic risks for cancers in African Americans, how these may be traceable to Africa, via "African" genetic background, especially in relation to DNA variants at the genomic position 8q24 that he and others have found to be statistically associated with prostate cancer. For more than two decades this quest has animated his research life. Personally, and also scientifically, Kittles describes himself as avowedly "pro-Black" and, moreover, "pro–Black male." Working alongside him during my field stays, seeing him explaining the need for genetic research to people in Chicago's South Side, from parks and festivals to Churches and museums, he clearly wanted Black people to absorb the information that some aspect of their biological and societal lives might be putting them at risk for specific cancers. If he cared about Black people as a group, he cared even more about Black men as a particular kind of kin linked to his own life and body—as men in a world that hostilely marks them as threats, as lesser, as scary, as pathological.

Even as a bright kid who grew up in a middle-class family in Long Island, his father made sure that he was exposed to Black men who had it harder than he did. Kittles Sr. made sure that his son knew how people talked, lived, and struggled, and that he would understand that in the large scheme of things, the people's pressures were also his pressures.

Entering academia and the biological sciences as one of very few Black male graduate students in biological anthropology and genetics in the mid-1990s, Kittles often felt the weight of racism's burden. The professional world he entered was stressful. He grappled with his own moods and energies to get above micro- and not-so-micro aggressions holding him down or blighting his chances to thrive.

During three summer field stays at his labs, at the University of Chicago for two years, and the University of Illinois–Chicago for one, Kittles described his cultural politics and his science as set on the same goals: "helping Black people." This ethic was extended to men in particular, in what he called "the community," as well as to those he hoped to mentor through the world of academia which, he often lamented, was a difficult place for Black scientists. Yet men in his lab were a small minority. Most of his postdocs and graduate students were women.

Through multiple extended interviews I conducted with members of Kittles's research team, it became clear that scientists self-selected into his lab for reasons that were as much about scientific curiosity as they were the political stakes of health disparities that affect Black people. Their own personal identities figured into these discussions via various threads spontaneously, and off the cuff, in lab meetings or just as easily over juice in a campus café or over grilled food at one of Kittles's cookouts. Most of the people in the lab, with the exception of a postdoc from mainland China and a graduate student from Central America, were of African descent or of the African diaspora.

In recounting their prior experiences in other scientific settings, African American undergraduates, graduate students, and postdocs often reported that in prior labs they had not been taken seriously or were assumed to be less bright. African and Caribbean students did not experience this to the same degree. A thoughtful West Indian postdoc, whom I will call Shanice, explained that much of her graduate career in the United States was one where she was "hardly ever asked to offer ideas in the lab." I asked her how this negative gesture worked in practice. "My PI seemed to light up whenever anyone had ideas to offer, and she engaged them—however weird," Shanice told me. "But when I opened my mouth, it was like she would begin looking at her watch or something. She just wouldn't really pay attention." Time and again, the researchers explained that Kittles gave them the space to "be themselves" and encouraged them to inhabit the idea that they were scientists. This meant being able to theorize about genetics and ancestry as well as to openly talk about racism without any sense that these interests were distinct spheres or at odds.

The researchers often brought stories of their family's history, life experience, and tensions into the lab, and all were quick to diagnose or analyze systemic biases in the larger field of genetics that they were hoping to change. One woman who wanted answers told me: "The medical field still has no idea what causes fibroids, which disproportionately affect Black women all over the world. Most researchers haven't been bothered to ask the question!" Ebony, another woman who took up looking for possible genetic associations with fibroids for her own research a few years later, was quick to point out: "They wanna give Black women hysterectomies too easily as a treatment, but what's causin' them [the fibroids] in the first place?"

Kittles encouraged those under his mentorship to find their passion projects. He made no apologies for the fact that he himself was squarely focused on cancer of the prostate. The intellectual overlaps with specific health disparities that affected Black people that linked him and his mentees varied to different degrees. Although most of the women in the lab could be interested in some aspect of prostate cancer, what really drove them to be scientists were issues of breast cancer, HPV, which was largely seen as a female problem due to its traceability in women, fibroids again, and more general lifestyle and environmental issues that could breed carcinomas.

Despite the bodily differences that might segregate these cancers and other afflictions by sex, there was enough shared outrage about the gross margins of health, that US science and medicine had allowed to develop between Black people and others, to unite the lab as a group that was fashioning itself, individually and collectively, in the fight to reduce these disparities. Yet, even so, some of the women could feel that their passions and interests were being glossed over—a kind of erasure that was often patched over by a group focus on Black health more generally (as it pertained to men). This gendered dynamic of the tabula rasa/raza in the Kittles lab happened within a mostly Black social space. The women in the lab could easily support the considerable health justice goals that the lab represented in the larger field. Despite the erasures (*rasa*) of a fuller Black health, yoked to a re-instantiation of racialized (*raza*) prostate cancer genetics, a shared language of societal neglect, disregard, and America's inability to care about Black life and health became the communal lingua franca of the lab. Kittles cultivated this collective presentiment as a rallying point to motivate his team to work harder on the lab's projects—all of which were framed as using ancestry genetics to potentially rewrite Black futures by finally eradicating the burdens of injury, violence, abuse, and ignorance that have accumulated in people's bodies as perhaps preventable disease. Yet prostate cancer and issues related to it would take first order of the day.

Still today Kittles is still trying to figure out the connections between vitamin D, sun exposure, and dark skin as well as genetic background. His publications include admixture maps to analyze skin pigmentation—and melanin expression—in African Americans of all shades. He told me that he would prefer to create study designs that look at differences between Black people, not differences between "whites and Blacks" as is often assumed to be the correct dividing line in health disparities research. This ethos and outlook—to focus on a Black world that is diverse and important in its own right—structured the social science of his science—that is, the ways that the majority of people made their way into his lab. No matter their racial, ethnic, or national background, all of the researchers came to focus on Black health.[2]

That said, these scientists still had to interact in the larger field, and the larger world. Therefore it was not generally possible to act as if others shared a focus on Black-centered dynamics, or to rely on the culture of racial care that the lab could provide, even if this was imperfect and could blot from view other important dynamics like the women on the team wanting more attention on their research interests and more attention to gender bias in science, starting right there at the lab. But before we turn to some of those stories, it is important to get a better sense of the science politics that surrounded the lab from the outside and how Kittles as a Black male geneticist straddled research within the safer bounds of his laboratory and the outside world, where he felt his Blackness often preceded him—and by extension, his science.

5 Look, a Black Guy!

(WITH A GENETIC FINDING)

One evening after a long day with his team, a few disappointments, and a bit of soul searching about where his group should newly focus to figure out what might be going on with chromosomal region 8 at position q site 24 (8q24), Rick Kittles shared a story that had been upsetting him for over a year. His frustration, linked to his drive to explore the genetic region, was compounded by the very ways he felt marginalized from the larger mapping efforts in the field altogether. Other scientists, in his view, were dismissing and re-mapping areas his team had already staked out as their genomic territory. His lab's finding, he said, was "real-estate" that the group had already claimed through their discovery of it. At issue was a chromosomal focal point now called "region 4."

Part travel narrative, part conference proceeding, part science race, part racism in science, Kittles recounts a play of scenes where he could not escape the "fact of his blackness," in Frantz Fanon's words. Kittles realized that he could not always freely move (or publish) outside of the narrow confines within which others attempted to limit him. His lab's focus on African ancestry, genetic risk, and health disparities placed him in a professional world where the possibilities for recognition were often based in rigid ideas of race that he himself understood, negotiated within, and for reasons linked to his sense of social justice, also marshaled in his methods to address past racial harms Black people have suffered within medicine. His own racializing research on DNA linked to continent of origin, "African DNA" as opposed to "European DNA," for example, and the purification practices of actually creating genetic homogeneity in the lab for each, were specific efforts to redress past and present scientific mistreatment and neglect of Black people.

Kittles and his team focus on the genetics of health disparities that affect people of the African diaspora for reasons quite the opposite of scientists who sought to prove Black inferiority in the past. The Kittles lab's work was within the realm of racial science, but a racial science that they hoped would combat anti-Blackness. He proceeded with the confidence that with his subject position, his connections, his ability to garner the trust of potential DNA donors, his ease with "the community," things would work out differently. This focus led the lab to seek out genetic risk that might be tied to Black people's evolution in Africa. As mentioned in chapter 1, Ancestry Informative Markers (AIMs) can be thought of as genomic reservoirs of sorts—where alleles might be preserved for reasons beyond race-linked ancestry as most people conceive of it. Often chosen for their role in some superficial trait, like freckling or a clearly adaptive disease response, the markers can have some relationship to falciparum or vivax malaria, such as the Duffy-null allele or *G6PD*.

The group's results, they hoped, would pick up clues that other teams without such a focus on African biogenetic distinction would miss. Their findings worked to color their science as specific, highly particular, and in some cases as boutique or "tailored" to a population that is already understood to be different, other, and more disease-prone than white populations in the United States and elsewhere. Kittles thus embodied, both personally and professionally, multiple "facts of blackness." These pervaded how he became embroiled in a struggle over the validity of "Black" versus "white" scientific findings that in his case ran afoul of customary forms of scientific acknowledgment.

"Did I tell you about Estonia?" he asked me in one of our long after-hours debriefs. I shook my head no. As if remembering a forgotten mental note, he said to himself, "I should have told this story in lab meeting today." There had been a conversation that day about when genetic findings in non-white populations might not be taken as seriously if they weren't replicated in white people, or assumed "Europeans." There were new people in the lab since the experience he was about to share had taken place. Even if he had shared the event with his team before, it would benefit the recent hires if he recounted it again. Returning to the past as a present moment, Kittles leaned over to recount his story.

"So, I get invited to Tallinn. It's this small city. It's like Trans-yl-fuckin-vania! I mean . . ."

"Well," I offered, wondering about its remoteness, "they do have a biobank."

"In Estonia? Oh yeah, they have a biobank. There was a session on the biobank." Kittles was referring to the high-profile conference that had drawn him halfway around the world to a place he had never been and would find uninviting for reasons tied to how he was received out in the streets of Tallinn as well as inside the scientific circle of the conference walls. Some of his compatriots—his American colleagues—would make him feel just as out of place as the people who were staring and sometimes pointing at him as he passed by. He briefly came out of memory mode and took in the relatively safe space of the South Side of Chicago. I noticed him relax into his chair. He revved up his tone and began to chuckle. Kittles was clearly eager to share what he was about to tell me.

". . . Like Trans-yl-fuckin-vania! It was like *Ooold* World." I took this in and wondered about his own relationship to modeling the so-called Old World with AIMs in his lab and how neat these constructions always seemed to be. Eastern Europe never figured in the models. Recall from chapter 2 that what is termed "Old World" DNA was set up as the comparative reference to read, interpret, and fundamentally understand the disease risk and ancestral "compositions" of "New World" populations. Estonia and places outside the set schema of AIMs had no place in this American genomic world building. By tripling the *ooo* of "Old," Kittles's Estonia occupied a space beyond—outside of the relations of time and continental landmasses that configure much of American scientists' focus on ancestry and our (us Americans) DNA's linkages to the DNA of others on the planet. In his retelling, the spatial distance Kittles invoked went off the map he has helped to build as genomically relevant. He exited familiar territory, but he would be brought back soon enough.

"You've been to Eastern Europe, right?" he asked.

"I've been to Hungary."

"Yeah, okay. Well, it was just like Hungary." I shook my head and gave him a questioning look. He knew he was overgeneralizing, but this was not a retelling that was meant to be nuanced. I went along.

"It's funny," he said. "I walked around the night I arrived, and it was wild because I think I was the only African American in the *entire* city that day. Actually, the whole weekend—I mean, that week. I was there for a week.

"People were . . . looking?" I asked.

"Yeah. And there were some people who were polite and nice, but then there were others who would act like they didn't see me, but as soon as I would pass, say, 'There's a Black guy!'"

He jutted out his finger abruptly, then dropped his hand and turned his head away to act as if he was playing it off, like the strangers in Tallinn. He became the multiple persons pointing, the spooked Estonians who dramatically reacted to his simple presence and immediately feigned resuming what they were doing as soon the Black figure in the scene (Kittles himself) looked back to see them aiming forceful fingers toward him.

"And then they act like—" he froze in a poker face of innocence "—and then." His movement was sudden as he improvised the theater of their characters, one here, one there pointing, "There's a Black guy!"

His dramatization was comedic. His back and forth, his multiple embodiments of different people seeing him, getting caught seeing him, pointing, and then playing it off, all unfolded through Kittles's own body brusquely turning, pointing, and feigning a ho-hum normalcy.

Of course, many readers will have already been invited into similar scenes: those made familiar by Frantz Fanon in 1952 and James Baldwin in 1953. In a chapter of *Black Skin, White Masks* translated as "The Fact of Blackness" (*L'expérience vécue du noir*), Fanon recounts many instances of the finger taking aim, physically and psychically, directed at his Black self.

> "Look, a Negro!" It was an external stimulus that flicked over me as I passed by. I made a tight smile.
>
> "Look, a Negro!" It was true. It amused me.
>
> "Look, a Negro!" The circle was drawing a bit tighter. I made no secret of my amusement. . . . I made up my mind to laugh myself to tears.[1]

In his essay "A Stranger in the Village," James Baldwin gave the children in the Swiss mountain hamlet, where he would retreat to write, the benefit of the doubt compared to the American children who shouted the same word at him on the street: "There is a dreadful abyss between the streets of this village and the streets of the city in which I was born, between the children who shout *Neger!* today and those who shouted *Nigger!* yesterday. The abyss is experience, the American experience. The syllable hurled behind me today expresses, above all, wonder: I am a stranger here."[2]

In both instances the epidermal fact of Blackness, in Fanon's terminology, creates a hypervisibility these men were subjected to but could not control even if they tried to through reasoning, contextualizing, or a desire to be saved by humor. Both writers take up the larger themes that created the possibility of them being called out in the street—that is, the structural schema of whiteness, European heritage, culture, politics, and thought that established whiteness as simultaneously normal, supreme, and a kind of universal human standard that garners its power, for Baldwin in any case, through its own projection of Blackness. This projection, in Baldwin's words, constitutes a white supremist "personality," which he calls a form of madness, a misrecognition, that has little to do with actual Black life at all.[3] Fanon was also concerned with the measure of the white European human as the putative gauge to see how people of color fair and fail in their assumed aspirations to, in his case, French language, culture, and intellectual life in the Antilles. At the end of *Black Skin, White Masks*, he wonders about the possibility for a true recognition. That is, when and how might Black people actually be seen beyond measures and standards of whiteness, and actually get free?

.   .   .   .   .

"But anyway, I go to the meeting. And I present on region 4. And this is a week after"—Kittles interrupted himself. He named a US government scientist who is big in the field. "And he's leading that session. He's *the moderator* for the session. A week before, he submitted a paper where he named a *new* region, in whites, and he calls it 'region 4.' Now remember, up until ours, there were three regions, right?"

"So he didn't read your paper?" I say, confused.

"No that's not it." His tone gets higher. "They just ignore it." Kittles saw me puzzled. "I'm dead serious." His face slackened to convey its truth. He was speaking in the present tense, perhaps to signal that the scientific world is *still* ignoring him.

I looked at him in disbelief that scientists at that level would simply not cite a key paper that reported a discovery that they too were interested in. I only had one word to offer: "No."

His voice got emphatic: *"Ignores our region 4!* We published it in *Genome Research* and named it region 4."

"In 2007, right?" I say.

"Right! Right. He publishes a paper in 2009 and says this is 'region 4,' right? So, I present my work and I say, *'This is region 4.* So I don't understand why a recent paper calls another region 4 when ours is *already* published!'"

The "recent paper" was not merely one the moderator had just published but was, moreover, one he collaborated on with an international team of heavyweights in the larger prostate cancer field. They were joined by well-known geneticists more broadly.

Kittles continued: "So after I give my talk, he comes up to me after the session and says, 'Good talk. But just so you know, we meant *region 4 for whites.'"

Kittles paused. I too was silent. "I swear to God. *'Region 4 for whites.'* So I was like, wait a minute, so regions 1, 2, and 3 were *just for whites too?"*

He told me how he and other Black scientists and physicians often feel marginalized. They never know if their research and publications will be seen as valuable in a more general sense, or if they will be taken to only be serving their community, which can get cast as a "special population." This kind of niche thinking within genetic science—that emphasizes research on what might be thought of as the particular human—as opposed to the universal—has become acceptable with regard to personalized medicine and has created rationales that validate a focus on genetic health disparities. Both aspects of looking at particular groups can make scientists of color feel that today they have more professional opportunities overall. These emphases also permit them possibilities to be included in broader scientific conversations, including garnering the interest of journal editors and being invited to conferences, like the one in Tallinn. There is an assured sense that they will have a reserved seat at the table when professionals in the broader field want diverse teams contributing to the idea that diseases and genetic responses to therapeutics may differ in people from diverse racial or ancestral backgrounds.

Yet the frustration Kittles describes here is simply that these supposed opportunities in science slowly constrict and can actually result in being locked out of the center of the field, relegated to the sidelines. The recognition of Kittles's team's science was reneged on once the session moderator characterized his own larger study as the general "region 4." Period. In this exchange the numerical designation of the region as 4, when the Kittles team had already mapped a region called 4, was a clear usurpation of their finding—seen as their stake, or flag, planted on this genomic territory. That the session moderator's team did not call their own more recent discovery "region 5," for instance, excluded the work of Kittles and his team, which, in the way it was minimized because of its supposed racial focus and thus limitation, performed an erasure. The team's finding was overwritten, effaced even, by a strange kind of whiteness.

Kittles concluded his Estonia misadventures inside the conference by embracing the absurdity of what he experiences as a Black man, a Black scientist, an outsider even in his own field. He was simultaneously exasperated, willing to share, needing to vent, yet still able to see an aspect of farce in it all.

"The issue is, if we would have had whites, then ours would have been 'region 4.' But he says 'this is region 4 for whites' like he's done something above ours. . . ."

"I mean, your paper didn't say 'region 4 in Blacks.' And you didn't call it 'region 4 in Blacks.'"

"No!" Kittles replied, his voice rising. He was bothered recalling these scenes. But then he laughed. "You see, in genetics—and this is something I learned from David Altshuler—that this is about real estate. And I claimed something that they didn't want to give up. And in fact, they took it away, in a sense, with that paper. They said, 'Look, *this* is region 4.'"

One of his most prominent female doctoral students went into defense mode to protect their finding. "Even Wenndy got upset, she actually submitted an abstract for a meeting she attended that said 'THE REAL REGION 4.' I mean, everybody in the lab was mad. I'll send you the paper. . . . Why couldn't you just go 5, 6, 7? And it wasn't like our paper was in a fucked-up journal."

"Yeah, it was in *Genome Research*," I offered. "Not only that, but you also replicated the other stuff. You were playing ball. You were basically saying you guys did good science. We see it. Here it is. Here's what it's like in our population and we found something new."

"Right, exactly. And we replicated it in the multi-ethnic cohort." He simply shook his head.

When I read the paper that was aggrieving Kittles, I was surprised that it presented more like a report, with only two and a half pages of text. Examining almost twenty thousand individuals of "European Ancestry" from Europe and the United States, the data itself *performs*. It speaks its power and clout through its large numbers. More striking, however, is the performative work of the whole first page, which consists entirely of well-positioned scientists in the author section, which takes up half of the space, followed by a wide-ranging long list of European and American prestigious institutions, which comprise the second half. As for its argument, indeed, the second and last line of the two-sentence abstract reads: "This defines a new prostate locus on 8q24, Region 4. . . ."

The authors do, however, cite the fact that this locus had previously been associated with cancer—just not prostate cancer. Rather, they note that the genetic change in question had been associated with cancer of the breast. In the breast cancer literature, they would not have called it region 4. This naming of it as region 4 was to follow the pattern of discovery in prostate cancer that Kittles was respecting by naming the locus he discovered in a logical succession, as 4, after three other loci were found to be associated with prostate cancer by other teams. The convention of enumerating specific findings at these different locations is a tacit scientific social agreement for people in the field to collectively follow the order of discovery in how they name their findings. This is to acknowledge what has been done, and to validate others' science as they report their results for a key area as it may relate to, or correspond to, risk for a specific disease.

Before I actually read the paper, during Kittles's initial telling of his time in Tallinn, I asked him about the authors' wording and if they explicitly published the distinction the scientist was making for him that day in Estonia. In other words, did the authors actually write "This is region 4 for whites?"

"No," he said. He raised his voice again, yet somehow still took a softer tone as if he was intent on explaining a joke to someone who was simply not getting it. "They don't say 'this is for whites.' They would *never* say that—because then it's not *generalizable*." He slowed down and let me in on the punchline: "This for the *humans*," he said, and laughed. "This is for the folks who *matter*." He laughed even harder.

"But did they replicate it in Blacks?"

"Nooo! That's why he said, 'This is region 4 for *whites*!'"

Kittles and the scientist in question were in a game of "the obvious" in Kittles's mind, and my question—although also obvious—was seeking a different kind of sense, meaning where did the racial tagging start and stop, and where did the infractions lie for him and the lab?

Finally, Kittles said: "It was almost like he was saying, 'That was a good talk, son. Don't get upset. You got your lil' Black thing.'" He squinted. "You know?"

We both took that in. Then Kittles came back to a sentiment that I heard him say in many different ways but never quite this directly. "I don't understand why the hell would anyone study prostate cancer if you don't want to understand what the hell is going on in Blacks. Because that's a major issue. I mean, to be proud *just because you looked at whites*?"

.   .   .   .   .

This event and Kittles's frustration demonstrates the fine line that scientists walk when they are working to find genetic markers in populations designated by "ancestral" or continental groups that easily overlap with race in the American psyche. This said overlap is not just fantastical or an image construct. Peoples' places of origin as continentally based, where land and territory figure into ideas of what constitutes the "original" source from where a group hailed, are key elements of how race is made in the American context. We have seen how Kittles himself frames his lab's interest as one centered on Black people's broader health in the form of tackling health disparities that disproportionately affect those of the African diaspora, especially US Black men. The vast majority of the studies that have come out of his lab have this focus. The DNA markers, the larger databases, his lab's study designs, and their recruitment efforts zero in on Black men in the United States.

So what, one may ask, was so troublesome, so illogical, so wrong about another scientist deploying racial niche reasoning to cast Kittles's and his team's work as focused on a specific group? For Kittles the issue was that his region 4 suddenly didn't seem to count and could be passed over because it was initially found in Black people even though his team had replicated it, but in a multi-ethnic cohort that did not focus on white populations.[4] This implied that Black people cannot serve as a human norm in the way that people of European descent can. The latter have so easily functioned as stand-ins for others. Where he goes with this is that the humans whom he most cares about and identifies with (socially, ancestrally, politically, racially) do not matter. Following this, he later lamented that his science is also "devalued."

Kittles did not mention the number of people in his study or in the study "on whites." Neither did the scientist-panel moderator, as Kittles recounted their exchange. This is not a question of quantification, statistical power, or even robust science. It is a fundamental clash about where the lines of inclusion are etched, the demarcation of when one is in versus out, but still possibly in if content to stay within the sector allotted them—even if that means accepting that their fuller life's work gets reduced to marginalized bits. Kittles was also arguing that health disparities in prostate cancer need addressing by everyone in the field, and that with higher rates in Black populations, the participants whose DNA led to his finding within 8q24 should not be seen as a downgraded niche population but should be central to the field's mapping efforts.

On a more procedural register, there was also a breach of understanding that included the naming convention of using cardinal numbers in an ordinal sense, but that undergirded a particular politic of science when it comes to using distinct populations to search for genetic risks. The main scientific teams doing this work thus far had conceptually drawn upon highly particular aspects of their population data—people's ethnicity, their subjects' discrete nationalities, or the power of participants' societal emphasis on race at the political and cultural level (self-report). These very much structured how scientists in the field framed their positions and opportunities to be well situated to detect genomic gems buried in culturally inflected aggregate data within geopolitical boundaries, like Iceland, the United States, or even the Black world as a "diaspora."

These discourses and descriptions were obvious in how lead scientists detailed their research and their specific advantages to locate genetic risk in the people, or subjects, they were studying. Kittles's lab was surely enacting its own version of this patterned scientific approach. He, like other researchers working on prostate cancer, pitched that the particulars of his populations, and his own insightful methods to map genetic risk within them, would yield answers to disease causation that could be of value to a broader humanity at risk. To better situate his reaction in Estonia and his dismay at a US scientist's refusal to recognize him fully, we need to understand the stakes of these larger enterprises and the possibilities that they open for racialized chronicles in the current Age of (genetic) Exploration—its adventures and misadventures. This is the backstory, which must now be told.

THE SAGA OF MAPPING PROSTATE CANCER AND RACIALIZED RISK

African American men, taken as a vast and diverse social population, are burdened with some of the highest rates of prostate cancer in the world.[5] One of the first attempts to map and understand the possible genetic underpinnings of this cancer, however, came from a small island nation with only *one* known "African" person in its history before 1900.[6] The country in question was the Nordic Isle of Iceland, which launched a massive genomic and genealogical effort to genetically catalog every Icelander who agreed to give DNA after a series of public discussions starting in the mid-1990s. With such data, scientists could perform feats like unearth the "African" person whose traces live on in many of her descendants in the country today. The woman, Emilia Regina, and her son, Hans Jónatan, were enslaved in the once-Danish colony of Saint Croix but were eventually brought to Copenhagen when their master's influential aristocratic family resettled in Denmark. Hans Jónatan was an educated mixed-race person who fought in the 1801 Battle of Copenhagen, after which the Crown Prince of Denmark himself declared Jónatan free. Yet the mistress who owned him since his birth pursued her rights of property. Although slavery was illegal in Denmark proper, Jónatan lost the case in court. After the trial he "stole himself" and sailed to Iceland, settled in Djupivogur, married, had two children, and lived out his life in freedom through his own force of will.[7] Feats like finding evidence of Hans Jónatan's ancestry in many of his descendants, and tying them to his mother's genetic patterns of African diversity, could be carried out in 2018 through the national genomic and genealogical enterprise that the country has become known for.

Yet Iceland's vast research tool would lead to much more general science—having little to do with Hans Jónatan—one of which was a focus on prostate cancer in racialized terms that would implicate many people with African heritage in places like Chicago. Despite the fact that Icelandic researchers' genome talents were first focused on their own predominantly "Nordic" population, their discovery of a key prostate cancer risk region would lead to a transformation in thinking, according to some American geneticists I spoke with. If specialists in the US had started out not believing in using race as a general proxy for ancestry, the key area mapped by the Icelandic team, working within the company deCODE, permitted a picture of race, geography, biology, and risk to converge before their eyes "in the data."

In 1996, Icelandic geneticist Kári Stefánsson gave up tenure at Harvard Medical School and returned home to pursue a dream. His plan was to execute what was then an unconventional idea: to amass and exploit the relatively homogenous "isolated" population of Icelanders, which he and his funders saw as a large, recent human family of sorts, to find genetic variation that might be linked to diseases and drug responses, and also to provide other clues about human relatedness. Mapping disease genes through linking phenotype and tracing the patterns of genotypes in those affected versus those who were healthy was an early detection method for gene discovery in family pedigrees. Stefánsson wanted to see if the entwined social and biological history of Iceland as a relative population "isolate" would provide links that might be more informative than could be found in societies, and populations, elsewhere. If Iceland's population of 250,000 (at the time) could be seen (and function genomically) as a massive family, the country would have a mighty comparative advantage to eke out a highly privileged place in the world of genomics. Their work on themselves, as northern Europeans with a well-documented health and cultural history, could then help others around the world find rare or complicated traits in people of various origins since the majority of the human genome is shared.

This narrowing down the size of the proverbial genomic haystack to facilitate localizing areas where plausible needles might point to illness causation was compelling to many. That same year, with $12 million in venture capital, Stefánsson founded the company deCODE genomics. The unique appeal of deCODE on the global stage, then as well as now, was furthermore solidified when Stefánsson and his team of scientists explained that they would combine information from the country's meticulous church records (such as births, christenings, and deaths), centuries of census data (Iceland has the oldest nominal census in the world dating to 1703), and historical accounts reaching back to the ninth century— some rendered on calf skins. The oral and written histories were compiled in the early 1100s by the priest Ari Þorgilsson in a document that would come to inspire elements of the Icelandic Sagas in what is now a national treasure called *íslendingabók* (or The book of Icelanders).[8] With this detailed data on Icelanders' historical relationships, political ties, kin networks, and lines of descent, Stefánsson convinced the country's parliament and up to two-thirds of its citizenry of his big science vision.

Since 2003, more than 160,000 Icelanders have consented to submit their blood, DNA, and health information to deCODE's database. Through various genome mapping technologies, starting with microsatellite markers of the early days, then "SNP chips" and Genome Wide Association Studies (GWAS), and most recently full genome sequences, Icelandic scientists have been at the forefront of narrowing down specific regions of the genome that may contain genetic disease risk alleles. These issues were not without controversy, however. Concerned Icelandic geneticists, physicians, and ethicists protested that a private venture should be given the power to essentially monopolize the "population's" DNA, while disquiet over how every day people's medical records might be accessed by the company also drew critique. Nonetheless, the majority of the populace polled supported the idea with leading politicians citing it as an economic boon.[9] Drilling down on a key genomic region involved in prostate cancer would be one of deCode's first clear successes.

In 2006 the Reykjavík-based team published a watershed paper. They discovered a strong association between the incidence of the disease and risk markers on 8q24), the general region where Kittles wanted to join in the claims of genomic turf discussed earlier. After reviewing the statistics in Europe, the United States, and comparing survival rates of "African

Americans" with "European Americans," since African American risk would become a highlight of the utility of their method, the deCODE scientists described that their work drew on cancer registries and research databases for prostate cancer specifically in Iceland, Sweden, Chicago, and Michigan. The Icelandic population recruited by deCODE provided the team with healthy controls and provided key genealogical information on relatedness for men affected as well as those without cancer. This resulted in findings on heritable prostate cancer patterns in Icelandic families and the larger population, as well as markers that differentiated cases from unrelated Icelandic men not affected by the disease.

Using the linkage analysis from the Icelandic population, Stefánsson's group could see if the genomic signals they located remained when they examined the genetic markers of cases and controls in Sweden, as well as in "European American" populations from Northwestern University in Evanston, Illinois, and "African Americans" from the Flint, Michigan, Men's Health Study and the University of Michigan Prostate Cancer Genetics Project. The 8q24 region was repeatedly validated as a hotspot in each group. Two markers emerged as key players in the seminal paper, which appeared in the globally revered journal *Nature Genetics*. These were a microsatellite with a "-8 allele at DG8S737" and a single nucleotide polymorphism, "rs1447295 with an A allele."

A few months later, a team of researchers from Harvard University and the Broad Institute in Cambridge, Massachusetts, confirmed that something was indeed worth paying attention to in the 8q24 region. The US team's paper presented itself as a validation of the Icelandic group's finding, but only in part. "We strongly replicated this association," they wrote. But the paper also struck a competitive tone, with a litany of issues that the deCODE group did *not* demonstrate or discover. They continued: "The effect of the -8 allele was observed in European and African Americans, whereas the A allele effect was detected only in European-derived populations. The [deCODE] authors did not show, however, that either allele was causally involved in disease but instead suggested that they were both in linkage disequilibrium with an as-yet-unidentified causal variant. They also did not identify which gene in the region might be responsible for prostate cancer risk."[10] In other words, yes, two indicators or signals were found but no causal mechanism, and certainly no gene was discovered. There was still much more work to be done. The paper then laid out the American researchers' distinct methods for understanding the region on a deeper level, which opened the gates for a whole new mode of work on prostate genetic studies as others in the United States would try to replicate and go beyond their findings.

Using their own genome assessment tools that deployed genetic admixture mapping to suss out "African-derived chromosomes" against "European-derived" ones, the Harvard-Broad group put on display specific methods and techniques that they had been honing. Like the Icelandic geneticists, the Americans relied on the specificity and identity constructs of their home nation—the diversity of the US population—which was not seen as pure or "isolated." Quite the opposite. They consciously drew upon the potential functionality of the famed "melting pot" mixity of the United States as their own comparative advantage. For this they looked to genetic information from 215,000 people who took part in the University of Hawaii Cancer Center's "Multi-Ethnic Cohort" study (MEC). In the MEC's project description, and in the literature, study participants are comprised "primarily of five ethnic groups (Caucasians, Japanese Americans, Native Hawaiians, African Americans and Latinos)."[11]

Very few in the genomics world would argue against the boon of the identity apparatus deployed for the Icelandic biobank. Thus it would not be a matter of trying to beat Iceland at its own game. But what if others could brandish their own nation-based population features as genomic assets? For the Harvard-Broad Americans the shiny new tool they were perfecting in their own way was that of what is called "admixture mapping." Taking linguistic and historical forms from the US colonial, enslaving, and imperial history of conquering, naming, and categorizing territories and people in the Age of Exploration, admixture mapping is at base a highly specific cultural construct of genomic architecture where scientists go to great lengths to find markers that are not shared widely at a population level between people who have come from different regions of the world—most often Africa, Asia, Europe, and pre-Columbian Native America (as laid out more extensively in chapter 1). Even though these particular scientists tried to emphasize geography and stay away from language of racial identities, the result of such analyses still end up looking very much like race.

In this way, via a scaffolding of markers whose frequencies were reassembled by the geneticists to exemplify differences between these predefined continental groups, the Americans established their high-stakes technologies for gene discovery and demonstrated their efficacy by publishing a series of new markers they found in the 8q24 region. This tool and the findings it yielded were based on their own national and historical sagas that allowed them to mark and stake out a portion of the 8q24 genomic territory in the pages of the flagship journal *PNAS* in what would be the second seminal study on the larger region.

The principal complaint of the Cambridge-based researchers concerning the Reykjavík team centered on the idea that the informative -8 allele pointing to risk for African Americans in the deCODE study could simply be an effect of African ancestry. When chromosomes of Black Americans were said to be "African-derived" at 8q24, a host of markers, including the deCODE microsatellite, were on average two times as common. This, they concluded, was the confusion, or confounding, called "population stratification" that failed to take into account the myriad ways in which people who have lived and evolved in different geographical environments come to conserve similar genetic traits that often travel together. Something in the 8q24 region was highly associated with prostate cancer, while the many sites of genomic variation shared among those whom the scaffolding tool constructed as possessing "African Ancestry" in those bits on chromosome 8, seemed to make Black men more likely to get the disease, and at a younger age.

The Harvard-Broad paper tells the story of "admixture," race, prostate cancer risk, and genetics that Kittles, who was at the University of Chicago at the time, and others would set about detailing for the next decade.[12] It focused on admixture not solely as heritage but also as a practical tool. It was a way to see the architecture of genetic patterning that might be detected through the literal body—only now compiled of genetic markers tagged with identity/continental terms like "African" and "European." Thus the Cambridge group established their own watershed moment in the write-up of their findings:

> First, this study shows that admixture mapping can be a powerful and practical way to map genetic variants for complex disease. The results motivate

the application of admixture mapping to other disorders, especially those like prostate cancer in which incidence varies across populations. These results also highlight the scientific value of studies to find disease genes in specific ethnic groups, such as African Americans.

Second, we show that the 8q24 locus contributes to a major increased risk for prostate cancer in African Americans with African ancestry at 8q24. The difference between these individuals and African Americans with European ancestry at 8q24 explains a large proportion of prostate cancer in younger African Americans. If one could intervene medically to reduce the risk for prostate cancer in African Americans less than 72 years of age to what would be expected if all African Americans had European ancestry at the locus, the incidence in men less than 72 years of age would decrease by approximately 49%.[13]

Their story ended with the authors bringing another protagonist onto the racialized ancestral prostate cancer scene: the onco-gene *c-MYC* (pronounced "mick" for short). *MYC* overexpression during one's lifetime, in the soma, compared to germline conferral of disease, seemed to be at issue in animal and cell models. How might understanding this region, which could provide at least a partial answer to the mystery of why Black men have more prostate cancer, and also get sick and younger ages, advance when no proof seemed to actually inculpate *MYC* directly? In fact, no regulatory variants or structural genetic agents could be found in the region at all. In their words: "Somatic genetic data independently highlight the 8q24 region as one of the most frequently amplified regions in prostate cancer tumors. The c-MYC oncogene, a key regulator in cellular proliferation, lies within the peak. Overexpression of c-MYC has been shown to induce tumors in mice and to create a cancer phenotype in benign prostatic epithelium. It is possible that c-MYC could be the gene responsible for the prostate cancer risk, but no structural or regulatory variant has yet been identified."[14]

It is through both a sense of the obvious and a specific political rationality tied to place and history that the geneticists in Reykjavík and then in Cambridge illustrated how seamless the gestures to imbue genetic data with clear social characteristics could be. The Icelandic scientists and the Harvard-Broad team in Massachusetts each invoked the political strictures of their countries' populaces as idiosyncratically beneficial for revealing clues for prostate cancer disease risk. Similarly, Kittles and his team also equipped themselves with a database that featured their own cultural and biosocial commitment to a pan-Black, supra realm of diasporic people, which would not be bound by any singularly named national border. It included African Americans as well as people from the Caribbean and West Africa. More specifically, the data was largely comprised of samples collected from Black men in Washington, DC, and Chicago as well as from people whose DNA Kittles gathered or acquired in Cameroon, Jamaica, Senegal, Sierra Leone, and beyond.

In what may seem like a strange turn, white Americans, specifically those from Utah who gave DNA (see chapter 2) that was taken from Mormons sampled for some of the first French gene mapping studies in the 1980s and repurposed for the HapMap, were also included. Their presence actually makes sense on some level, however, as European descendants living on US soil. White people who at some point ended up in Utah lived within the same larger society as Black Americans and were now functionally positioned in the Kittles lab's datasets to assess "admixture" ancestry estimates for "European" alongside "West African" heritage within African American bodies that the violence of slavery forced upon Black people living in the New World.

If this geographical shuffling, with regard to labels, time frames, and convenient proxy switch-ups can feel dizzying, that's because it is. Especially so if one remains on the outside of the world-building enterprise where the AIMs modelers orchestrated flexible architectures for ordering human difference through geographies and peoples that require suspension of disbelief with regard to the long histories of mixing that happened in each locale cast "Old." Yet the political and emotional stakes of being part of the 8q24 research effort were nonetheless all too real, a thing of the present world, of health disparities and scientific inclusion, where disparities also marked how people had to fight for validation and recognition. This left members of the Kittles lab frustrated, as the opening scenes of this chapter lay out.

The paper that Kittles's team published to announce their "region 4" discovery, which appeared in the Cold Spring Harbor journal *Genome Research*, was earnestly titled "Confirmation Study of Prostate Cancer Risk Variants at 8q24 in African Americans Identifies a New Risk Locus." The authors diligently confirmed the premier Icelandic findings, then the subsequent Harvard-Broad results, before going on to review several successive studies that detailed even more variants and a third region. The tone of Kittles and colleagues' paper strives for inclusiveness. They did not seem interested in competitively undercutting others as they offered a modest description of their novel finding among the many that those before them had already located. The tactic was one that effectively validated the big and smaller players in the larger field, while marking their lab's entry into the club of teams that had begun to detail a genomic region with multiple risk variants for prostate cancer.[15]

The Harvard-Broad team for their part immediately raised an issue with the initial Icelandic paper. The strong association of a microsatellite (-8 allele at DG8S737) with prostate cancer that the Icelandic group characterized as being a higher risk factor for African Americans, compared to the "European" population of Icelanders, Swedes, and white Americans, had not taken into account the phenomena of "genetic admixture." The Icelandic and European population samples in the deCODE study were seen to be more or less pure and were not subjected to the same kind of assumptions (that they too might be considered "mixed") and scrutiny by the Massachusetts-based Americans and their collaborators.

Kittles and his colleagues addressed the first issue by evaluating the major markers reported by the Icelandic team in their Washington, DC, African American samples. The initial cause célèbre in the deCODE paper centered on two results. One was a single nucleotide change, rs1447295, which was not found to be associated with prostate cancer in African Americans, and the other was the microsatellite mentioned above. It was the microsatellite that was said to put African Americans at a higher population risk and therefore might account for the disparity in their prevalence rates compared to the European groups in the larger Icelandic study, and by extension white Americans.[16] The Kittles team did not find that either conferred a higher risk or were associated with prostate cancer in their DC Black men.[17] When they stratified their data by age, they did find that the rs1447275 SNP had an effect in their population, but when they compared cases and controls for markers said to reveal African ancestry at this locus the finding was less statistically significant. The Broad-Harvard group and others reported that this SNP seemed to be associated with early onset disease in Black men.[18]

After reviewing the major findings to date in regions 1, 2, and 3, Kittles and colleagues reported on two associations in their population, one of which, rs16901979, was also seen by teams who reported variants in region 2. The other, which was an even stronger and more robust signal, was the Kittles lab's altogether new finding of a strong association with prostate cancer in the DC samples of Black men. They detailed the findings as follows: "In our study, genotyping of additional 8q24 SNPs that show ancestral allele frequency differences implicated two separate risk loci on 8q24. Two SNPs provide evidence for 126.8 Mb and 128.4 Mb regions influencing PCa [prostate cancer] risk. SNPs rs7008482 and rs1001979 [*sic*] revealed a highly significant association with disease even after correction for age, and local and global individual ancestry. Therefore, the association we observe is unlikely to be biased due to admixture. Our most significantly associated SNP, rs7008482, represents a new region of independent risk (region 4), which is mapped to 8q24.13."[19]

Curiously, prostate cancer cases in the US samples had higher frequencies of ancestry informative markers shared with West Africans than the controls. It was not by much, as both had high levels, but it was enough to eke out statistical significance. These markers are not thought to confer or cause prostate cancer but to signal one's level of ancestry that could be a confusing misnamed driver (i.e., confounding). Kittles's team would conduct many studies later on skin pigmentation as well as melanin genotypes, while trying to assess the role of vitamin D and the shield of dark skin from the health-promoting ultraviolet rays that make vitamin D in the body. For now, this signal of more "African ancestry" was a crucible wherein genetic specificity shared among many Africans at this locus was not meant to be unpacked or detailed. Rather, the tool of AIMs created a surface scaffolding in order to hold the "ancestral" aspect of the comparative data stable so that a candidate risk variant might emerge.

And there it was: a strong association with SNP rs7008482 and prostate cancer in Black American men. This prize finding allowed Kittles, for the time being, to stake a claim on 8q24 and call it "region 4." And this was where he drove his flag into 8q24 genomic territory—the "real estate" he would later fight to have acknowledged as the "real region 4" in Tallinn, Estonia, faced with highly positioned white American colleagues.

RUNNING IN CIRCLES

Kittles could be despondent and often seemed to take our interviews as a place to work out and process some of the problems and frustrations he had with his field. In those conversations I would often bring him back to his participation in it and my concern that some of his work might enable racist ideas to which he was morally opposed. One such moment came up in a conversation about admixture studies that use Black Americans and Latinos. My apprehension was that those designs relied on notions of purifying these groups to resemble homogenous racial entities. His comeback centered on the fact of a purge rather than a purification. He wanted to purge the research community of the idea that Blacks and Latinos represented, in his words, "dirty data not worth touching."

When I asked him to explain, he said that Black potential recruits were seen as too genetically heterogeneous in many scientists' minds and that it was simply "cleaner" to study white men (such as the population of Iceland). In the hands of researchers like himself or Morales, with admixture tools these "mixed" understudied groups could be made respectable for research. I pressed him on his thinking with regard to African ancestry at 8q24 and prostate cancer and wondered with him if putting things in the language of "African ancestry" does little to change the ways that people conflate race and African heritage. Wasn't it too easy of a slippage, or invitation, to see biogenetic "race" as the reason for a particular predisposition to prostate cancer?

In a subsequent conversation we got a bit further into some of the vexing issues at hand. Kittles's situation was one that literary scholar Laurent Berlant might describe in terms of "splitting," with regard to attachments to objects and habits of thought—race in this context—that are cruelly enabling and simultaneously disabling.[20] Being caught in the race split is not something particular to Kittles. It is, in my view, a larger problem plaguing American academia and political life more broadly. During fieldwork I was interested in how the enabling/disabling comeback of racial thinking entrenched in a science focused on genes and race worked for Kittles and others in the areas where he has staked his career path.

"Again, I guess the reason I'm asking about going beyond ancestry," I began, "is that in a lot of people's minds African ancestry just means Blacker people."

"Riiight . . . ," Kittles drew out the word while thinking.

"So if you leave it at African ancestry being the culprit, or the cause for the phenotype, it could do the same damage potentially as people saying that is race genetic."

"It could . . ."

"And I think in a lot of people's minds genetic ancestry *being African ancestry* is a more robust idea about race—"

"But it's still race," he responded, finishing my sentence.

"Yeah."

"Yeah, I've always had a problem with it, and I've written about that, and I talk about that in my talks. But I also say that these estimates of ancestry allow us to deconstruct race. I'm very clear on that. I think that's one of the advantages of using it. It allows you to deconstruct race." He paused to think, then continued. "But it's weird."

He looked at me and turned his head up to the side slightly to share the insight as it was forming. "Because, in a sense, it's like a circle. You can deconstruct race, but you can also reconstruct race. [. . .] We're not using these homogenous-bound racial groups anymore, we're going to use these more fluid ancestral-bound groups. So yeah, I think it is problematic. I think scientists have to be *very, very* careful . . ."

"And so . . . when you say that . . . AIMs deconstruct race, you're thinking . . . ?"

"So, right, it deconstructs the one-drop rule."

"It deconstructs the one-drop rule," I repeated to mentally jot it down. "So then how would you say it *re*-constructs it? As a scientist, you're saying you're trying to be careful, what are you trying to avoid?"

"Because it's just another—it could be—" he interrupted himself to get to the point: "It's used by many scientists as another way of saying *race*. They still think in that racial—they still have that racial thinking. But they're using genetic ancestry in its place."

THE ETHICS OF HOW WE LEARN TO SEE

At the end of the summer in 2010, I asked Kittles to help me understand why he had made 8q24 a risk area of such high focus for his team in the first place. Why were he and his lab so committed to continuously studying this chromosomal area when they seemed to be coming up empty-handed after that first initial finding that was not being widely recognized? I told him that I wanted to do a focused follow-up interview to clear up a few things in "lay language" since the 8q24 story was getting admittedly complicated even for many in the lab to imagine their next steps for study. Despite the four regions mapped, no one had a clear sense of how to detail pathways about the higher-order cellular and biological mechanisms and why the designated populations they studied might exhibit different responses to the base-pair changes that were being documented in the field.

Kittles was at his computer, visibly tired, scanning a paper, and re-reading an email he would later discuss with me about an invitation he had received to comment on a journal article on African ancestry and asthma—an invitation that would be rescinded when the journal editor discovered that Kittles owned a private company that sold products marketed as "African ancestry." Despite his academic research on ancestry informative markers that were not the same kind that he used for the company's products at the time (mtDNA and Y-chromosome markers), the editor cited a "conflict of interest" for her disinvitation. Kittles was confused about why the editor was conflating his medical research expertise on ancestry for disease and ancestry for personal heritage quests.

He scrunched his face. "Let's get out of here. You hungry?" We decided to grab a bite in Hyde Park, not too far from the lab but seemingly a world away from it—for the simple fact that the waitress sat us outside. Small reminders of summer floated past in the form of ornate cocktails delivered to the table next to ours. Faces lit up with smiles and a sense of relaxation when the chilled glasses arrived, ice cubes swimming in curious colors.

Life on the patio, the warm summer night, made me recall that some people were on vacation. Our shared turquoise 7:00 p.m. sky contrasted strangely with the fact before me: that the scientist doesn't easily clock out. Most often, he or she can merely decamp to a new space. There were white umbrellas overhead and starched white napkins on the table, reminding me that the crisp lab coats were left behind, but not the obsession with the lab's stories, its dramas, what was working, what was *not* working, and what might be going on. I was now part of this clock. I was part of this pursuit to understand the science as it was unfolding alongside these committed geneticists. We would read papers on the chromosomal regions in question and discuss them in lab meeting or talk methods and statistical assumptions at the bench.

In the Kittles lab, as in every lab I've studied, the scientists themselves are members—humans, social beings, aspirants—in fields that are constantly on the move, where learning on the job is the quintessential nature of the job. No one knows everything about any one study, methods, theories, mechanisms, much less the entire field. They are self-summoned to constantly read, to perpetually absorb, to relentlessly try to figure out the bigger picture, even though they are usually focused on the minutiae—in somewhat constrained terms. Constraint allowed for a tighter concentration on findings, however, and their constraints included US racial groupings as an easy and somewhat practical lens through which to see genetic risk.

There came a time in 2010, after a few summers in the Kittles lab, where racial thinking also got deep into my own senses and perception. The more time I spent away from my habitual socialization, from home, friends, the intimacy and consistency of my own larger life, I could take on scientists' visions of people as "genetic admixtures" in the stark continental terms the lab worked with.

On one particular occasion, when I was having lunch with the clinical coordinator for one of Kittles's outreach initiatives for recruitment, I noticed that I felt exposed, a bit naked to be out in the world. It was as if I had stepped beyond the shield I was coming to experience as a form of social protection. The lab was primarily a Black diasporic space of African American, Caribbean, and African researchers (with the exception of two Latinas and a very committed male researcher from China during the time I spent there). The familiar banter, the open talk of race, an ethos of helping Black people—but also a practice of constantly measuring African ancestral DNA, in contradistinction to other ancestries—started to feel normal even if I knew that thinking of people in terms of genetic ancestry compositions wasn't *my* normal.

The lab as a home hub for these commitments made its confines a micro Black world led by Kittles and maintained by the others. As the team's anthropological observer, who was sympathetic but also concerned that their notions of ancestry might run too close to genetic reductivism about race, as well as the potentially extractive aspects of genetic science more generally, I established a healthy mode of questioning while remaining committed to actually learn and understand the work they had set out to do. So it was jarring when something in my own consciousness began to narrow and sort racial differences in a hyperactive way.

It happened as the outreach coordinator was recounting stories of mistrust in the community. "Yes, Kittles is Black but he's working for the U of C, and the U of C has not had the best reputation with the community. You even have Black staff who work here telling me of family members talking about bodies in the basement. Experimentation, you know." It was as she spoke that my feeling of nakedness, of being in the vast space outside the insulated one of the lab, began. It made me notice with a strangeness the many professionals and students around us in the bookstore café . . . the majority of whom were *not* Black.

There was a white man, with animated tufts of dishwater blond hair, wearing chino slacks and a lab coat. Then another with a frizzed, thinning Afro (could I even call it an Afro?) with a pen in his mouth while he mused the menu. Over by the wall of cold drinks, a group of white students were pulling out juice bottles. They were confident, with loud laughs and strong muscular bodies. They spread out as they sat among several tables. An empty space in the café separated us from an Asian couple engrossed in a hushed conversation. It was as if all of a sudden race was on the face. Traits and skin colors overran my perception to become the dominant filters. These pieces of people, in hues, textures, and physiques, rushed forth and ballooned into hallucinatory inflations of themselves in my visual center stage. Soon it was races that were walking by, standing in line, paying for their iced tea, Odwalla juice, or chicken salad.

It was both awakening and disturbing to see my thinking do something so foreign at this intensity—suspect in its indulgence, if not flat-out wrong. I felt I was temporarily in someone else's mind, or on an uncontrolled substance that I could still come back from. But I decided to ride out my fieldwork-induced psychonautical voyage. I knew that by noticing it, I could end it. But I could also let it take me to see more—to learn, from my own submission to it, how easily I too could be primed for *extreme* racial thinking. Now it was not only the lab that could not be left behind, even though the lab coats were, but the compulsion to map race itself was in the café. And it was in me. Admixture worldviews had the potential, in their often-drab routinization, to simultaneously seduce. This seduction was an aspect of their potential to colonize, or should I say re-colonize, my mind.

Growing up in Oregon in the 1970s and 1980s taught me to see race, in skin color and hair texture mostly, but also not to see race as an absolute. The language of color that my interracial parents taught us about their skin in the English language didn't match anyone. No one was truly "white" or "Black." We knew early on, or rather sensed, that these were bad poetics for a much more complicated reality. Following what the adults said, we used the language of being "mixed," however, and since the rare mixed kids in my Portland public school were adopted in those days, I was therefore told by other children that my pale-skinned, red-headed mother "couldn't be my real mom." This was the beginning of my understanding that fundamental racial thinking could impose cleavages that were not just about fractions and parts but also about relationships of family that others could try to sever if these did not make racial sense to the outside world.

The difference with "admixture" thinking in the lab was that it took that early childhood racializing vision as we lived with it, and within it, to a whole new level. The scientific visualization of admixture percentages in the hands of the geneticists attempted to purify itself of those lived relational negotiations of the said and unsaid, the felt and deeply sensed that often had no words. Instead, admixture testing promised precision—percentages, clear-cut lines, and few questions or doubts even in "the margins of error" that were usually minimized, metaphorically airbrushed, or textually absent in scientists' PowerPoint presentations.

Admixture genetics performed "a false transcendence," in Donna Haraway's words, of objectivity and distance from the old racial sight that feels wrong to so many of us precisely because it is subjective, personal, and too often prejudicial. With admixture genetics the subjective is rendered objective via mediated statistics and data organized on tables and charts. It is "a view from nowhere," as Haraway would have it, but simultaneously also always a view from many *somewheres*. This imperfect "God trick" performs through the conceit that one can actually split subject from object, which can be crazy-making. And, I might add, America-making. Only a partial, situated perspective allows for holistic objectivity because it acknowledges the body, life, and mind that are responsible for "the eye." There is a person whose vision focuses in; a person who also possesses blind spots. In this Haraway convincingly invites us "to become *answerable for what we learn how to see.*"[21]

Rather than forcing myself not to look, to reject what had just transpired in the bookstore café, I wanted to become answerable for my own racial sight and to recognize how it could also be activated given how I, myself, had learned to see. In the years of doing this fieldwork, I wondered about the need to search, questioned the obsession to visually organize aspects of people in such fragmenting ways. What do we do with our own thoughts of fellow humans as mosaics of phenotypes? When do such moments arise? What are the racializing structures of separation that our consciousness is either seduced by, compelled to call out, or both? Working in certain labs, such thoughts, uninvited, could usher forth in a matter of minutes. Time and again, I was forced to contemplate the spatial, formal structures that make people stand out—or be caught up within racialized terms at all.

What I didn't know growing up in Portland, Oregon, or when I entered the genetics labs all those years later as an anthropologist, was that my freckly-skinned mother who had zero tanning ability—and who had the strange fortune of having brothers who converted to Mormonism, and who created genealogies that revealed in detail their Scotch, Irish, and English heritage—would turn out to be "Asian." In various discussions about the heritage of our families, I shared some of what I knew about my parents' forebears with the scientists. Given my own skin color and traits, like freckles, they usually assumed that I had a large portion of European ancestry. I told them that, yes, my mother was "white." I don't exactly remember how it came up in the particular context of the Kittles lab, but I do not remember it being a topic that we lingered on.

Nonetheless, this fact of what could be in my genes did come in handy on one occasion. It was a slow Friday morning. The team began discussing how they "needed a control" for samples in Colombia, South America, that they thought would most likely have Indigenous mtDNA lineages. Similar to cases mentioned earlier, the Kittles team often used their own DNA for various purposes, and almost everyone knew something about their own ancestry. Given that they assumed the Columbian study participants would have Indigenous or African mtDNA patterns, a "European" sample could serve as a helpful control. But no one in the lab had European mtDNA—no one except (perhaps) the anthropologist who had at that time arrived a few months earlier to study them.

At one point during the discussions about the need for a control, they all simultaneously looked at me. A silence replaced the earlier chatter. At first, the obvious thought on everyone's mind did not seem totally serious because when they started to approach me with the idea some were smiling. There was soft laughter and some nondirect language about me giving them my DNA. Then finally one of them just went ahead and asked for it. After all, wasn't I expecting them to share their time, stories, and work lives with me? Surely I could reciprocate something. The assumption was that I would have mutations in a "hypervariable" region of the mitochondrial genome that would likely be classed as haplogroup H, one of the most common patterns throughout Europe.

After realizing that they were indeed serious, I told them why I might be hesitant. When I was doing fieldwork at a different site some years earlier, the head scientist there asked for my DNA to use in a commercial ancestry product that also had a forensic component. In addition to my genes, he wanted a photograph of my face, which he hoped to correlate with my genetic markers. Uncomfortable and unsure about the future of the whole endeavor, I declined on that first occasion.

Nefertiti assured me that my sample was safe with them. There would be no future or commercial uses. "I don't think Dr. Kittles would do anything bad with your sample," Nef assured me. Shanice let me know that there was nothing to worry about and that they would discard it if I wanted. I consented and they sat with me as I rubbed a long Q-tip with little teeth against the inside of my cheek for the ritual swab. I went back home at the end of the summer and thought little of it. After a few weeks, a text from Kittles chimed through the room as I was folding laundry.

We can't use your DNA.

                    Why, what happened?

It's Asian.

                    What do you mean?

You'll have to ask Shanice.

When I followed up with Shanice, who was a postdoc at the time and who had spent her doctoral years assessing ancestry in Caribbean populations, she explained what she could—that my mtDNA was haplogroup F—and then said that since her specialty was on African lineages, this was a bit outside of her scope. She referred me to Wikipedia to find out more.

mtDNA haplogroups consist of clonally inherited mutations in the mitochondrial DNA that are passed on within a single line of descent in women (i.e., one's mother, and her mother, and her mother, etc. back through time). The origin of haplogroup F is said to be forty-three thousand years old, much older than the "European" haplogroup H. With this new information, the women on the strictly maternal line of my maternal side (meaning none of my maternal relatives who are men) were now said to be ancestrally and geographically "East Asian" and "South East Asian." The highest frequencies of haplogroup F are found throughout mainland China.

I recounted this story to others with some amusement. I was even more amused when some people told me that they could now "see" that I *could be* "Asian." In these exchanges they would point out my eyes. I responded that there was no way to know how far back this ancestry actually occurred on my maternal line, but that it wasn't recent given what my Mormon maternal uncles had reconstructed in their genealogical obsessions. That didn't seem to faze these onlookers who took a new distant view of my face. Now some part of me was signaling "Asian." With this small detail they learned to look for race in my face beyond any timescale that made phenotypic sense. Becoming answerable for how they learned to see was not as compelling as having millennia-old DNA mutations provide them with a convincing, albeit questionable, revision—a prospect that gave DNA the power to answer for them.

6 A Family Affair

MORNING SHIFT

We rushed through the heavy steel doors that threatened to close almost as soon as they flung open. A stranger, a medical staff person, had just swiped his card for us to enter a new section of the building. The two researchers I was shadowing that summer, two years on from my first field stay, thanked him as we marched on. The maze of the county hospital was one they would take days to map.

Ines was an eighteen-year-old student whose family had relocated to Harlem from the Dominican Republic. A few weeks before we met, she had completed her freshman year across town at the University of Chicago, where she was on full scholarship. Dee, her colleague for the summer, who was Black American, had recently graduated and was excited to "make a difference" by doing what she called "real community health," which she hoped would help her gain admission into the U of C medical school. Both women were excellent students and had clearly been rewarded for getting tasks done well. How this would translate into clinical and community recruitment was less clear.

Dee and Ines's shared job was to recruit Hispanic and Black men who were seeking care off campus at a local hospital for a genetic study on prostate cancer and its possible links with genetic ancestry. Many men who come to the site in question were considered indigent and did not usually have regular health insurance. Kittles had added the Hispanic component in the summer of 2010, as he began collaborating more with scientists interested in this population. The study was designed to suss out the influence that African genetic ancestry, as well as skin color, and vitamin D levels had on prostate health. For this reason, in addition to taking blood for genetic analyses and demographic information, the women conducted a multipage food frequency questionnaire to assess people's dietary vitamin D intake since most did not take supplements.

Once into the main building, the two young researchers and I joined the sea of hospital professionals in motion. The structure was visibly old and in disrepair. Cracked cement along the ceiling line and dingy peeled paint contrasted with Dee's and Ines's starched white lab coats and the fresh youth they embodied. Ines's natural ringlets dangled in a mane down her back. In front, a miniscule part divided two sections of hair she straightened with the force of pinning them tightly to the sides. Dee wore a short ponytail but left her evenly pressed bangs free to frame her face. She applied a neutral pink lip color. They were girlish and professional all in one swirl. I would often see them circumambulating the ambiguous space between the two modes, feeling out new projections and understandings of themselves.

The sound of the human herd patted the flooring and carried us past occasional small clusters of patients sitting in pockets of waiting areas. Their quiet constellations interrupted the yellowy stream of what seemed like endless hallways. Dee looked over to Ines as we approached Urology and said: "I hope there are biopsies today." The researchers' potential study subjects were generally older, acquiescent men who often flirted with the young women. This was not lost on the women, nor on Kittles, their PI, and was even discussed at lab meeting as to why they—he called them "the girls"—were so "effective" at collecting samples. The young women also flirted, within bounds. But even with the distracting exchange they offered, sometimes the men awaiting the serious news about their prostates refused to join.

I was surprised at how seldom this was, however, since the collection of their blood, measurements of a light reflectivity capture on their inner arms and foreheads to assess skin color and tanning potential, as well as their detailed answers to the food frequency survey, were supposed to happen between the time that they completed their enema and the moment when the doctor was ready to take a biopsy of their prostate. The procedure, which had already been explained to them on a prior visit but would be explained again, was a protocol of sequential events listed to de-emphasize the crucial piece of information—squeezed in, in plain sight, between the many lines—that a biopsy device would be inserted into the anus. It would contain a spring-loaded needle, which would pass through the wall of the rectum into the prostate gland allowing the physician to take several samples. It would also be explained that the forceful needle could alternatively be inserted through the perineum (the area between the scrotum and the anus). I could never tell if the steely, measured medical tone, perhaps meant to dedramatize the procedure, made the men feel better or worse.

The timing of Kittles's research assistants' effort to recruit men into their study was generally one of anxiety, and of potential emasculating vulnerability for those awaiting their dreaded punctures. The young women's strategy was to turn this situation around. Ines and Dee both smiled quite naturally and used the language of offering to "entertain" the men, to take their "minds off of waiting" for what they imagined, and some returnees already knew, to be a painful, invasive procedure. Most men welcomed the distraction.

THE HUMAN DYNAMICS OF DATA COLLECTION

Our wait for the men at the urology clinic started out uneventfully. On this particular day, there were only four people who had appointments, and the two young research assistants' frequent sighs were audible as they hung out, trying to spy potentials.

"I hope they show," Dee said, making an obvious point to stave off the impending monotony.

"Me too," Ines murmured. There was a nondescript wall that had been designated as the women's "recruitment area" where they stood and anticipated men passing between points on their visit. The first person to appear today was a South Asian man in his late sixties. He checked in, did the initial part of the protocol, and sat in the exam room. The nurse invited us in to talk to him as he waited for the biopsy. The women began asking about consent, but before they got very far, the urologist came in—a bit confused.

"Sir, your PSA is normal. You're sixty-eight and it's 2.6. You don't need a biopsy. I'm not sure why your doctor referred you here." The man was also a bit confused and wanted to know what to do next. The urologist told him to get dressed. We exited the room and went back to the recruitment area at the wall. The doctor was a renowned specialist and seemed to be on a tight schedule. Indeed, before clinic was over that day, he would leave to join a meeting with Kittles and other local researchers to discuss a study comparing Black men who are financially successful with those who are poor to begin to understand why class does not do much to protect Black men with means from prevalent health disparities. I would later learn that these scientists also felt themselves to be at risk for common diseases like prostate cancer along with their less economically fortunate patients. They hypothesized that there was something about Blackness (as African genetic ancestry) coupled with anti-Blackness (global yet specific US racism) that preconditioned this condition.

Even after the doctor said that the man with the normal PSA could go home, he came back out to the waiting area and settled in one of the chairs. He appeared to have gone back to waiting. Dee and Ines wondered if he had understood what the doctor said. They decided not to interfere and to let the nurse take care of it when she reappeared. I asked Dee and Ines if this South Asian man with medium dark brown skin, yet not as dark as some people from the Indian subcontinent, might be able to enroll in the study based on the fact that he shared a skin tone similar to many Black people in the Americas. If the Kittles lab was indeed interested in race, not as a biological trait but as a malleable social one linked to pigmentation but not defined by it, then people of different backgrounds like this man could be interesting to include.

Dee and Ines mulled over the question. They weren't sure. It was clear they both initially had some hesitation about how to classify him. The Kittles study was supposed to be focused on people with African ancestry to some degree. Did South Asians possess it? "Maybe, but further back," Dee concluded. I wondered if one of the other doctors in the hospital network who was aware of the study might have referred him as a control. They didn't think so. We smiled at each other in silence to fill the void and waited some more. After a while, Ines began to sway her body back and forth. She bore down on her right leg, then the left, allowing the soft soles of her shoes to squish the air of impatience into the floor's old tiles. The women would visibly light up whenever anyone resembling a potential patient passed by. Human action in the clinic was their reason for being in those uneventful moments, and there was none for longer periods than comfortable.

Soon the sole squishing and body adjustments from standing were becoming unbearable. The women started to look at the clock. After another forty-five minutes and one false start, Dee announced: "Only twenty-five more minutes 'til clinic ends." Suddenly, as if on cue, a tall older Black man emerged from another room where he had just finished getting a biopsy. They had not seen him go in since he likely started the procedure and was processed by the nurse before the women arrived that morning, or when they were in the room with the man with the normal PSA, or maybe there was another entrance.

The nurse took him into another exam room and said, "These young ladies would like to talk to you." She left the door open.

Dee started up with the man in a playful demeanor, and he played right back. When she asked if he was interested in joining a study on vitamin D, genetics, and prostate cancer in Black men, he said, "I've been in another study for some three years—so, why not!?" Every sentence that each of them uttered was punctuated with a smile and a sweet laugh. Their breezy rapport was instant.

Dee's words flowed along in a singsong as she explained the procedure for what they would need from him. Following her training, she mini- mized the blood draw.

"We'll only take a few teaspoons. A red cap and a purple cap," referring to the tube volumes as if she and he shared a common language of vial metrics. She briefly explained why their lab wanted people like him in their study: a Black man with prostate cancer. She described vitamin D deficiency and the fact that the team wanted to see if this crucial vitamin, or its near absence at low levels, played a role in prostate cancer risk in men who, in her words, "have darker skin."

Before she got to link sun exposure and ultraviolet B rays in vitamin D production, he shot back. "Darker skin?!" Dee explained that the tool she was unearthing from her case to show him would measure his melanin index—comparing the hue of his inner arm where he doesn't get sun with the skin on his forehead. The strange contraption was the DermaSpectrometer ("derma-spec" for short) mentioned in chapter 4. He had already started to shake his head back and forth. Now, his eyebrows—still scrunched at the darker-skin comment—jumped up onto his forehead in perfect arcs as if to protect that sun-exposed area from Dee and the derma-spec. He drew his full body back on his chair.

"I'm not sure about this dark-skin thing," he said. Dee continued with her mellifluous explanations. A smile was her resting face, her voice still sweet. She slowly and confidently brought the machine over to his body to demonstrate its parts in an attempt to demystify its outlandishness.

"Nah, nah, nah. I don't think so," he half chuckled. "Measuring my skin color? I don't like things gettin' too racial? No, nooo. I'm gon' pass on that."

The nurse came in and informed everyone that we would have to put a hold on things because they needed to see him again in another exam room. The man seemed glad to get up from the chair where Dee was standing over him. For the few minutes prior, I felt myself wanting her to give him more space. She was physically close and leaning slightly over him. Now she finally pulled back. Ines and I were seated on the other side of the small room and smiled as he left, maintaining hope that he'd return. We all knew that the contraption and its purpose were weird. He no longer looked happy. Dee gently placed the machine into the mold of velvet lining that housed it in its case.

"Okay, don't worry," she said. "You can skip that part. We'll see you in a few minutes for the rest, okay?" He laughed, playing the good sport, and seemed to accept her amends with that simple gesture of latching the machine's case. In the split-second timing of body language, he looked partially relieved to follow the nurse to what must have been familiar territory and leave the strange situation that had just ensued. But as he approached the door, he put his hand to his forehead and saluted us to reanimate the fun he had been having with Dee—with me and Ines serving as a small but sufficiently absorbed audience.

When the man came back, he was ready to engage what Dee had in store for him. The initial banter of their interaction was a good warm-up. It appeared that he realized that he genuinely liked some aspect of this impromptu interaction. His connection with Dee was instantly familial. But he also didn't want to overoblige. There was now a second space of vulnerability for him beyond his status as someone with prostate cancer. He was an older man in the presence of a young woman researcher who seemed to know how to connect with him, in fun and seriousness all at once. He didn't want to disappoint and, on some level, seemed to want to please. This second space of vulnerability, for any of us, lies in the very act of accepting to enter a social engagement, of giving oneself over—however minimally or impermanently—to another. Degrees of obligation accompany most bonds, even fleeting ones. How do any of us navigate this tension of being both available for and exposed to the other when a kind of familial care arises in our interactions with strangers?

At points the man seemed to teeter on the threshold of an emotional point of entrée—where he could allow or deflect access—to inner truths about his life, his health, his tissue, his DNA. In the back and forth that followed, it was clear that he was watching his steps, careful not to freefall into the weakness of being human, and that eons-long edict of kin and clan, that yearns for connection. He played with putting up and dropping his guard, with sharing social codes like precious coin—and then flipping them on their heads to stuff them deep into his pockets out of Dee's reach. His avoidance of the interpersonal traps that prevent anyone from saying no were a barely perceptible adrenal dance that became clear to me as Dee asked him to sign the consent form to use his DNA for future studies.

One of the main issues that I discussed with Kittles that summer was the reuse of people's data and how participants are consented and informed about this possibility. The Havasupai Indian tribe's lawsuit against Arizona State University had sent ripples through Kittles's and most geneticists' labs in 2010, when tribal members found out that DNA they had given to ASU researchers for a study on diabetes ended up being used for studies on migration patterns and the tribe's origins as well as on the sensitive subject of schizophrenia and the tabooed subject of incest.

Dee handed the man his consent forms to sign after flipping through their pages and indicating the important areas where he would need to initial. One such area was a statement about the future use of his blood and DNA for studies. She explained it as: "You can initial here to allow us to use your blood if there is a little left over."

"*A little left over?!*" he asked incredulously, jutting his head back. "What little left over? Do you see how skinny I am? I get poked so much. Only take what you need—or put it back!"

He laughed a big laugh and so did we. He wrinkled his brow and pulled out a pair of thick, plastic-frame glasses. "I'd like to read it over, scan it, because I got really screwed one time." He left it at that and went into a silent meditation on the forms.

Even though she was enjoying his hilarity and wit, it appeared that Dee hadn't encountered this much spirited resistance from one person on two important counts. She couldn't argue with him ethically in her role as researcher nor instinctually as someone who was engaged in a lively tête-à-tête with a study subject who was immediately real with her, which seemed to allow her, too, to feel like family. Between them it was understood that both could jive and keep it light, even if the issue at hand was serious. Each of his retorts and refusals made sense, even if it was the opposite sense of the study recruitment effort. When he handed the forms back to Dee, she took on a soft caring tone and explained in clear terms the reassuring parts of the text about his rights as a subject and the benefits of joining the study.

He listened and replied, "That's fine. That's all fine." When she asked at the end if he had any questions, of course he did.

"Will you all call me if you find out something about my cancer? Or if my vitamin D is too low?"

"Well . . . ," Dee hesitated. "We don't handle the biopsy—we're not . . . well, the hospital will call you for anything related to your cancer. We don't usually call people and tell them about vitamin D either, but you can call this number on the form in three to six weeks."

The man looked away. This was a one-way deal with two distant, somewhat unsatisfying outcomes. There might be a true benefit somewhere, or there might not. She offered to give him a copy of the consent form. For the moment he could content himself that he would receive something tangible—an important form with a phone number. That was just the beginning of the paperwork.

SHOWIN' UP

Later that week at lab meeting, the team would give updates on each aspect of their work. Kittles would make it clear how pleased he had been with "the girls," their progress in recruiting patients at the clinic, how well they had organized the process. He would lament that prior to them coming onboard, things did not run as smoothly with the initial data intake. He would go around the table and check in with everyone, but mostly report on Dee and Ines's "stellar job." Both had joined the team earlier that summer and were somewhat shy to speak out, Ines more than Dee, when among the older more experienced researchers who might be working on their master's, PhDs, or, in the case of their urologist collaborator, already have an MD and be fully engaged in his professional life.

Lab meetings were often unpredictable, lively affairs. Kittles's style was to drill down on specifics. Each person usually had a task. Had they done it? If so, what did they find? The science was paramount, but that did not prevent him from playing around, making off-color jokes, or randomly straying from his loose agenda to digressions, which were often noteworthy stories that took the group on a ride to the place his mind had gone for the moment.

Minutes before the meeting started that week, the urologist and I were chatting among ourselves as people shuffled into the room. He shared that he had thought more about gender disparities in his field, a topic we had broached in my interview with him earlier that week. He reflected deeper on how the biases that led people to assume that urology was a field for men were part of the problem. He emphasized that there are urologists of the female body, which are separate from gynecologists, and then there are female urologists who are women who examine the male body—although they are rare.

Kittles called the meeting to a start simply by speaking. His voice is distinctive, and his tone was commanding, confident, and often ambiguously playful. Some lab members told me that they could never be sure when he was kidding around, or when he was seriously offended or angry. Sometimes his serious face was a farce, and he would erupt into laughter. The sudden change could be confusing until one spent enough time in the lab to trust that he had their best interest at heart and was a complex person working out a lot of stress and tension over the course of any one day or week. Several of the women simply diagnosed him as "moody."

"Yeah, I went to a female urologist once when I was younger," Kittles joined in. "I wanted to know if I was average size. The girls I was with would always end up cryin.' I didn't know what was going on."

The room broke out in laughter. Several of the women, although laughing, displayed a *WTF!?* expression. Some put their heads down or covered their faces with their hands. Dee dropped her gaze to escape into her notepad, where she stuck her ballpoint to the paper and kept it there. She stared at it with her eyes exaggerated wide—not even pretending to write—and slowly shook her head. Back and forth.

"What? *Whaaat?*" Kittles pleaded. "I really wanted to know. *Come on . . .*"

This was one of those moments where his focus on the male anatomy clearly went beyond his passion for prostate cancer. Yet some of the vulnerability that prostate cancer leaves men with—questions about their virility and desirability—manifested as wider cultural phenomena. This was not the context to assert oneself if one indeed had such inner concerns, however.

I looked at him incredulously, wondering if he had forgotten conversations his older graduate students had previously brought up—notably one of his PhD advisees, one of two Latinas in the lab, who had confronted him one Friday morning (two summers earlier) after deciding that he had gone too far with his "joking around." Clearly upset and razor sharp, she spoke for several research assistants and interns when she let Kittles know his comments were "inappropriate" and "unacceptable." On what were usually relaxed end-of-the-week mornings, that Friday she called a meeting, on the spot, when he entered the lab: "You're supposed to be our mentor, we should be able to look up to you!" He listened, apologized almost immediately, and said that there was a "reason" for his behavior, but that did not make it right. He promised to change and told them all that he didn't mean to offend anyone.

In October of 2008, when I was away from the lab fieldwork, I was attending a small invitation-only conference hosted by the National Human Genome Research Institute. After the morning session I was standing in line with a highly decorated young molecular geneticist who had recently given a talk on my campus and who would soon receive a MacArthur "genius" award. The lunch options were not great. Dividing my attention between a depressed chopped salad on display and the avalanche of detail he was regaling me with—"People do not realize how much genetic diversity there is in China," "Africa is important, but China has a billion plus people"—I listened with the ear closest to him. Then, as if out of nowhere, Kittles appeared right up close. I barely had half a second to register his presence. There was no time to even say hello before he belted: *She caused an explosion in my lab.*

I was of course caught off guard. "What are you talking about?"

I noticed him slightly smiling. I was still unsure if I could feel relief. I asked myself, in the split second that it takes to register that you are in an awkward situation, and you don't have the skills to exit on a dime, *Why would he announce to the lunch line that I blew up his lab?* "Explosion?" I repeated back to him.

He looked at the young star scientist. "She was in my lab—she's studying my lab every summer and after last summer everything exploded."

The scientist smiled but was clearly confused. He knew Kittles, but I was not sure if he knew him well enough to know that he was sort of playing. Kittles put on a serious face and pointed his finger at me in a matter-of-fact scold: "After you left ...," he said, before going on to recount how the women in the lab had not seen a woman of color with my "power" before. My simply being there as an observer with tough questions at times inspired them to demand more from him. "Your presence was huge. It empowered them, gave them a vision of what they could become. They saw a Black woman, a Harvard professor, pressing me about hard issues. They realized they weren't being mentored like they want."

"Okay, so it wasn't like there was something I left in the microwave with tinfoil on it."

"No! Worse!" He let out a laugh.

"Please, stop exaggerating." I relaxed a bit from his laughter and decided to redirect.

"Let's definitely catch-up at coffee break."

"Alright." He moved on at his usual busy pace. I ordered something with spinach for energy and found a table with the young scientist.

During the afternoon break Kittles and I picked up where we left off. He was antsy. "So, what happened?" I asked.

"I'm telling you—you caused an explosion."

"But they were already calling you out long before I left."

Only a few months before, the PhD student, who had agitatedly addressed him on that calm Friday when she let him know that some of his behaviors were out of line, shared with me that she had also complained to Kittles previously about his public lecture slides. He had a full-body picture of Jennifer Lopez to illustrate some of his talking points about Latino admixture. His takeaway (which I had heard when I was an audience member at different venues) was always that Puerto Ricans have substantial African ancestry.

While displaying a photo of the singer and actress in shape-revealing attire, he would performatively drop the line, as the student complained, "'So now we know why Jennifer Lopez has what she has!' There is no need for that." The PhD candidate continued: "Hispanic women are always sexualized enough as it is—and for this to be coming from a faculty member? So I've been approaching it, and I've been approaching it."

The student also expressed her gratitude for the group, for Kittles, and for the opportunities that were afforded her. She told me that earlier in her education, at a prior university, Kittles had taken her out of harm's way when a male faculty member who made her uncomfortable (with commentary that she could not have a boyfriend, while demanding that she eat lunch with him every day) repeatedly called Kittles to ask if she could do "training" in the man's lab.

She was relieved when Kittles listened to her fears about the faculty member's boundary crossing. She told me: "Rick was like, 'I had no idea that was going on,'" and refused his colleague's insistent request. Now as a graduate student, she was better equipped to speak up for herself and others. She emphasized how important Kittles's lab was for her, but also for its emphasis on African American health. She made a point to be clear that she actually agreed with Kittles's priority ethic of investing more in Black male students than in women. Yet she had developed her own ethics of looking out for vulnerable young scholars and now used her own experience to speak out to Kittles himself.

She continued: "I remember telling him that part of our culture, Hispanic culture, is to feel grateful for things—feeling grateful for these opportunities. And also, there're a lot of recent immigrants who feel that they can't say anything because they'll get deported. And I'm someone who is not in that situation. I became a US citizen. So I just think that if something is not right, I have the right to say it. And I have the right to take it from person to person until somebody listens to me."

I witnessed her calling meetings with Kittles to let him know that she needed more direction for her thesis, more time, more mentoring in addition to asking him to examine some of the ways that he made women uncomfortable. I asked: "Do you also feel that you have an *obligation* to speak out as someone who isn't constrained by the legal issues you mentioned?"

"No. I don't feel like I have an obligation." She paused. "But I feel that I *want* to do it. You know—I—it just feels good." She smiled, laughed a little, and sat up straight. "I do it, also, because I *know* there will be others who *can't* do it."

During that summer I noticed that Kittles dialed back some of his penchant for "joking around." He became more serious and wanted to talk about the dynamics with his students, which he was not always sure how to navigate. One afternoon, during one of our summary conversations about the week's happenings, he recalled his own days as a graduate student and young scientist. He looked across his desk—where a metallic globe of the earth floated in mid-air, manipulated by magnets—to a shelf full of heavy tombs. He pointed to a book among a row of gigantic genetics reference bibles and asked me to hand him one in particular. At first I didn't see the one he wanted me to retrieve.

"Not those," he said. He wanted the out-of-place, tattered one that hid among the shinier spines. It was a hardback, with no jacket. Wear had turned the once bright red cover to an aged maroon-gray. With its fabric fraying, the loose threads visibly decomposing, this relic, or perhaps heirloom, seemed to ask to stay half-buried.

"If you really want to understand me, read this." Although I could not have imagined what it turned out to be, I took it. In letters that were rubbed out by fingertips and time, I could still make out the gilded print of the title: *Message to the Blackman in America.*

Elijah Muhammad is not light reading. I was reluctant to "apply" the chapter-by-chapter detailed prescriptions and proscriptions for Black life in the pages to what I was seeing in Kittles's lab. He in no way lived by the book to the letter. Yet there were definitely signs of Muhammad's general "message": to build and cultivate Black spaces, economies, relationships to foster dignity for the Black man, to create a world that does not degrade him. Kittles told me that the latter was his life's work.

The role of "woman" in *Message to the Blackman* is largely complementary. Supportive. Her helping function is interwoven with calls for men to treat woman well in order to be regarded with respect by other races, and ultimately to be able to respect oneself. "Until we learn to love and protect our woman," Muhammad writes, "we will never be a fit and recognized people on the earth. The white people here among you will never recognize you until you protect your woman." He continues:

> The brown man will never recognize you until you protect your woman. The yellow man will never recognize you until you protect your woman. [And again] The white man will never recognize you until you protect your woman.
>
> You and I may go to Harvard, we may go to York of England, or go to Al Ahzar in Cairo and get degrees from all of these great seats of learning. But we will never be recognized until we recognize our women [. . .].
>
> She is your first nurse. She is your teacher. Your first lesson comes from your mother. If you don't protect your mother, how do you think you look in the eyes of other fellow human beings?[1]

Kittles would later tell me that he saw himself repeating what was done to him early on in his career, in the non-Black world. The question became how did he look in his own eyes now? He realized that in possibly making his scientific space uncomfortable for women that he was in essence alienating them, and potentially harming them, making them feel as others had made him feel within educational settings where he could be invisible and simultaneously hyper visible but never really felt *regarded.*

When I brought up his slide deck, Kittles explained that he was trying to keep his lectures lively, to help people like him, who might be in the audience, envision another way of approaching scientific questions—as less uptight, more *down*—to attract them to the sciences. In his words, he was just "being me," which he hoped would appeal to young Black men. That said, he heard the critique. The student and I both noticed that the image of J.Lo no longer appeared in his slide deck the next time we went to one of his talks.

Back at the conference, Kittles continued telling me about the "explosion." It was clear that the tinfoil in the microwave had been nuking independently of me. I was simply a woman, now interacting with all of them, who was more "senior" than most of the grad students and postdoc researchers. The optics of my presence there balanced out the professor gender dynamics, if only a bit. But they all knew that I was a temporary observer who was mostly trying not to "interfere." Yet, like the women he mentored, I let Kittles know when comments were inappropriate or when his "joking around" was not okay. I did this because I cared about the women, and I cared about him. Looking back now, it was also a way of caring for myself. I knew these dynamics well, which did not necessarily mean that I knew how to deal with them. I was battling a worse version of them, largely alone, at my home institution. At the risk of making a gross generalization, they are not good for anyone.

Kittles wanted me to know how much things in his lab had changed for the worse—which meant for the better—when I left after the second year. "It's different now," he said. "They keep bringing up the fact that you are there in the summers, saying they want female mentors. They want to see more senior women of color."

"That's not a bad idea. What do you think?"

"I want that too. But it's hard. I've tried. They often leave." He went into how they want to start families. And how he felt that it was hard to invest when he knew that someone would move on within a few years. We talked a bit about academics wanting to have children and the gender bias around this when those in question are women versus men in the early career stage. I reminded him of two incidents when I had witnessed the women "exploding" because they themselves wanted to address comments he made that they felt were inappropriate. Because he was so open and spoke his mind about racism and mistreatment, and often encouraged them to do so, they in turn trusted that they could bring up his unfair treatment of them *with him.* His ethic of helping young Black males ironically made the women in the lab also see Kittles himself as potentially attuned to the ways that they too needed *someone*—him!—to look out *for them.* Nevertheless, he was the first to admit that he could be sexist.

During one of these honest moments, he turned to me and said: "You should interview me about my sexism."

"Okaay . . . ," I wondered if it was just a fleeting retort. *Is he serious? I'm sure he'll back out.* But I detected that he was not speaking on a whim. From my prior experiences, I came to believe that no man in his position would talk openly and honestly about even the slightest possibility of misconduct. Here was a lab director and renowned geneticist with responsibilities to young women scientists who, at the moment, were letting him know that he was failing them on a very serious count. Rather than denying it, as so many do, he was in fact volunteering to open up about it.

Kittles explained why he wanted to do the interview, that he didn't want to repeat what had been done to him as a young scientist. He had felt isolated and disregarded by some of the people who might have mentored him. He was often left alone to "navigate the maze of academia" where there were hardly any Black geneticists or scientists. There had been the occasional off-color joke, the hurtful offhand comment, and too many microaggressions to recount or even recall. And now he feared that he might be doing that to others. Not about race but about "sexist stuff."

When we sat down for the formal interview, Kittles tied these issues to his personal life as well. "I'm passionate about what I do because I bring with it who I am, and my experiences. And look at me, I am *a Black male.*"

"Uhm mm."

"And so, that is the worldview—if you want to use that term—or the centrality which I operate from. That's what keeps me balanced, as a Black male, in this crazy world of academia, right?—and business."

"Right."

". . . I don't really see myself as disrespectful to women. And when I say *really,* I know that in the past I might have treated women . . . ," he paused, "in relationships . . . not . . . the best. So, I can't say that I'm the *best* at treating women well. But I *do* recognize that women are important and should be cherished—and bring expertise to the table just like a man can. I understand that."

I later brought up the comments: "The women in your lab are always giving you cues or calling you out . . . the comments . . ."

"They should." He paused. "No, it's interesting that they feel it. It's probably *there* subconsciously. I don't hate women. I don't. I don't disregard them. . . ." It was clearly not easy to talk about these issues. He was being careful, which almost anyone in his position would be, understandably. It seemed that he was also trying to work out and work through behaviors that he wanted me to witness him processing.

Back at the weekly lab meeting, almost two years had passed since Kittles asked me to interview him about his sexism. His behavior had improved but there were lapses. Now there were new women hires in the lab hoping to bring their interests and expertise to the table, and they would need his support.

I refocused on the urologist: "This is actually one of the things we were discussing the other day about sex education, and young men needing to understand when a woman is ready, and to be aware of her needs as well."

The urologist looked around the room as if giving a sex ed lesson: "Yes. They were probably not ready."

We moved on.

·   ·   ·   ·   ·

"Okay, okay. So let's hear from you," Kittles said, as he turned to Ebony. Born and raised in Chicago, Ebony was unlike anyone else in the lab. Growing up, her family was part of the Nation of Islam (NOI). She witnessed first-hand its emphasis on Black people helping each other and trying to create social and economic possibilities to live better within American society. The Nation was a soft presence among the scientists when they moved from Ohio to Chicago. Louis Farrakhan's house is in the same Hyde Park neighborhood where the University of Chicago sits. When the lab first set up there, before moving onto UIC, my bike route would take me past Farrakhan's home to get to the lab or to get to one of our postwork meeting spots. (It would also take me past men in black suits with earpieces connected to mics in their sleeves to communicate any suspicious activity outside the gates of the Obama residence, about a half mile down.)

The cultural imprint of NOI came up again and again, besides the time Kittles let me borrow his book: like when lab members were doing community outreach and would get questions about melanin research, or when Kittles donated slightly outdated scientific equipment to an NOI school and gave them lessons on genetics, when he tried to correct some of their ideas about human speciation as the kids tried to grapple with why white people had done so much violence to Black people. He took it upon himself to carefully address their embodied, centuries-long exasperation with anti-Black racism that found its defense in judgments about devilish genes for blue eyes. "I couldn't let them go on believing that." He shook his head. "Nonsense." That said, he also understood them.

Ebony's relationship to the Nation was likewise nuanced. She cherished the sense of community and care that she always felt growing up, but she was also able to question it when divisions among people were emphasized. "I was such a loving child, so that part never stuck for me." She explained how her neighborhood next to Hyde Park, called Bronzeville, was a place where people could aspire. Her father had a good job as a physician's assistant. "I always thought he was a doctor. I would see him in his white coat and say, 'Ooh, *I* want to be a doctor.'" She later learned that he hadn't gone to school for his formation but was simply trained by the doctor he worked for. "That white coat!" and the pride in his work inspired young Ebony to imagine herself donning the powerful iconic vestment one day.

As an adult, with much more life behind her, she was drawn to Kittles's research because she wanted to learn more about the diversity of African ancestry and the range of ways that people could appear Black, or not, but identify with being Black, which was evident in her own family and friend group. Her interest in ancestry, as linked directly to Africa in ways that she did not have the details on, also made the lab appealing.

Ebony shared with me that her mother had died from breast cancer. She memorialized her mother's death in her renewed desire to become a medical scientist of some sort. Before her mother passed, she had spent time in Senegal, West Africa, where, as she told Ebony, she had epiphany after epiphany about her connection to the land and its people. Her mother didn't need an ancestry test to tell her where her African origins lay. She sensed it. In recounting her deep feeling to Ebony, the daughter adopted her most immediate ancestor's presentiment. She came into the lab wanting to learn about breast cancer, and also about fibroids, the afflictions that she knew affected Black women in West Africa and the United States in particular ways. Her first project was on obesity, however, because so little information existed on fibroids and genetic links, or on fibroid causation at all. Also, breast cancer was too large and overwhelming for the time being. An obesity project for the master's thesis was not only more feasible, it also had the potential to make a wider impact.

I would find Ebony in the lab some afternoons as she was preparing her proposal. On several occasions she was ecstatic that she had found this or that single nucleotide polymorphism associated with obesity in Black people. Ebony wanted her master's ASAP! but, as it turned out, she wanted a child even more. She was single at the time. This proved to be an obstacle on some days, and not so much on others. Increasingly she began to imagine a decision to have a child on her own. It did not take long for her to learn how her desire made Kittles nervous, but she mostly did not let his anxiety derail her deep longing. The clock was ticking. She said that when she talked to her family and friends about in vitro fertilization, artificial insemination, or just finding a donor, the people in whom she confided said things like, "We don't do that, Black people don't do that. . . . Just find a man."

Ebony felt alone but also said that she understood their reaction. The medicalized way of baby making was not her first choice, but she saw no other option.

·   ·   ·   ·   ·

"Her master's thesis proposal is the bomb," Kittles announced to the group before Ebony could offer what she had been focused on that week. He was impressed by the preliminary research she had done, emphasizing it a few times over. She got slightly embarrassed but smiled. "It felt good to be acknowledged," she said, especially since she had taken much longer than planned to finish her bachelor's.

Ebony had started out her undergraduate education with high hopes to embark on a pre-med track. She was even on scholarship at Grambling State University, a public HBCU in Louisiana. But, as she told me: "I wanted to party and connect with all the beautiful Black people on campus. [. . .] Erykah Badu was there my freshman year. There was a powerful poetry scene, and a lot of love. I lost interest in my classes and moved off campus. [. . .] By the time my mom came down to try and see what was wrong and straighten me out, it was too late." Ebony lost her scholarship and had to drop out of school for a while. Before she could get back on track, she had to return to Chicago when her mother developed aggressive cancer due to what she thought could have been medical oversight. Years later, in her mid-thirties, when Ebony entered the Kittles lab, she felt empowered by a new sort of family to pursue the higher degrees that she had envisioned for herself as a child. She wanted to be a doctor then, but now "*research* was calling."

"Yep, you guys should read it. It's the bomb," Kittles repeated. He tapped his index finger on Ebony's proposal as it lay on the table in front of him.

Ines noticed something that no one else did at that moment. "Is that why you're using it as a coaster?" she asked. The group erupted in laughter. His cup sat still, in innocence. Ebony waved the air away from her with a limp hand, as if shooing Kittles. The gesture was a playful he-can't-be-serious swipe.

Kittles put on his stern face and pinched his eyebrows together. He leaned over dramatically and zeroed in on Ines. "I have to be careful with you." He squinted as if trying to read a detail that was too fine to see. Ines blushed a bit, held her own, and smiled. Kittles looked at everyone seated around the table, then confided: "I look at her, and all I see is a flower child. I have to be careful."

I was impressed with Ines for speaking up. She was an honors student on scholarship and in a prestigious undergraduate science program. Whenever she spoke, she revealed that there was more going on inside than the older adults might have assumed. I offered one of my observations: "A flower child? The other day she was wearing a T-shirt with an ice-cream cone—but the strawberry and chocolate scoops were lit grenades."

Ines nodded, and looked at Kittles assuredly. He made a comment about the wars going on and turned his chair toward Ebony, keeping his eye focused on Ines for dramatic effect. "So yes, this proposal . . ."

"Um hmm," Ebony said and cleared her throat. "You were saying . . ." She brought everyone back to focus.

.   .   .   .   .

One early evening as she was packing up to leave work, Ebony asked me if I would go to see the film *The Switch* with her. She explained the premise: "It's about this woman who wants to have a child on her own because her relationships haven't worked out."

"Okay . . ."

"It's with Jennifer Anniston. She's determined to have a baby, so she does IVF. But the guy who's in love with her switches—oh, let me not *ruin* it."

The film, a romantic comedy where almost everyone in New York city is a white person, nonetheless promised some deeper philosophical moments in the trailer about the modern human race being in a rushed course against time. The main protagonist agonizes about her biological clock and is on the hunt for a nonanonymous sperm donor so that the man might be in the child's life and father her, or him, outside of the confines of a traditional couple. Ebony told me that she herself had discussed a similar prospect with men over the years—to help her in this way—but it never felt right.

As she talked about it more and began to have consults with specialists, she found that even soliciting support from her girlfriends to help her through the process proved difficult.

"I don't want to ask anyone I know to go see the film with me. None of them want me to do this. It's like they think in vitro is for white people."

I told her that I had heard that before, too. "It's probably that it's so expensive that it becomes a class issue, and so might actually mostly be for white people." She agreed but was still frustrated at her friends' faith in what seemed to be a hopeless traditional route of "man + woman = baby" kind of math that was not working out for her. We planned to see the film later that week.

When Ebony talked to me about her desire to have a child, she put it in terms of the Black family and what she saw as Black people's sex/gender dynamics. The hope that she might be part of a unit, that the child might have a father, and that she might have a partner to raise her, or him, was a fantasy—a phantasmic specter that threw into relief the reality all around her. She would tell me that Black men were not showing up for Black women, even though she witnessed women trying to make it work, committing, supporting, and being let down. It wasn't only the individual men. It was social. It was about opportunities, or the lack thereof, about incarceration rates, about educational disparities.

I could not help but think of how this larger family dynamic expressed itself in the lab in some ways. The basic sex and gender politics of the lab were rooted in the structure of Kittles's main focus on a cancer of the male anatomy, his desire to map its risk in Black men as a vulnerable group in the United States and in the larger diaspora, while a team made up mostly of women helped him to do so. However, these young women scientists also wanted to focus on women's health and cancers that affect the female body, especially the Black female body. They wanted the Black man in their scientific lives to "show up for them."

This all-too-familiar, lopsided support was a topic that Ebony broached given her quest to be a mother (yet others in the lab also described an unevenness between Black men and women that manifested in their lives in different ways). When explaining why she felt so discouraged about relationships and waiting for the right person to have a child, Ebony offered me a whole paragraph without pause:

"I'll say that it is because of the dynamics of the African American family. Unfortunately, there are a lot of educated single Black women, or just a lot of single Black women period. Because of our Black men—there are a lot of them in jail, or they are not working, or they're on the streets. So the pool is small in terms of marriage. And this is not just what somebody told me, this is *what I see*. I'm not just going on what everybody is sayin'. This is what I see with my sisters, who are beautiful. They have children, but the men are not stepping up and wanting to get married. This is with my friends. I have friends who look like Halle Berry. So, you can't just say that they aren't pretty enough. I have friends with PhDs and others with no education. The men are just not marrying us. I have decided that I am going to do in vitro fertilization. I don't know how this is going to play out. . . . I'm really ready to be a mother now. When I go to my doctor's appointment, they'll let me know how viable my eggs are. And if everything is everything, then I'll wait until a couple of years. But if they tell me, 'Girl, you gotta do this now,' I'm gonna be pregnant next week." She laughed and swiveled gleefully in her chair.

"You are?" I said surprised (she really was so excited about those nucleotide polymorphisms she had just found in the HapMap database).

"Yeah, it's gonna happen—and I'm very excited about it."

"Good," I responded, as my mind flashed to what I imagined to be Kittles's expression when he'd find out.

"I'm ready to be a mother. I am. And that's another reason why I want to complete the master's. . . . If I decide to have a child, I'll have a master's, you know, and I'll feel proud of that—and in the sciences? *Whaaat?*"

Her voice went high as she sang the last word, the proud astonishment at what would indeed become her successful achievement. As for the ensuing IVF and child, as so many women know, that journey was uncertain and years-long. When Ebony did eventually have a child, she had given up on artificial means. She had undeniably spent way too much effort on the not-so-magical technology. She eventually did meet someone and embarked on a relationship. Although she had abandoned the costly fertility route, she conceived naturally shortly after her and her husband's final IVF attempt.

Throughout this long process Ebony stayed with Kittles and the team at UIC and also when they moved out of her city of Chicago to the desertlands, where the team relocated to the University of Arizona. Over the years Kittles watched her struggles, felt her frustrations, and identified with her determination. In the end, it was Kittles who proudly shared the news with me when she did give birth. He was giddy—and genuinely happy for her. And it was Ebony who was quick to tell me when Kittles would expand his focus to pursue collaborations outside the lab on breast cancer as well as adverse drug responses to the commonly used anticoagulant Warfarin, which they worked on together. Eventually, in 2021, Ebony earned her PhD.

7 Sci Non-Fi

CELLS, GENES, AND THE FUTURE TENSE OF "DIVERSITY"

As part of my ongoing research to understand how genomic scientists deployed concepts of race and human diversity, I began doing field work on the Personal Genome Project (PGP) and the laboratory of George Church at Harvard Medical School starting in 2008.[1] Church and his team wanted to look at full genome sequences, not just a slew of hundreds or thousands of biomarkers that so many other projects, direct-to-consumer (DTC) companies, and large-scale studies were focused on. Personal genome scans, which focus on inter-individual variation, would move away from racializing methods that often start out with group-based identifiers as categories for organizing DNA variants by design. Full genome sequences, at this point, were still technically not the entire thing, however. Church was primarily focused on the coding regions, called exomes, which would yield somewhere around 1 percent of the six billion DNA base pairs that comprise the entire human diploid genome (forty-six chromosomes). That math comes out to sixty million bases, which is still a whole lot.

George Church is a towering yet gentle figure, with thick graying hair, an equally impressive beard, and a mind full of visionary plans for science that he often narrates in dulcet prophetic tones. Over the years I have heard him compared to Santa Claus, George Clooney, and Jesus Christ. I came to realize that he might just possess some of their most alluring personality and phenotypic traits—and these attract his own group of future-seeking followers.

SCIENTIFIC SHIT STARTER

An integral player who actively participated in the rise of computing and the creation of automated methods to read DNA, Church has had his eye to "doing things very differently," as he told me, for most of his life. When I entered his lab, in one of our discussions about the PGP and his philosophy on just "putting everything out there," because "whoever wants to will probably figure it out anyway," Church pointed me to his website where he had posted for all the world to see what is usually cherished personal information. This included his medical conditions, among them heart disease and narcolepsy, the coordinates of his home address, his date of birth, a letter expelling him from graduate school for flunking out of coursework, and up until recently, his social security number. His wife asked that he not make it so easy for people to misuse or steal his data, and to please redact that last bit of prized information. He complied.

The PGP is Church's brainchild, which, as he recounted, stemmed from his "disappointment that the goal for the Human Genome Project was just *one* genome—and that it would cost $3 billion." As he envisioned the PGP, he advocated for full sequences for a cohort of many. He was committed to bringing down the eventual cost to $1,000, so that everyone might enjoy the gift of knowing what lies in their genes. The PGP differs from most other genetic studies in the vast amount of data that it gathers and produces on multiple fronts: genome sequences, trait/phenotypic data, the collection of blood and skin tissue to make immortalized cell lines (then synthesizing participants' biology to make workable tissues), and its ongoing series of "ethics experiments" to see if people enrolled would consent to allowing the PGP to make their data public. The participant-driven science they imagined was a highly specific form of what medical anthropologists Deborah Heath, Rayna Rapp, and Karen-Sue Taussig have called "genetic citizenship."[2]

The scientists on the Personal Genome Project have adopted the philosophy of the open-source movement in computing and applied it to DNA, tissues, and more on multiple levels, regarding what they call "open hardware," "open wetware," and "open genomes" that will rely on "open consent." The latter rests on the idea that participants should be aware that they could be *re-identified* by known or unknown actors via their *de-identified* data and cell lines with little effort. In a key PGP article on these issues, Church and his colleagues positioned their concept of open consent against other genomic study consent protocols that promised to anonymize donors' data. When studies like the large and influential HapMap project did broach re-identification in their consent protocols, deceptively the wording made the possibility appear to be "vanishingly small."[3] The PGP team called this out, writing: "For example, the consent form of the International HapMap Project assures the participants in the following way: '[. . .] it will be very hard for anyone to learn anything about you personally from any of this research because none of the samples, the database, or the HapMap will include your name or any other information that could identify you or your family.'"[4]

Church and his team also cited the acknowledgment on the part of the American Society of Human Genetics (ASHG) that despite the fact that huge numbers of people are needed in studies to find risk alleles for complex disease in large-scale Genome Wide Association Studies (GWAS), a possibility lurks that donors may be re-identified because of the unique quality of DNA to reveal aspects of the person. In the same paper where they introduced open consent, they wrote: "The ASHG declares the following in a statement on genome-wide association studies: '[the ASHG is] acutely aware that the most accurate individual identifier is the DNA sequence itself or its surrogate here, genotypes across the genome. It is clear that these available genotypes alone, available on tens to hundreds of thousands of individuals in the repository, are more accurate identifiers than demographic variables alone; the combination is an accurate and unique identifier.'"[5]

In general, there exists an agreement in the genomics world that data anonymity should remain a best practice—despite the technical loopholes that can undermine it. When data is anonymized, most research institutions including the US government's health and genomics arms, the

National Institutes of Health (NIH) and National Human Genome Research Institute (NHGRI), have established professional norms that it is more or less safe to share donors' information and in some instances their samples. The PGP researchers who penned their open consent article, which included a bioethicist and a lawyer, argued that such a guarantee does *not* deal in the "truth." In their words, the human genetics research enterprise as a whole, whose efforts at base always depend on the trust of humans to offer up their DNA, was often peddling in what they called "false promises." They argued that with the business-as-usual approach, research subjects would be allowed to continue with "wrong expectations" about genetic privacy.[6] They ended the piece by stating that human genetic research "requires veracity on the part of the researchers, as a primary moral obligation."[7] How would this truth-telling be possible going forward as genomic studies and huge collections were ramping up in the 2010s?

Putting central players in the genomics world on blast for assuring a nearly impossible-to-assure feature of privacy was a key characteristic of Church's genomic counterculture ethos even as he simultaneously enjoyed the position of being a recognized leader in the field.[8] The PGP's proposition, from the moment of its launch in 2005, was that since a person's genome is highly specific and individualizing, it would be possible to uncover a supposed anonymous sample's real identity even amid large swathes of data. That is, if someone with the right skills attempted to do so. It turns out that such skills in this day and age are more prevalent and stretch far beyond the imagined boundary lines where government-funded lab scientists and university researchers conduct science today.[9]

PROOF OF PERILOUS CONCEPT

Starting in 2008, computational biologists started to publish on holes in what many had assumed to be secure genetic databases. They began to expose the weak premise of anonymity often offered to people who participate in genetic studies or who give DNA to companies through home test kits. One of the first instances of a sample being re-identified was when a computational biologist in Arizona named David Craig and his team showed that if an individual participated in what were then promising

GWAS, which can number into the tens and even hundreds of thousands of individuals, then some participants could be discerned from the data. This was the case even when the data was aggregated, and allele frequencies were presented in summary.[10] A few years later in 2013, even more destabilizing to the work of the National Human Genome Research Institute at NIH, a computational biologist and security expert named Yaniv Erlich re-identified samples that were said to be anonymized by the US government's NIH in its first large dataset to provide "full" sequences of diverse people's genomes called the 1000 Genomes Project. In a 2013 report in *Nature*, journalist Ericka Hayden narrates the key events so succinctly that I quote her at length.

> In 2011, as Erlich was setting up his first independent lab as a Whitehead Fellow, he met a Colorado-based woman, Wendy Kramer, whose son had managed to track down his father—an anonymous sperm donor—by searching a consumer-focused genetic-genealogy database for people with DNA similar to his own.
> [...]
>
> Erlich wondered whether a computer program that he had been working on with an undergraduate student, Melissa Gymrek, might enable a similar trick using de-identified genome data from human research studies.
> [...]
>
> His team extended its analysis to men whose genomes had been sequenced as part of the international 1000 Genomes Project. Extensive information about these men, including their ages and detailed family pedigrees, was available on the website of the Coriell Institute for Medical Research in Camden, New Jersey, which distributes cell lines made from their tissues to researchers.
>
> Erlich's team used [their software invention called] lobSTR to infer the men's STRs [short tandem repeats] from their 1000 Genomes data, and then searched Y-chromosome databases to find linked last names. After that, it was relatively easy to search public records databases to find men with those last names who were the right age, came from the right place and had similar family trees. The team identified nearly 50 people, including DNA donors and their relatives. When he first saw the results, Erlich said later, he was so shocked at how easily the method worked that he had to go outside and take a walk.[11]

As prophesied by Church himself, the identities of PGP participants who had decided to try to play it safe and not put their data on the web, as Church and the other first ten participants had done, were also re-identified in 2013.[12] The not so difficult feat was carried out by then MIT data scientist Latanya Sweeney and also subsequently featured as part of the PGP experiment on being comfortable with the discomfort of genomics but proceeding ahead anyway. Some potential participants might slough off, but others would get on board.

RICH IN IDEAS, POOR ON PAPER

As concerns the molecular biology and cellular bioengineering aspects of the PGP, its open-source genetics was not necessarily restricted to an emphasis on crowdsourcing to improve upon the genome, although this could be possible with synthetic biology and CRISPR gene editing, eventually. Rather, the Church lab envisioned the openness of PGP data as "a resource" to researchers to pursue questions of their own and to promote "inclusive" science without an old-fashioned proprietary caginess affecting everything. But this way of doing things would pose other problems, as Jason Bobe, the PGP director of community (and frankly the principal organizer holding things together much of the time), told me, such as "getting frozen out of the NIH for funding opportunities."

Bobe was Church's second-in-command, and Church often referred to him as "an army of one" because of his "great ideas." One of those lightbulb moments was when Jason established the project as a nonprofit organization 501(3)c very early on. Interest in getting one's sequence and being in a project led by George Church brought in thousands of people at certain points in a matter of months. Jason was primarily responsible for PGP recruitment, conference organization, publicity, and finding funding so he witnessed firsthand the lack of backing for the scale the rush was bringing on. He filled me in on the project's cash woes over and again. I soon recognized that he ran on hope, perhaps even more than Church. As the lab tried to scale up the project beyond the first ten participants in 2009, he walked me through all of the places where capital was *not* appearing for them. Most of the places and people they had anticipated giving had not followed through. Church had royalties from patents, some money from the NIH for technology development more generally, and then there were some private donors, but the hodgepodge of money pots did not cover the projected costs.

"Without this money the organization just won't be effective, you know," Jason lamented. But he quickly picked up his habitual optimistic thread again. "So, but—we *can* squeak by with the PGP-100 without that money, but we could use it, like, *yesterday*. So that's a really, very stressful thing right now. Just trying to do all this work on no budget—just pulling it out of the air. But we've managed—we've really managed to pull it off. *We're not there yet*. But we're almost there." *Phenotypic hope*, I thought to myself in silence as Jason whirled through the ups and downs.[13]

But on a broader societal level, the issue came down to who would feel most comfortable and trusting having their genetic information on the Web for anyone to see, use, and potentially manipulate.[14] The researchers knew, but didn't know exactly how to work around, the fact that most minorities, the poor, or anyone concerned with police, state, or corporate surveillance in the slightest would likely not be joining anytime soon. More important, anyone who didn't have health insurance guaranteed for life, or who wasn't independently wealthy, would possibly forgo the so-called "moonshot" opportunity.[15] These barriers went beyond the usual ones that, as concerns many African Americans, manifest as early as the seventh grade. These include poor secondary education in math and the sciences, a general apathy toward science as useful for everyday life, an inability to see themselves as competent "doers" in a scientific field assumed to be a "white" vocation, a lack of exposure to scientific role models, and a general distrust of scientific research as a result of the history of race relations in America.[16]

Church paused when I asked him about diversity in the PGP. He admitted that he and his team would not be able to approach the idea as conventional multicultural inclusion in the United States, which, as mentioned in chapter 1, has been the NIH-mandated model for biomedical research since 1993. He acknowledged that the science pipeline is disturbingly non-diverse (as was his lab) and that the PGP's make-up of those eager to become participants—and who at the time sat in a queue of nearly seventeen thousand people, when the project was supposed to get sequencing in earnest in 2012—were 70 percent male and 90 percent "white." The team tried to address this when they could.

In one attempt early on they invited Rick Kittles to join the PGP-10, but he declined. Kittles later told me that he "would never" put his genome data on the Web. He cited people's ability to link traits with those he might choose to redact, which could be stigmatizing or worse. As a Black man, he was deeply concerned about the use of PGP data for forensic applications, which he felt were societally biased. (Forensics and potential misuse figured among the many risks detailed in the consent form, more on that below.) The initial thinking for the PGP team was that high-profile people would signal to everyday folks that the adventure was something they should do.[17] Then there was the hope that the more people who got involved, the greater the possibility that some general risks concerning data privacy might be diluted. This was basically the idea that if we are all at risk for something, all flawed by "nature," then carrying scary traits might not be so scary as mutations became normalized. At one point they drew up a list of African Americans connected to the world of science who might make good candidates to serve as examples. Finally they recruited Harvard professor Henry Louis Gates Jr. as well as his father, Henry Gates Sr. (Gates featured his own participation in one of his widely viewed PBS films).

Practically and philosophically, however, "genomic diversity" for the PGP was, as mentioned at the outset, to be explored where the wellspring of it resides—that is, "inter-individually" between any two given unrelated persons on the planet, compared to one another. Shifting the focus from groups, who are often racialized, to individuals, who would be characterized by the power of full genomes, would teach us "a whole lot more," in Church's view, about what we all carry. In other words, the PGP would unearth untold numbers of newly, not yet discovered, variants that everyday people are walking around with in their genes. Beyond the possibility of re-identification, Church wanted to make sure that people understood what could be done with human genomes. For this reason, he included a slew of highly undesirable "risks" that participants were asked to contemplate before signing onto the study. On page 13 of the twenty-four-page, single-spaced consent form listed under "Risks and Discomforts," participants would need to agree to the following possibilities:

(ii) Anyone with sufficient knowledge and resources could take your DNA sequence data and/or posted trait information and use that data, with or without changes, to:

(A) accurately or inaccurately reveal to you or a member of your family the possibility of a disease or other trait or propensity for a disease or other trait;

(B) claim statistical evidence, including with respect to your genetic predisposition to certain diseases or other traits, that could affect the ability of you and/or your family to obtain or maintain employment, insurance or financial services;

(C) claim relatedness to criminals or other notorious figures or groups on the part of you and/or your family;

(D) correctly or incorrectly associate you and/or your relatives with ongoing or unsolved criminal investigations on the basis of your publicly available genetic data; or

(E) make synthetic DNA and plant it at a crime scene, or otherwise use it to falsely identify or implicate you and/or your family.

The most shocking for me was that with one's full genome sequence locatable on the web, people with know-how could synthetically recreate a participant's DNA sequence and plant it at a crime scene. Synthetic biology was an area of Church's expertise, so he knew the ease with which this imagined, not so futuristic, scenario could materialize. He wanted to responsibly forewarn people of the real risks of just putting one's sequence out there: risks that few others in the field dared to divulge.

EARTHLY PROBLEMS AMID THE HOPE TO MOONWALK

The highly visible first ten participants were all white people, except for one. This man, the original sole racial minority in the effort, was christened "PGP-10."[18] On the PGP website his important enrollment number, the "10th," would finalize the initial smaller cohort of people (most of whom were scientists or professionals in the medical field themselves) who, as Church imagined, would be the "heroes and human guinea-pigs paving the way" for the general public.[19] These ten would be the ones to put all of their identifying data on the Web and undergo all of the trait and sequencing plans that Church had for an eventual one hundred thousand people. That is, if Church and his team could just get the funding. But "PGP-10" also had a name. He also had a race (African American). And like the other participants, he had a photo (both a frontal shot as well as a profile, staged with a ruled piece of adhesive measuring tape on his forehead for potential forensics research). He also had a vision.

I met James Sherley at his office on a frigid Massachusetts winter morning in a little place called Watertown, which, for someone without a car, is a brisk walk or a bike ride through the better known town of Cambridge where I worked. I rode my bike to avoid the slick February sidewalks of centuries-old brick (individuals are supposed to shovel their own snow), preferring the more reliable city-cleared streets. I arrived for our 10 a.m. interview at his office invigorated by the cold. As I settled in, we talked a bit about where we were from, an unspoken first assumption that we were both transplants. He was from Memphis. So was my father, I told him. I would come to notice the optimism and high notes of cheer in Sherley's voice as he recounted his story—notes that my older cousins Poncho and Lil' Bob had also kept in their diction, despite their many trials with racial discrimination and harassment in both the US South and North.

As a child, Sherley had always wanted to be a scientist. Now, with his crisp light-blue starched shirt, thick-rimmed glasses, and curiosity, he led a life far more complicated than any aspiring child might imagine. At the time, Sherley was part of a team at a nonprofit biomedical research center specializing in a range of diseases including Alzheimer's, the dystrophies, cancers, and heart disease. His role was senior scientist in their regenerative biology as well as cancer biology programs. I asked about why he joined the PGP, a hyped and controversial project, which might bring more trouble for him since he was already embroiled in debates about whether scientists should use embryonic stem cells. This, in addition to racism, may have negatively affected his tenure prospects at MIT. Sherley enthusiastically responded: "To be the first amongst the firsts."[20]

Despite his woes with fighting MIT over his promotion and later the US government over embryonic stem cell research, he was excited by the prospect and asked Church—who had been an integral supporter in his tenure case and who, with his wife, even set up the protest website to make Sherley's tenure fight and hunger strike more public—"Why he hadn't asked me sooner?" Others of those first ten enrolled in the PGP, collectively called "*the* PGP-10," (Sherley is simply "PGP-10"), wanted to join the project largely to get their full genomes sequenced at a time when such a feat still cost hundreds of thousands of dollars and was reserved for the über wealthy.

Sherley continued. "I joined not just to have my DNA sequence—but there is a vain component to this—which I'll admit to in a moment. But the DNA sequencing was not a part of that for me." I nodded for him to continue, as I took a few notes. "I've been interested in the social issues around technology, and how it impacts people. And how nonspecialists process the information, the various ways it is presented to them. Some of my MIT protest was all caught up with my position that embryonic stem cell research is not ethically, scientif—is not something we should be doing. And so I think George realized I was interested in these issues. How do you conduct what you hope is going to be beneficial science, in an ethical fashion?"

We talked for a bit about what Sherley saw as the inevitability of not just the $1000 genome, but eventually the $100 genome and how scientists, politicians, and the lay public would have to be informed about the technology that, as he said confidently but calmly, is "gonna happen." Sherley returned to the other motive that was shared among the participants. This came up as a semi-serious joke, in the summer of 2007, at one of the initial gatherings of the first ten participants on the Harvard Medical School campus. One of them forecasted that there was one trait that would likely be found in all of their genomes. This was, perhaps all too predictably, "the narcissism gene."

THE FIRST ADOPTERS

PGP-2, the second enrollee in among the first ten, was then chief information officer (CIO) and a dean at the Harvard Medical School (HMS).[21] He oversaw several hospitals tied to the HMS system (including Beth Israel, Deaconess, and Mount Auburn) and had other predictions for his fellow cohort to consider. In his introduction he told them that in his capacity as CIO, he was responsible for the security and privacy of sensitive information on over three million people, "including all the data for the Red Sox players." He emphasized his interest in the "moral, philosophical, and medical implications of the PGP." Adding, "as a corollary to that, I have an implanted RFID that has my medical data, my identity on it." He cautioned them that the societal reaction that he had received when different publics learned about his chip might be a harbinger of what the PGP-10 cohort would confront for flouting cherished notions of privacy. PGP-2 went on:

> So I've had experience going public with such a thing, and there's been a huge reaction to that—ranging from "You've lost your anonymity" to "You've got the mark of the beast!"
>
> And so, I certainly will be happy to share with you the thousands of email responses that I've received from the general public about something that's just trying to push the envelope a little bit.

Another participant, PGP-5, had already been getting lots of correspondence from the public—an abnormally large subset of which just happened to be his own biological children. In the 1980s and 1990s, PGP-5 had donated his sperm regularly to a clinic in southern Michigan. He was a smart young medical student—one with blond hair and sky blue eyes. The clinic realized the American market gold value of such a specimen and used his sperm "far in excess," he offered, of what should have been the case. One day he ran the math and discovered that he had fathered more than four hundred children.[22] In 2008, at the PGP public data release event, he told me that many among this mass of his unknown progeny had contacted him via Facebook. Through him, they were henceforth in touch as half-siblings. Although he and his wife had borne their own biological children, he found ways to have relationships with some of his once anonymous kids as well. In his introduction to the larger group of the PGP-10, PGP-5 stated his reason for joining the project: "I became really interested in my own personal genome—just because there were so many copies of it."

In a discussion on all of the questionnaires and recurrent updates that the PGP wanted its participants to fill out over the course of several decades, the small group of early adopters mulled over whether years of updating one's lifestyle and phenotype information would be a turn off.

Figure 5. A still from the documentary film *Genome: The Future Is Now* showing the first ten PGP participants in mugshot-style facial photos. Both frontal views and side profiles were placed on the PGP website, along with their initial genome and trait information. Courtesy of Marilyn Ness.

PGP-3, who was an entrepreneur and angel investor serving on the board of the then nascent 23andMe, and also training with Russian cosmonauts for space travel, didn't think so. "Let's face it," she reminded the group, "people think they're fascinating." And the meeting went on. Scenes from this event, which happened on the same day that the PGP-10 had their faces photographed with a strip of adhesive measuring tape on their foreheads, had their saliva collected, and blood drawn, were filmed for the documentary *Genome: The Future Is Now*, available on YouTube (figure 5).[23]

IMMORTALITY IN VIVO

Back in his office with me on that wintery day in Watertown, Sherley continued his thought. "But there's another piece here, and the other piece is that—and this is the vanity of it I guess—it's that *I am now an immortalized person.*" He smiled. A clear giddiness rang in his voice. "One of the challenges in a project when you want to have genome information and [you want to] allow other laboratories and scientists to evaluate pheno-type versus genome, you need a source of that genome in a cell. So the project has immortalized B cells. And every one of the PGP-10 right now has cell lines that are immortalized. They're immortal, which means that they're not *quite* normal. But the method that George chose to do this actually keeps them pretty close. They are not like HeLa cells. They are not aberrant tumor forming cells."

I jotted down a note to follow up on that thought, as his voice shot up.

"And anyone can get them—including me! Although I haven't bothered to order my own cells. . . ."[24] We both laughed at this weird capitalism.

"How would you describe that, in lay terms?" I asked, going back to his comparison of his cells to HeLa. "Why are they *not aberrant*, or 'close' to normal?"

"They are closer to normal because the viral proteins that are used makes the cells proliferate continuously, but if you take these cells and put them into an animal, they won't proliferate as a tumor. So they have longer growth potential without taking on the properties of forming tumors. And in terms of gene regulation—" He interrupted himself to cut over to a broader thought. "So remember, what we are trying to study, or what the project wants to study, [is] how does genomic sequence predict the behav-ior of cells?" Sherley already had some ideas about how cells worked in the finite body, but proliferation—the ongoing nature of life at this basic level—fascinated him.

I thought deeper about why a scientist whose life work concerned the vitality and behavior of cells would have a slightly different obsession with life than most people. One of the mysteries in his work was to figure out how to keep cells from proliferating at the hyper-boosted reproductive pace seen in cancers, but that was also tied to an interest in regeneration. The life character of proliferation run amok, therefore, coursed strangely close to the controlled process of immortalization where cells are made to constantly exponentiate and are estimated to live well beyond the wildest human lifespan predicted goal of 150 years.

In the meeting where the PGP-10 group raised the possibility of the narcissism gene, one other statement stood out for its strangeness. When discussing the possibilities for the immortalization of their cells, PGP-2 mused about the sheer amount of him, measured in tons, that would or could exist in the future. His cells, allowed to reproduce over time, could far outweigh the corporeal mass of the 165 pounds that constituted his mortal middle-aged body. Sherley came back around to this idea, and the prospect of living so long in a line of cells that might at one point wrap around the circumference of the earth, perhaps several times, like the HeLa cells that he had worked with his whole life. And amazingly, like the HeLa cell line, Sherley's would be here for as "long as people were on this planet." It was during these musings when he uttered his strangely articulated new belief: "It's better than *children*."

.   .   .   .   .

Sherley is the son of a Baptist minister from a religious family. He is also the father of two daughters as well as a scientist who believes that life begins at conception. He assured me that it was his scientific understanding of human fertilization at the cellular level that made him opposed to using embryos for research. He was smart, reasoned, and gracious in discussing these matters with me.

At the time in 2010, he had brought a lawsuit against Kathleen Sebelius, the Health and Human Services secretary in President Barack Obama's first administration, and Francis Collins, the head of NIH, when Obama issued Executive Order 13505, which expanded federally funded research on human embryonic stem cells (hESCs) in 2009.[25] Sherley and co-plaintiff Theresa Deisher of Nightlight Christian Adoptions, also a stem cell biologist, suffered a setback in one court, had success in another, and finally a loss, which prompted them to take the matter to the Supreme Court, which refused to hear the case in 2013. At issue for Sherley et al. was the reach of Obama's order. They claimed that the government was in breach of a rule that federal funds could not be used for research on destroyed embryos.[26]

Sherley has spent many of his life's days on what are called induced pluripotent stem cells (IPS cells), which are adult cells that could be reprogramed back into an embryonic like state and become lots of different tissue cell types. These include motor neurons, blood cells, liver cells, bone precursor cells, and even egg and sperm precursors.[27] Therefore, he found it unethical to experiment on living beings that he considered to be human. In fact, he likened embryos' lack of rights to enslaved Africans in the Americas who had no rights and were often experimented on, telling reporters that he would not cease his efforts "to emancipate human embryos from research slavery sponsored by the NIH."[28] These embryonic subjects without rights were beings that he was defending in the Capitol by taking the US government, under its first Black President, to court.

Given this, I wondered what exactly could someone with his commitments consider to be *better than children*? "So, talk a little bit more about this immortality idea and why that's important for you. Is it to be like Henrietta Lacks and contribute to research? Is it more spiritual? Is it just coo—"

He started to answer before I could fully exit the last word from my mouth. "What was the last thing you said?" His voice perked up.

"Just cool."

"Just cool," he offered. We both laughed at the silly truth. "That is just what I was getting ready to say," he said with a big smile. "I can't explain it really. . . . It's just this idea . . . that, you know, what makes us alive, living organisms, is having the information that defines us captured in something different than what we are—a coding system." Enthralled with the prospect for the future of cell science and his own life within it, he continued: "The idea that is *really* exciting for me is that right now as long as somebody either has my cells in a freezer, or they're culturing them—they're probably better in a freezer—in some sense, *I exist!*"

He was careful to include the human experience that his cells *also* experienced by just being part of his lived life on the planet.

"Right? 'Cause . . . these integrative things that happen as an organism is developing, there are processes which will happen at some frequency but they are hard to predict. It's this chaotic kind of process of gene expression. All that stuff is a part of me as well."

He returned to the exciting prospect of his immortalized cellular (perhaps) future self. "We have no idea how likely it is that it will be reproduced if we turn that same genome on again. So there is that element. But the *basic* core information is there. So, you know, maybe [in] some time. I don't know, maybe in one hundred years or five hundred years people may be able to pull out genomes and—*resurrect people*."

"Hmm."

"And we sort of think, 'Well, you know, that brain is really our experience, that's what makes me who I am.' It's what this body has encountered over time, but actually we really *don't* know that. We really *don't* understand memory that well *at all*. We *don't* really know how much of our structure—our chemical structure—relates to how we think about the world, for instance. So it may be the case that this genome reestablished, again, has very similar features so that the person who would be sitting in front of you wouldn't have the *same* experiences [as me], but they would have *reacted* to the experiences they have in the same way that I have. There are some really, really deep questions about humanity there."

I nod, taking in the imaginative future he's envisioned.

"And what I like about the immortality issue is that although we can debate whether what I just described is a good thing to do, ultimately we are still talking about how society treats *people*, however they come into being." Yes, the social and human tendencies to mistreat others were always there. But, he continued, barely skipping a beat, to the excitement inside when he pondered immortalization: "But that's kind of intriguing, ya know, to think that you might wake up in 3050 or something and have some recollection of yourself, because it's in the DNA."

GENESES

There is much to say about the PGP beyond the scope of this chapter, and like other labs and scientists featured throughout this book, Church and his collaborators could easily fill the pages of a monograph all on their own. In fact, they have. One of the PGP-10, known as PGP-4, or Misha Angrist, who holds a PhD in genetics as well as an MFA, wrote his own personal and highly detailed account about the PGP's birth called *Here Is a Human Being*.[29] When I interviewed him for my book, he interviewed me for his. Afterward, we even compared notes. Sociologist Jenny Reardon, relying heavily on interactions with the PGP director of community and recruitment, Jason Bobe, as well as Angrist, also included many of the PGP's concerns with open consent and open data in her book on a much larger set of issues that we are confronted with in what she calls "the postgenomic condition."[30] Others within the PGP and in the science media have written articles and thought pieces that add to the project's ever-growing archive.[31]

Given this, I feel even more compelled to dive deeper into parts of the project that I observed ethnographically. One theme in particular that never came to fruition, but that highlights the futuristic imaginings brewing at the time, concerned developing cell lines to safeguard against the risk of death itself and what one PGP researcher called the "inadequate reality" that human beings were proceeding without "backups" of their genetic material. The notion of a human analog that is so prevalent in the sci-fi genre was a common feature of PGP scientists' discussions of immortal cell lines. Some technologists in the Church lab wanted people to be aware of the myriad scenarios where full humans, or human parts, might be cloned for what they imagined to be good reasons. Therapies that could involve aspects of cloning range from addressing diseases that compromise an organ to accidents or wars resulting in the loss of limbs, from neurodegenerative disorders that might kill brain tissues or limit function to the possibility of death for individuals, or the destruction (and regeneration) of the whole human race. There were several in the lab who thought it would be a good idea for the scientific community (they often thought in terms of a "commons") to have a reserve of humankind.

I am using the terms "backup" and "reserve" to signify an assurance and also to highlight the aspects of the PGP scientific imaginary that draw from visions of life tied to human engineering. An ethos of bioengineering may not seamlessly gel with many people's idea of human organic uniqueness. The scientific vision I heard discussed in the Church lab was to create the possibility for PGP participants to benefit from the source of their own biological material, while also advancing science and the plight of the species through a buildout of that same source's potential. The cloneable reserve, potentially living in immortalized cells, could present the possibility of constructing new organs to prolong one's life, if needed, and also the opportunity to create a "seed bank" to one day reanimate human matter to reproduce disappeared Homo sapiens. Societal acceptance of cloning in light of such impending "risk" scenarios would have to move beyond the current ethical barriers against human cloning, or at least begin to chip away at them.

A futuristic junior scientist, whom I will call 00MAT.Rx since he would at times consciously reference *The Matrix* (albeit with more therapeutic ideas for any source human bioenergy), was one lab member who was particularly forthright on talk of a human reserve. He recounted how he envisioned designing cellular regeneration in ways that could, nevertheless, sometimes shock others. Church told him not to reference growing human organs or body-esque forms as "babies," "spares," or any other potentially triggering image that could create more ethical issues for the project. Despite these researchers' neoteric hopes at molecular wizardry, they realized that science must be done within limits, all the while questioning those limits with their own in-house ethicist, Jeantine Lunshof.

00MAT.Rx, who was an MD/PhD, spent an afternoon walking me through these future scenarios and the technology he was hoping to develop to achieve them. He started by describing the importance of a "human model organism that could be dissected" for large-scale projects beyond cellular studies without alarming medical, legal, and ethical committees. It was late morning when I briefly met 00MAT.Rx for the first time in Church's office. He was in an upbeat mood, proudly dangling a wishbone-shaped tool that he had fabricated himself. The contraption joined a large needle attached to a device that resembled a mechanical pencil. He wanted a less invasive tool than the standard biopsy puncture needle to use for the PGP. Cells taken from skin tissue biopsied from participants' arms were being used to create cell lines of the PGP volunteers. At that point, in 2010, four immortalizations were in process.

"I think it will be much better than the traditional skin biopsy needle," MAT said as he jutted the object out into the space between us. He had just come from a meeting with Jason, who oversaw recruitment for the PGP, to assess the ease of the new tool, which, to MAT's surprise, only took a few minutes to get Bobe's thumbs-up. "The rest was basically about table settings and name tags . . . to tell you how smoothly that went!" he gloated.

·   ·   ·   ·   ·

MAT told me that throughout his life he hated authority figures. He rebelled against his parents and others in his home country before immigrating to the United States.

"I've been cocky, thinking that 'the MAT way' is the right way. But that was until I met George. George is right 98 percent of the time—and *that*

is a first." He continued: "I really respect and admire him, but I don't necessarily want to be friends with him because that might taint the way he
sees my work. I want people to be able to dislike me, and still see the merit
in what I do. So many people judge your work by how much they like you.
I want George to judge my work by how good it is." MAT said that he
wanted an independently gathered, candid take on the PGP, and frank
feedback from the participants. "I want to know if the tool hurts, if they
hated it, what the problems were. I don't just want to hear that 'This
project is great, and the skin biopsy didn't hurt at all.'"

He wanted participants to have a meaningful yet honest experience in
this project that most of the scientists likened to "moon walking." Half-
joking, I offered the idea of making the skin pincher into a keychain for
them, to materialize some aspect of the PGP's emphasis on "veracity." I
added: "There are risks. And they aren't always minimal. They can come
with agonizing costs."

MAT loved it. "A keychain! They could wash it off—or better yet, we
could leave a little blood on it." He imagined PGP volunteers' desire for
connective memorabilia of this bodily (present and future) investment in
the project. Here was a skin-puncturing wishbone to memorialize the
naissance of their own immortalization.

GENDER AND EMBEDDED "EXCELLENCE"

On the day of our interview, I met MAT in his dry bench workspace, which
resembled the general aesthetic of the lab. The room was stark, modern,
and commercial clean, with floor-to-ceiling glass walls and gray fabrics
wherever anything soft, like carpet or an ergonomic chair, figured.
The entire building design was minimal but once in a while small details
stood out, like the color-striated chromosome pairs of XX and XY desig-
nating the sexing of bathrooms. This logo of cytogenetic staining to signal
sexes on the restroom doors—which on the one hand went beyond
the block pictograms of silhouetted stick figures in dresses or pants,
while on the other were just as deterministic, only now deep inside the
cell—reminded me that despite so much forward-thinking "open science"
being done in the lab, there was still some old-fashioned "biology as

ideology" aestheticized throughout the modern design of the scientific work-grounds.[32]

Some of the researchers who walked through the space daily, and who noted the old ways of thinking fashioned in chic new forms, questioned other instances of passive biological power structures embedded in the lab environs. For instance, a female postdoc pointed out the dismal numbers of women in the lab. She took me to their website portal where people sought information on how to apply to join Church's projects as researchers or postdocs. The instructions there started with a very high bar: applicants, it said, should consider themselves to be "the best" in their fields. The postdoc looked at me plainly, as she began to pull back the digital curtain to the lab. "That's basically a filter—a deterrent. Very few women in science I know have that kind of attitude."

"The lab does seem to attract highly confident men," I responded.

"We haven't studied the effect of the website, but I found it intimidating, and I'm sure other women do too." She moved on to talking about the Wikipedia pages that she was busy writing on key concepts in the field. One highly important page was full of her revisions and edits. It was the wiki page titled "Genetics."

A few days later, I encountered Church in the late afternoon outside of his office. He asked me how things were going. We chatted for a bit, and I brought up the possibility of a gendered reading of his lab's recruitment call on the website. He listened intently, then shook his head a little as he looked down at his feet. He fidgeted for a second, to aid in thinking, before responding. "Well, that's really interesting. I would have never thought about it like that."

He was clearly bothered that his emphasis on excellence could be read as code to select for hyper-male confidence, a potential contributor to gender bias in the sciences (particularly at Harvard, where he launched the PGP in 2005—the same year that the school's president at the time, Larry Summers, commented that women's brains had less scientific aptitude). But Church agreed that any gender bias on his website should be called out. He seemed to regret not catching the error before someone had to point it out to him. "Well, we need to take that wording down," he said. "That's definitely not the message we want to be sending." We walked over to the lab of Ting Wu, Church's spouse. The three of us talked a bit more about "blind spots" as he shared the story with her. She, too, was glad to hear it (being called out, that is).

Wu, for her part, works on chromosomal behavior as well as the development of "oligopaint" technologies to render aspects of the genome visible. She researches how genetic technologies can be applied to the exploration of space and climate change. Her lab also runs the Personal Genomics Education Project (PG-ED). PG-ED was born alongside Church's PGP as an education and awareness program so that genomic scientists like she and Church would be reminded to leave the "halls of science" and enter the public sphere in an effort "to meet people where they are—in classrooms and community centers, libraries and museums, places of worship, movie theaters and government offices, and online."[33]

THE GENERATIVE BODY IN A SCI NON-FI WORLD

The beaming confidence of some men in the lab did seem to bear out the postdoc's diagnosis. But, then again, it was hard to tease apart certain male scientists' confidence from a broader culture of vanguardist, optimistic risk-taking and bold entrepreneurism that Church cultivated and infectiously spread in a cultural well where such attitudes have wide and rippling appeal. MAT was admittedly deep in the Church waters. He and I talked for only a few minutes before the conversation pulled my eyes to one instantiation of his dedication: the wounds on his forearm. "For me, it was important that when I'm proposing to biopsy one million people— rather than writing a paper about why it should be pain free—I should just test it on myself to see . . . what are the complications? And *these* are the complications I actually saw . . ."[34]

Bold self-experimentation was an act of expediting the science and demonstrating that to be a specimen was not to be feared, or that if it was, then Church, and in this case MAT, would encounter any dangers first. Study subjects could rest at least somewhat assured knowing that the scientists themselves were willing to undergo what they were asking of participants. This act of specimen pioneering was a gesture George Church demonstrated from the very inception of the PGP. He was participant number one in the study—alternatively known as PGP-1. What followed

from this was that PGP-1 was also perceived to be among the most inno-
vative and influential thinkers in the field of genomics. He was the only
participant that the Harvard research ethics and review committee would
allow to join the unconventional project of DNA data sharing for the first
few years. No one else, they argued, could truly be said to be informed, and
therefore protected, as a human subject. Eventually the Harvard IRB lowered the bar for entry to anyone with a master's degree in genetics, or the
equivalent thereof.

My eyes were still on MAT's arm and its scarified track marks. Ordered
as if arranged on a grid, perhaps in the ways of science, the scars were
nonetheless horrifically large—a reality that the needle that he had
deployed, and just downsized from, was massive compared to those most
people are already averse to. His embattled markings, the result of "self-
puncture," were offset by a fifth messy wound of dark brown on his lighter
tan skin. MAT looked down, then back at me, twisted his arm slightly to
get the right light and pointed to the dark shambolic splotch he knew I
was going to ask about next.

"This is the scar from the Band-Aid I had, actually."

"You had a scar from the *Band-Aid*?" I was expecting a more dramatic
mishap due to the seriousness of the skin damage. The scar looked the
most violent but the others, even in their neatness, looked worrisome too.
"And the four little dots you have there, these are *all* biopsies?" His attitude was light yet resolved, a mix of amusement and nonchalance. He
clearly did not want to suggest that any of this was too serious. He returned
to his straightforward explanations.

"Yeah, they're all biopsies. So there was skin discoloration. I had welts
forming . . ."

His countenance was calm as he talked. A clear contrast with his beat-
up forearm. "You had welts?"

"Just from the Band-Aid, taking a shower, you know."

MAT laid out the scientific ground for his work, a tone that echoed his
insouciance about his wounds. "I just wanted to experience all of the possible consequences of a skin biopsy. So after I explored that, I designed a
very tiny, like, *click!*, pen-like biopsy tool. You just go in, pull up, and
click—you have a little core biopsy of bulk tissue."

He mentioned having some trouble doing any kind of puncture on African Americans because, he said, "They keloid." Then he thought about the only Black person in the lab, a West African man of athletic build. "It's also hard to do it on someone like him—because he has *no* body fat!" MAT exclaimed, with a bit of envy.

PGP participants were promised a full genome sequence and the immortalization of their cells, which would be deposited at the NIH's Coriell Institute. As we talked about getting to their goal of one thousand participants for phase 1 and then to ten thousand and on to—what Church calls the movement of "Genomes for All"—MAT went into detail about the PGP creating a repository for a future human race. "We need to consider if this one gets wiped out," he explained. But in the short run, cell lines transformed into organs would make for more robust in vivo trials for drug reactions and would expedite the timeline for R&D.

To bring me up to speed on his projects, MAT described methods for growing cells from different individuals collectively and giving them bar-codes. These codes would be able to distinguish the genomes of what he called individual "synthetic humans" from the larger pool in which they would be made to grow in colonies. These "synthetic" beings, he told me, could eventually be "be farmed," as he said, for research or therapies. Methodically, careful to find the right words for this sensitive terrain, he explained his vision: "One thing I've really been thinking about is that scene from *The Matrix*—I don't know if you remember it, where people grow in pods?"

"Yes."

"Cells are fine, but obviously we need an even better disease model. Obviously, we can't grow pods of humans, right?"

"Mmhmm."

"But I think there are ways of growing differentiated . . ." He paused, before enunciating every phrase carefully. ". . . of growing very early, non-humanoid-looking, organism-looking-things, without any nervous—without any functional nervous network."

I let MAT know I was still with him, even though I was saying very little. He was deep in getting the details right, going from his native tongue to English in his mind and also tip-toeing on sensitive terrain. As for my mind, it was now in film mode, picturing the images of synthetic life he was describing. I was trying to follow his lead, to see the human origin, but to stop short of imaging a person, or even a body. My verbal responses lessened more as the imagery took over. I was now in the theater of his scene and offered the musical replacement of language, in my semi-speechlessness. In sounding *mmhmms*, people often communicate non-lexical affirmations. My reduced language of *mmhmms* was meant to say "Yes, continue, I think . . ." And MAT did.

"So I talked to George about the idea of high-throughput micro-placentas. So these are essentially sort of like based on high-throughput colony sequencing, but with two layers of metal with thousands of holes—arrayed holes. In between is a mixture, it's an oil [. . .]. The bottom layer essentially secretes little droplets of growth media with micro-carriers or cells mixed in there. So we know that we can grow embryo bodies that way."

"Mm hmm—"

"So I want to insert biosensors, like photoactive biosensors, into these cells so that their gene expression can be controlled with light. So, using a flow, light, and growth factors, right? I think we can develop very large numbers of little artificial, human assemblies that hopefully will be ethically acceptable."

With this, I sensed the scene was finished, in a way, but there was so much left unanswered. "So there's no nervous system?" I asked.

"Right."

"So, what is it, like a mass of—like a cell on the eighth day of gestation?"

He paced his words as if enunciating a list. "It's something that contains a functioning heart. Primitive kidney. Primitive liver. Primitive skin. Epithelial cells. Primitive neuronal streak, but no signs of a brain." He paused.

"Okay," I offered, feeling how minimal my reaction was as I tried to take it all in without stopping his flow.

"The reason I want the photo-activity is, if we know that there is a brain tissue that has a chance of developing, we can insert a gene, which is neuron-specific, to kill that cell type."

"So you would want to suppress the existence of a brain?"[35]

"Right. Right, right . . ."

"Meaning, that—a brain would make it human? And alive?"

"Right, right. I don't know how reproducible this idea is, but knowing—" He interrupted himself, suddenly pulling back a bit to reign the idea toward its scientific function. "And there are people who want to make real babies, artificially. We don't want that. We just want a robust and reproducible way of generating these 'micro-bodies' from a large population."

I returned to the cell lines. "So those tissue bodies would be taken from an individual's cell line?"

"Yeah."

"It would be a reproduction of the human in a kind of body form that wouldn't be alive? Is that a good way to put it?" I felt like I was getting lost in the oily medium of the micro-placenta idea and looking for some language to hold onto. I circled back to the basic question: "How would you put it?"

"I would say—" He stopped short to find the right words. "Mmm. I would say . . ."

"What would you say this mass would be?" I rephrased.

"I would say—it would be a living, non-sentient surrogate of you."

I slowly repeated his measured phrasing. "A living, non-sentient surrogate of you."

"Yeah."

"I could see the project being a headline on *The Drudge Report.*"

He laughed lightly. "There is a danger, no matter how safe you make it. The idea of cloning a surrogate and making a farm out of it is kind of a scary one. We'll see how society's attitude evolves. We'll see whether or not this ever becomes ethically acceptable."

He again reminded himself of Church's caution. "I have to be careful about the language. George never wants me to talk about this in terms of 'babies,' or 'clones.'"

A few days later I was scheduled to check in with Church. Among other updates on recruitment and selection of the PGP-100 for the next phase of the project, I wanted to get his take on whether the genome sequencing machine that he had designed, called The Polonator, and that could also aid work in synthetic biology with iPS cells, might be likened to a placenta in the way that 00MAT.Rx had imagined.[36] I asked: "So the synthetic biology aspect of the PGP would be the creation—not just the creation of cell lines but potentially mutating those cells to have certain phenotypes?"

"Right," Church responded. "And we're *enabling* people, in the research community, to mutate them."

"Okay, and so MAT had also mentioned creating—potentially—something he was calling a placenta. Does that ring any bells? Sort of a synthetic placenta that could be farmed, or . . ."

He hesitated a bit and, like MAT, seemed to be searching for the right words. "Yeeaaah," he said slowly, drawing out a hesitant affirmative. "I think what he's talking about are ways of getting—" He interrupted himself, and went broader. "So in the Wyss Institute we're making, we're making—so we're making all kinds of artificial tissues, like lungs, kidneys, livers, that do exchange the blood. And in tissue cultures one of the things that makes it highly *manual* is that you have to change the solutions by hand when the color starts to change, or after a certain number of days."

"Hmm. Okay."

"Sometimes you have to trypsinize—hit the cells with enzymes that release them from the surface, dilute them the right amount, put them on fresh plates, all this sort of stuff. And it's costly in terms of consumables and labor. So if you have a way of doing the exchange of gasses and fluids in many of these body tissues that are highly vascularized, including placenta, lung, liver, kidney, that becomes more automated."

"Okay."

"I *think* that's what he meant by that. . . . We're not making little babies. That's for sure."

"Right."

"But we *are* making tissues."

"Tissues," I repeat. "Yes, MAT said that it's not the same thing as a *baby* but that it *could* also have some ethical questions arise around it."

"Well, it's good of him to take the leap, to, you know—it's not that—" He interrupted himself to focus on explaining how he saw the technology. "Well, the kind of fluid exchange, air coming in with the blood vessels and arteries and, you know, arteries and veins, is similar to what you have with a mother's circulatory system interdigitating with the baby's."

I let him know I was still following him, as I nodded for him to continue.

"And so, he's trying—maybe either trying to engage you, or alert you to where this could go."

"And so," I asked, "is the placenta idea something that would vascularize these different organ tissues?

"I think it's more like an artificial kidney, or heart, lung instrument," he explained. "It's a placenta in the sense—well, he's thinking of it like the mother is cleaning up the baby's waste and giving it fresh food and oxygen. And so, you've got these interdigitating blood systems. And the same thing would be true of these artificial tissues, which is—what you've got is that the *mother is now a machine* that is cleaning everything up and is providing fresh fluids and that's interfacing with the tissue which is personalized [from the PGP cell lines]. Each, each—" He was at a loss for words for a moment and, again like OOMAT.Rx, took time to search for the best phrasing.

"*I* would reserve the word *placenta*, myself, for actually developing embryos. That's what I would reserve the term for."

"Yeah, the language . . ."

"And there *is some* possibility that the field will go in that direction as well."

"But that's not what you're doing here per se?" I asked.

"That's not what we're—" He cut himself off again and thought about the possible future. He seemed to not want to rule out the possibility of one day conceivably making new people. Finally he offered, "I would think that would be an advanced concept after we had gotten quite a bit more confidence in making any kind of tissue that is semi-normal."

.   .   .   .   .

In his caution George Church acknowledged societal and ethical landmines that certain aspects of the work in his lab might set off. MAT's projects and his willingness to blur the lines of living human and *living non-human but human-like* life forms were the unconventional material hybrids that neither researcher was sure would see the light of day—at least for the time being. Neither wanted to peddle in a contemporary *Matrix* of an illusionary present where the masses were not aware of where the science was going, or how it might be used and fueled by their DNA and tissues.

EUTHENICS

Church held other ideas about children in the present, however. He focused on what we as a society are doing to stamp out diversity in the general sense of the word. There was a time when the PGP got a small amount of funding from the now nonexistent Counsyl, a DNA screening company that was acquired by Myriad Genetics in 2018. The small fund was granted to a collaborating Harvard graduate student who was developing a potential new test for IQ that they wanted to use for the PGP participants' cognitive traits.

At one point, early in 2010, Counsyl launched an ad banner that proclaimed they were on a "campaign" to eradicate genetic disease. Ting Wu and her PG-ED team flagged it and brought to Counsyl's attention similar campaign slogans linked to photos of "fitter family" contests from the American eugenics movement. The company leadership removed certain aspects of the strong messaging from their website. Because the PGP entrance exam that participants had to pass in order to enroll in the PGP was so extensive and required basic knowledge of genetics (disclaimer: there was no limit on how many times one could take the exam, and PGP staff would tutor people if necessary), there was, Church admitted, already some bias toward individuals with a certain level of cognition.

When I asked him if, in light of this bias and their prior brief relationship with Counsyl, there could be perceptions that the PGP was an elite study with eugenic components, he doubled down on his commitment to diversity. His notion of diversity, however, had to do with what he called "the brain pool" as well as the gene pool.

"I think our project *embraces* diversity," he told me. "I've gone on record, a number of times, of saying that we already have a *euthenics* problem where people are trying to make everybody the same by changing their environment, and in particular changing their drugs."

"For instance, Prozac and this kind of thing?"

"Exactly. Well, in high school is where you see it."

"With ADHD and Ritalin," I offered.

"Where kids are making up their mind as to whether they're gonna be the way they are or whether they're gonna be the way their teachers want them to be, or who their classmates want them to be." I nodded, indicating for him to continue.

"And so, people who are fidgety, you know, are given Ritalin, or what-ever. People that are depressed are given Prozac and people that have other distracting traits are basically asked to leave, or there are just a whole bunch of things that happen. You know, if you fall asleep you are given uppers. If you're out of sync with the school starting early, which is actually a huge fraction of high school, again you're encouraged to take stimulants of all sorts. So basically, they don't accept bell curve. They try to truncate it or squeeze it into, you know, one standard deviation. So that everybody sits still, is attentive, and works hard."

All of this concerned Church to the point where he concluded: "I just think that's an example of—in a way—that's *much* more threatening than eugenics. Or at least the way it would play out, you know. A few wealthy individuals choose a non–cystic fibrosis child. That's unlikely to change [. . .] our brain pool, to change our social and cultural environment nearly as much as programming kids in high school."

He went on to talk about the far more widespread "special" neurodiver-gent traits, such as the narcolepsy that he, his father, and his daughter all possessed. This trait, as well as so many of his entrepreneurial phenotypes and stances on science, formed aspects of Church's own identity politics, even if his were not explicitly racialized in the ways that Kittles's or Morales's were, given how they identified with their projects, their poten-tial recruits, and ultimately with their science.

Connecting a line from his own and his immediate family experience, Church wanted to make clear that there are positive and negative aspects to special phenotypes like his. The same genotypes, if seen in enough individu-als, often do not lead to the most extreme or dire expressions of how these might play out. This was one of the deeper goals of the PGP as he saw it: to highlight the ways that the traits that all of us possess, and that we are con-ditioned to "squeeze out" of our behaviors and ultimately out of *ourselves*, might be validated and accepted as part of our remarkably diverse species.

RACIAL CODA

In the Church lab, experimentation was both scientific and social. As technology developers, they were extremely future-oriented and saw themselves and those in the PGP study as iconoclastic "first adopters." Not only did people in the lab refer to the project as a moonshot but the "first of the firsts" were also seen to be moonwalkers themselves and in at least one of the cases quite literally. At the same time the project was, in Church's own words, exceedingly "white," "male," and "wealthy." These have remained principal biases that the project grapples with, even as Church told me: "All studies have bias." Yet he still wanted to balance out some of the highly skewed "representational issues." The team realized that people who might lose health insurance, who were not independently wealthy, or who might have to worry about government, border control, or police surveillance did not have the same "freedoms" to walk the moon of the PGP that those who first enrolled might have. The consequences of all of this returned them to the terrain of race.

Although the PGP scientists initially imagined that the project—with its focus on inter-individual genomic variation—would not be structured by race and ethnicity in the ways that other genetics work in pharmacogenomics and the search for disease genes have been, in 2012 they were forced to change course. The PGP was just too glaringly homogeneous and socially insular since involvement was often based on entrepreneurial interest, word of mouth, "being in the know," or invitation. On some level, studying people of European descent could be useful for certain questions. After all, a small portion of western Europeans carry a mutation on the *CCR5* gene that protects against HIV infection. Yet non-European individuals have clued researchers into viral immunity on a whole other level. Perhaps less well known than the *CCR5* story is someone described in the medical literature as an African American gay man, known as "donor 45," whose body produces several effective neutralizing antibodies for HIV, especially one now called VRCO1.[37] Donor 45 has lived a healthy life for years without the need for antiretroviral drugs. Information from his cells has been at the crux of the search for an HIV vaccine. One of the hopes of the PGP was to see what in fact any of us mere mortals as individuals, like donor 45, carry within.

After five years of watching the same patterns of who felt comfortable to enroll in such an endeavor, the PGP team decided that it was time to launch a call that read "Seeking Diversity" on their blog. They sent it through their broad networks, and their networks' networks. In it, they

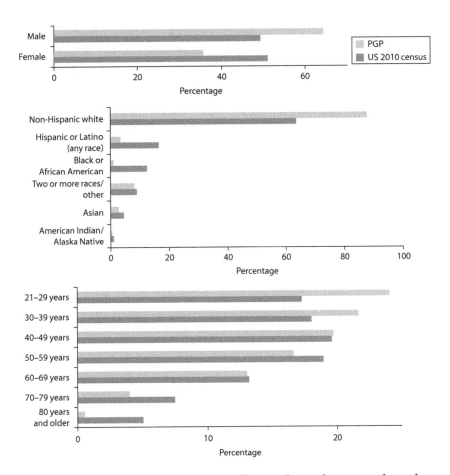

Figure 6. PGP participants and potential enrollees as of 2012, by race, gender, and age. From Mad Price Ball, "Seeking Diversity (Especially Families)," Personal Genome Project blog, November 29, 2012, https://personalgenomes.wordpress.com/2012/11/29/seeking-diversity/.

provided the public with a graph of US census categories of the racial, gender, and age demographics of the United States compared to their own demographics of PGP participants and seventeen thousand "PGP hopefuls" (those queued up to join the PGP when more resources to fund the project would become available) (figure 6). Scientifically, a study on inter-individual genomic difference that did not have enough data on people from the rest of the world, or at least a subset of a broader swath of humanity—and was instead made up of self-assessed healthy, relatively young, white male Americans with European heritage—would be incredibly narrow at best. In the years since they have expanded only slightly via PGP offshoots in Australia, Canada, and South Korea.

In all of the labs where I conducted fieldwork, scientists cited an ethos of inclusion. This was the idea that racially diverse teams as well as study populations would lead to better science or, at the very least, that people would not be left out of what some called the genetic revolution. This optimism circulated at the level of discourse in these spaces but was often limited in action. Basic nods to inclusion both feed and allow for the disciplinary state of genetics in its present, acceptable forms. Despite their constraints, the PGP scientists held future-oriented hopes that their work would advance the social good. They toddled the new terrains that they were bringing into the world, while acknowledging their trepidation around being part of an institution of science that perpetuated systemic social ills and inequality. They remained confident that the future they were helping to usher in would provide humanity new degrees of freedom, rather than veer off into renewed forms of scientific myopia—or cloistered visions that spawn newfangled raced, classed, gendered, and acceptable dystopias.

The next chapter begins with scientists who look to the past, rather than the future, to automate "inclusion" for the people of Africa regarding ancient DNA in their genomes. Full genomic sequences, racialized continental groupings, and algorithmic artfulness continue to usher in new hybrids of technological feats to reconstruct stories of human life.

8 Seeing Ghosts

FROM THE EXCAVATED PAST TO THE
HAUNTINGS OF THE PRESENT

In 2020 two geneticists in Los Angeles, California, published a paper that started with their journey to download DNA data taken from West Africans who were labeled "Esan in Nigeria," "Gambian in Western Divisions, the Gambia," "Mende in Sierra Leone," and "Yoruba in Ibadan, Nigeria." The bundled information traveled from the genetic repository that houses DNA from people described by their ethnicity and geographic placement (regions, cities, states, and countries) that span major oceans and certain continents. This repository, or genetic database, known as the 1000 Genomes Project, contains the "deeply sequenced" genomes of 697 people.[1] The Esan ("ESN"), Gambian ("GWD"), Sierra Leonean ("MSL"), and Yoruba ("YRI") genetic files trekked along a series of telephone wires, fiber-optic cables, wireless transmissions, and satellite links known as the modern internet. They voyaged not just as human but as bits of digital data that would be explored and excavated for information via algorithms that were programmed to suss out points of genomic difference in the people from whence they came—for various ends.

Once the West Africans' genomes were safely stored on the two men's computers, their UCLA team got to work to begin answering a research question for which scientists in the larger field had some clues but did not yet have a clear answer. The question? Did people living in sub-Saharan Africa possess *ancient* DNA? Several teams had already used various modes of simulation and, later, machine learning tools to predict that Africans did show evidence of genomic pieces from a long-ago ancestor who was not yet the *Homo sapiens* who would thrive in much of the continent and then, at different points, would migrate to what is now the Middle East, then Australia, eastern Asia, Europe, and later to the Americas.

This query, of modern Africans' quotient of ancient DNA, like most scientific ones, did not just appear out of the blue. A brief history of recent paleogenetics and the search to map ancient hominin genomes will shed light on its emergence. But before we venture there, the journey of this chapter itself requires a bit of a road map. As the starting point, it is important to note that the seasonally rainy, humid, and hot material conditions of many sub-Saharan African climates do not favor the conservation of organic material over millennia. Yet these ecological limiting factors have themselves actively conditioned how scientists are able go about assessing ancient DNA in different global populations—populations demarcated by continent since the ancient DNA researchers, like so many working in genomics, compared people who were defined in part by continental origin as vital to their framework.

There is much to analyze in paleogenetic studies from an anthropology of science perspective, but what interests me concerns how some researchers explicitly concede the limits and prior assumptions that often shape models of "genetic admixture." They articulate an acknowledgment about a regular feature of this work, which is the fact of *missing data*. A wonderful early example on this count comes from the Swiss evolutionary biologist Peter Beerli who writes: "It is commonly assumed that samples from all populations are in the dataset. This is rarely the case when researchers work with natural populations; some populations are not sampled because of logistic difficulties, or because one is not aware that additional populations exist."[2]

I am interested in how, when pressed by the confines or "logistical difficulties" of having *no* material fossils or bones from which to extract DNA, the UCLA researchers confidently deployed artificial intelligence, simulations, and a heavy reliance on inferred predictions to calculate an answer, or their "finding," that African populations carry archaic hominin

DNA in their genomes. What is most remarkable to me, because of what it reveals, is that they named this unknown West African ancestral line "a ghost" lineage.

As we will see, the UCLA scientists *excavated* this ghost by mining the DNA of contemporary, living people who identify as Esan, Gambian, Mende, and Yoruba to those of other Africans seen to be ancestral to them. They then compared these original four groups' DNA to the genomes of the Neanderthal and the Denisovan. In the absence of actual physical ancient DNA from the hominid in question, the researchers used algorithmic machinery to puzzle together predictions of "putative archaic segments" whose source did not appear to derive from any of the comparative populations under study (select southern and central Africans, Neanderthals, and Denisovans). Since the source of these segments could not be attributed to a known human species predecessor, coupled with the feat that their algorithms uncovered *something*, the team offered up *the ghost*.

How geneticists and bioinformaticians think of the ghost, as an extinct ancestor who nonetheless lives on as an elusive trace, may not be so very different from many people's everyday conceptions of the spectral qualities of what can be done with our DNA and our digital information more broadly. These range from attempts to represent us, to tools that reductively characterize us, to technologies that negate or render unintelligible some aspect of who we understand ourselves to be. We can take these scientists' acknowledgment of ghost DNA, and their admission of working around the fact that they are essentially missing data, as an instructive point of departure.

It is helpful to consider the ground covered in this book to invite the ghosts out of the crevices and black boxes where they have often receded from our sight. We can excavate them from the shadows, or from the wings of the world stage where the people and populations who have been most featured in representations of the human take front and center in the built world of genomic admixture models. In chapter 2, I showed how scientists deployed methods of algorithmic sieving and sifting that create a sense of continental racialized homogeneity—African, European, Native American, and so forth. There, the ghosts of Muslims residing in Spain of Middle Eastern, North African, and West African descent as well as people throughout Europe (Slavic people from the Black Sea region or descendants of Vikings from what is now Iceland) were some of the few who disappeared from the outset of AIMs panel designs. The historical record contains documents of some of their lives and family genealogies where their genes "admixed" in what was known as al-Andalus, or Islamic Iberia. This area extended to the upper middle of what is now present-day Spain and contained the future country's major cities in the long period between 711 and 1491.

How these "ancestors" have disappeared from view and now reside below a nominal surface reading that describes New World Latinos as largely descended from a population simply called "Spaniards from Spain" is just one example of how diverse humans go missing in genomic algorithms that are attempting to simulate a certain version of the world. These acts of selective population spotlighting of some (alongside the selective forgetting of others) inform and draw from racialized visions of the globe. These moves are repeated time and again and end up conveying the picture that stark continental divides equate to marked human "ancestral" (often read racial) lines.

Lastly, I want to explore when other instances of missing populations reveal telling absences in large-scale studies. I am interested here in when groups have taken themselves off the table for study, knowing full well the erasures inherent in the tabula raza that would reify and preserve their so-called population-specific genes often at the expense of them as *actual people.* I discuss such bold responses on the part of many Indigenous groups in the second part of this chapter. Here we go beyond paleogenetics, while expanding the concept of ghost DNA. In this section the presence of the ghost, and the haunted past, can help drive home the necessity of probing scientific projects for legacies of injury and harm that live on within them as consequences of the choices that colonial governments made and in some instances continue to make.

Human genetic science heavily relies on people, no matter the degree of computer simulations one might imagine. Therefore, it is critical to find ways to open the field's frameworks to anthropological and social inquiry more broadly, rather than assuming its sewn seams result from a natural threading of an inevitable needle. My final query on this count will be to ask: How do those missing data—which in contemporary admixture analyses are past and present human lives—furthermore evoke a historically pervasive and continuous haunting for many Indigenous people and African Americans today?

MAPPING *HOMO NEANDERTHALENSIS* AND DENISOVAN HOMINID GENOMES

A decade before the UCLA paper on the West African ghost lineage, in 2010 a team led by renowned paleogeneticist Svante Pääbo in Leipzig, Germany, at the Max Planck Institute, together with an international team including key players at the University of California, Santa Cruz, the Harvard-MIT Broad Institute in Massachusetts as well as the US National Institutes of Health, published the first draft of the Neanderthal genome and would go on to release a much more complete map later that same year. This was after an earlier targeted gene sequencing effort in 2006 (on what turned out to be a Neanderthal sample that was overrun with contaminating DNA from the various humans who had handled the bone as well as microflora bacteria from the environment that had also colonized it). For the 2010 low-resolution genome (not quite full but still a feat) the Max Planck "clean lab" was able to finally sequence the Neanderthal from well-preserved bones excavated from the cool climes of the Vindija cave in Croatia. These prized usable bones—selected from a collection of many that were unearthed but, like the samples used in the 2006 study, were too saturated with other human and environmental bacterial DNA to be of use—were from three females who had lived at the Vindija site over thirty-eight thousand years prior.

The Neanderthal genome mapping was a formidable accomplishment at the time. Getting DNA from ancient remains is an arduous and multistep process that requires a range of methods, with contamination verification protocols at nearly every stage. But the reason that compelled scientists who undertook what are now several detailed ancient hominin genomes was that they wanted to know how these "cousins" (sometimes called "sisters") of modern *Homo sapiens* might shed light on how humans evolved in ways that diverged from their hominin kin. Some of the hypothetical questions that paleogeneticists posed concerning Neanderthals and humans centered on comparative genomic and evolutionary adaptations that might affect cognition, metabolism, and immunity that could have contributed to our survival, and to Neanderthal extinction.

In these scientific explorations the team wanted to investigate questions of genetic variation, but there was a larger inquiry at stake. Prior archeological evidence had made it clear that these early hominids shared the planet, overlapping in terms of time and in some cases location, with early *Homo sapiens*. Thus one of the field's main questions was to see if these two early humans had actually interbred. To home in on the answer, geneticists would strategically use "diverse" contemporary human population samples from distinct parts of the globe. Their focus would be on separate continental areas as well as on people who approximate ideas of ethnic and racial differentiation. Specifically, the thinking went: "If Neandertals are, on average across many independent regions of the genome, more closely related to present-day humans *in certain parts of the world than in others*, this would strongly suggest that Neandertals exchanged parts of their genome with the ancestors of these groups."[3]

To test this idea, the researchers compared the Neanderthals' autosomal DNA to the genomes of "one San from Southern Africa, one Yoruba from West Africa, one Papua New Guinean, one Han Chinese, and one French from Western Europe."[4] They then indeed found evidence of Neanderthal DNA segments in the genomes of *some* of these individuals. To be precise, the ancient DNA appeared to live on in humans who were, as they put it, "non-African." The contribution from interbreeding was found to be 1 percent to 4 percent of Neanderthal DNA in those classed as "Eurasian," meaning people who live in present-day Asia and Europe.

In the public sphere, the news that only some people on the planet possessed Neanderthal DNA launched discussions in certain circles about the validity of the "Out of Africa theory," also known as the "Recent Single Origin" account, which explains how the human species evolved in Africa and, through various successive waves, began to populate the rest of the planet. People everywhere—in Asia, Europe, North and South America, and all the less-featured places in between—are descended from a subset of African populations who migrated out and adapted surface traits to respond to the dramatically diverse climates and environments where they have lived for what are now tens of thousands of years. Yet the common refrain that "we are all African" does not sit well with many people for different reasons. For some, migrations from a human homeland that happened between one hundred thousand and twenty thousand years ago are too far back to claim a meaningful identity. They could also contradict people's own stories of their origins. For others, they simply do not want to be biologically, genetically, or socially associated with Black people. Period.

The notion that the ancestors of today's Asian and European populations interbred with Neanderthals, which could influence how that resultant heritage manifests in the expression of traits that were conserved through positive adaptation, got some of those in the latter camp thinking about the limits of our human genetic sharedness. When I first heard about geneticists mapping the Neanderthal genome, it was through the American media. As I was reading the comments section on one news site, several posts led me to a white nationalist webpage called stormfront.org. I was both amazed and amused to see its adherents gleefully refuting liberal pronouncements about humans being 99.9 percent genetically the same. Several posters were all too glad to embrace the "biological difference" that would separate them *evolutionarily* from Black people—even if that difference stemmed from the fact that they were "part Neanderthal." Their superiority claims were a hodgepodge of references to wording that scientists on the Neanderthal genome project had mentioned regarding "cognition" and "skull size" as areas of research in humans that the Neanderthal genome could illuminate. Other posters chimed in with a bizarre dose of optimism that even though they were basically talking about a "caveman," maybe the Neanderthal DNA would be found to confer some intellectual advantages to those who would become their white offspring—that is, them.

I shared the blogpost with historian of religion Terence Keel, who was interested in how eighteenth-century monogenism (the Christian conviction that humans were all one race) and polygenism (the belief that God created the races differently) were not in fact so intellectually separate when he examined key natural historians' ideas in this era. In the twenty-first century, some of what the stormfront community had posted, and possibly what the Neanderthal admixture studies could offer for interpretation, was that, yes, the *timing* was important for any kind of "mono" theory about shared human origins. Even if we humans were at one time of one origin, or "stock," interbreeding with other hominids in other parts of the world *could* lead to a separate evolution for Asians and Europeans with Neanderthal DNA—or so the thinking went.[5]

In conversations with Keel, I wagered that the then fast-growing DNA testing company 23andMe would soon offer a direct-to-consumer genetic testing product for "Neanderthal ancestry." Yet I also wondered how such results would fit into anyone's admixture analyses, results that were delivered as proportions of continental ancestry at the time. How would Neanderthal ancestry fit into what were often racialized pie charts? It did not take long for the company to seize the moment. The consumer test for Neanderthal DNA launched in 2011. Customers were given "their Neanderthal Ancestry percentile," were offered a Neanderthal relatedness "connections leaderboard," and "ancient history stories" as part of the package. In 2020 the suite on offer was revised if consumers updated to 23andMe's "new V5chip." Those who would pay for the new update were promised newly discovered Neanderthal "trait associations," such as "having difficulty discarding possessions you may never use," "not feeling irritable or angry when hungry (hangry) very often," "having a worse sense of direction," and the list went on.[6]

Meanwhile, as some were hatching plans to unleash the first Neanderthal ancestry test for sale on the open market, researchers on the Max Planck collaboration at the National Institutes of Health focused their press announcement on the evolutionary history out of Africa *and* on the feat of recovering and detailing such an old *Homo* species's genome. Almost as a rebuttal to those who had immediately ran with the finding that Africans and "non-Africans" could now be understood to possess different genetic heritages, evinced by Neanderthal inbreeding, Vence L. Bonham, the senior adviser to the NHGRI director on Societal Implications of Genomics, was quick to point out that "the findings do not change our basic understanding that humans originated in Africa and dispersed around the world in a migration out of that continent."[7] Back in the spaces of the Max Planck and its collaborating labs, the scientists proceeded apace. The race continued to see what other ancients could be found in caves, bones, and genes.

In December of 2010 scientists from this same team published yet another hominin genome that they mapped using a fragment of a prehistoric child's finger bone—smaller than the size of a US penny. The piece was found in the Denisova cave in the Altai Mountains. This unique massif range, spanning Central and East Asia where China, Kazakhstan, Mongolia, and Russia meet, has been made a UN World Heritage Site and is rare in its climate and ecology that feature weather dynamics and animal species remnant of the last ice age. The Denisova cave, located in the southern Siberian area of the Altai, was also the site of many other hominid bones, of both Neanderthals and the prehistoric humans who came to be called the "Denisovans" once the international team distinguished them genetically as a new type of ancestral contributor to modern humans. Eventually, in 2018, an "admixed" offspring of the Denisovans and Neanderthals—who was genomically reconstructed and affectionately named "Denny" by the Max Planck research group—would be found there as well.[8]

In their publication of the first Denisova genome, the researchers undertook techniques similar to those of the first effort to sequence the Neanderthal draft map, which was to suss out a human and archaic human comparison. Their work revealed that of all the contemporary human genomic sequences taken from different parts of the globe for their study, the people of Melanesia alone possessed notable Denisovan DNA—around 4 percent to 6 percent. A more detailed genome draft of this same Denisovan child appeared two years later at much higher coverage. "Coverage" refers to a process where a sample's genome is sequenced multiple times (thirty in this instance) to verify that observed DNA changes are real and not simply sequencing errors.[9]

Here the Max Planck team and their international collaborators set out to assess the relationship of the Denisovan hominid to modern humans by testing whether they shared more specific allelic variants with any of the eleven present-day human genomes in their comparison. These were labeled "Dai," "Dinka," "French," "Han," "Karitiana," "Mandenka [sic]," "Mbuti," "Papuan," "San," "Sardinian," and "Yoruba." This analysis was restricted to the genetic variants that were not present in thirty-five African referent genomes, which would give researchers confidence that any DNA differences were therefore more likely to come from archaic humans.[10] Again, the Melanesian samples showed evidence that people from Southeast Asia had almost uniquely integrated the rare Denisovan genetic ancestry when compared to other groups.

Throughout this period scientists began to offer the possibility that Africans could have small traces of Neanderthal and Denisova genetic signals, if more populations were eventually examined. The thinking went that if found, these small DNA bits would had to have come from Europeans or Asians who journeyed *back* to Africa thousands of years after the out-of-Africa migration waves. Around the same time that the UCLA researchers published their paper on the ghost lineage, another paper also based on computer simulations appeared where scientists found some evidence of Neanderthal mixing in African populations. Indeed, they explained this as a "back-to-Africa migration" by non-African groups.[11] They also argued that the estimates of Neanderthal DNA in Europeans and Asians had been overstated. Even though the field was now finding low levels of this particular close cousin's DNA throughout the three continents that were the focus of so much of this work, the question still remained about the possibility of other types of archaic contributions to African people.

Yet it must be said that several researchers in several other labs had already pointed out that archaic hominins might be living on in the genomes of contemporary Africans. Mostly this work involved looking at specific genetic elements, not whole genomes, or if whole genomes were examined, the teams researched just a few individuals.[12] It was actually in 2016, four years before the UCLA scientists' paper, that geneticists at the University of Arizona and the University of Pennsylvania performed their own simulations (based on readings of comparative alleles in different groups of people and ancient hominins), "thus providing the first whole genome-level evidence of African archaic admixture."[13] Their focus was on two Pygmy populations, Central African Biaka and Baka peoples. In their review of the science to date, coupled with their own findings, the Arizona-Penn team posited that interbreeding between archaic forebears and anatomically modern humans in Africa happened several times and over a long period of evolutionary history. In this way the UCLA researchers were not actually the first to find an archaic hominin presence in the genomes of people in Africa, nor were they even the first to attempt it without reference to real human biomatter. But they were the first to bring into being the somewhat abstract "evidence" born of simulations that projected an inferred presence almost a million years from the past into a virtually graspable image in the present. They were the first to give form to this simulated presence as Africans' "ghost ancestry."[14]

Traces of a murky unnamed branch on the hominin tree that stretches perhaps hundreds of thousands or even a million years into the past motivated the Los Angeles researchers who ordered the ESN, GWD, MSL and YRI genomes to begin their own journey to complement the work of teams that had given them their lead. Yet, as stated at the start of this chapter, this science on African populations is limited by technical issues that go beyond the obstacles of contamination. And here we return to the difficulty of DNA extraction from bone fossil remains on the continent when the genetic material likely would not have withstood the climate for millennia. As one team put it, "We note that unambiguous evidence of introgression is difficult to obtain in the absence of an archaic reference sequence, which currently does not exist and may never be feasible given the rapid decay of fossils in Africa."[15]

The earlier sequencing efforts for the Neanderthal and the Denisovan came from a simple fact, albeit not so simple technique, that researchers had the boon of being able to work with small bits of a fifty-thousand-year-old toe bone here, or a thirty-eight-thousand-year-old finger bone there. Having these prehistoric human remains was of utmost importance, even if early genotyping was of low coverage and not always complete, which required the geneticists working on the 2010 studies to also sometimes statistically infer derived sequences to fill in the gaps.

As mentioned earlier, modern humans began migrating out of Africa in several different exoduses, starting about 130,000 years ago and as recent as 20,000. But long before that, there is much archaeological evidence that the very first bipedal hominin called *Homo erectus* left somewhere between 1.8 million and 1.9 million years ago and continued to spawn descendants who would create their own branches of prehistoric hominids in different parts of the world. Another more recent subspecies of *Homo erectus* was *Homo heidelbergensis* (although classification varies since some see the latter as its own species), whose remains have been found in Africa as well as in Asia.[16] If *Homo heidelbergensis* spawned *Homo sapiens* in Africa and *Homo neanderthalensis* and Denisovans in Europe and Asia, it would seem possible that the archaic "ghost" in the African genomes might indeed have an identity that could be narrowed down.

Archeological evidence from these and still other *Homo* lines (such as *Homo floresiensis* found in Indonesia) indicate that these groups lived in various settings—in Europe, the Middle East, and Southeast Asia—even though in several instances they died out altogether. As researchers in the Max Planck collaboration compared and examined the genomes of Europeans and Africans to what are now multiple Neanderthal and Denisova genomes, they and other researchers still found gaps that they attributed to "missing" ancient DNA in Africans that has not yet been linked to a known ancient group.[17] The possibility of unearthing a long ago predecessor has been a particularly pressing question for researchers in the field concerning African populations since, before the 2020 paper out of UCLA, the substantial ancient DNA findings were, for the most part, in continental "non-Africans."

NEWS OF THE GHOST

As the world was learning about the possibility of the COVID-19 virus spreading in ways most would not fully grasp until a few months later, on February 12, 2020, the UCLA researchers' paper on the Esan, Gambian, Mende, and Yoruba genomes began making waves globally. Titled "Recovering Signals of Ghost Archaic Introgression in African Populations," the article appeared in the journal *Science Advances*. The language of "introgression," which is a highly specific mix of a word meaning "the transfer of genetic material from one species into the gene pool of another by the repeated backcrossing of an interspecific hybrid with one of its parent species," made perfect sense.[18] Here was a process of describing a species that had left Africa, evolved descendent lines who would remain outside of Africa and then later mate with a species (*Homo sapiens*) with whom they shared a common ancestor many unknown years prior. The term "introgression" was also used frequently in the paleogenetic field regarding the Neanderthal and the Denisovans in 2010. Sonically, the word rings similar to "transgression." It was as if these encounters, in all of their mystery and sexual intrigue, were in fact a crossing of a line that many people were nonetheless captivated by.

What was most interesting, however, was that the ancient DNA "found" in the 2020 discovery came from the genetic sequence patterns of the West Africans themselves through simulations. The focus was on points in their genome that were statistically inferred to indicate areas of evolution-ary "divergence." These would be regions where the West Africans in ques-tion did not exhibit contemporary DNA patterns when compared with Southern and Central Africans, nor did they share sequences with the Neanderthal or the Denisovan in these regions. To understand the distri-bution of archaic ghost ancestry across the genomes of present-day Africans, the researchers first focused on the Mende and Yoruba samples. Importantly, they did this *without* the unknown ancestor's bodily referent material (bone or DNA). In their words: "We used a recently developed statistical method (named ArchIE) that combines multiple population genetic statistics to identify segments of diverged ancestry in 50 YRI and 50 MSL genomes *without the need for an archaic reference genome.*"[19]

This phantasmic presence, a specter marked by its physical absence, would need help to appear as the apparition of the ghost lineage. It would be ushered into existence with the aid of the researchers' invented tool, ArchIE—an assemblage of algorithmically based AI features that the same team developed and published on in 2019. ArchIE used simulations and predictions to compute archaic population segments in the genome from present-day people's sequences. The latter served as the input data to esti-mate the probability that a substantial stretch of fifty thousand bases of DNA in select contemporary individuals would likely be archaic. In the researchers' words: "Inference of segments of archaic ancestry was per-formed with ArchIE. ArchIE proceeds by simulating data under a model of archaic introgression, calculating population genetic summary statis-tics, and training a model to predict the probability that a 50-kb window in an individual comes from an archaic population. We apply the resulting predictor to genome sequences from the Yoruba and Mende populations."

Once they found a signal of what could be archaic ancestry, they wanted to see if any other contemporary humans taken either from a range of African populations or from a known archaic hominin could be the source. They wrote: "We sought to test whether the putatively archaic segments identified in YRI and MSL traced their primary ancestry to other African populations or to known archaic hominins such as the Neanderthals or Denisovans. We computed the divergence of these segments to a genome sequence from each of six populations: southern African KhoeSan [*sic*], Jul'hoan; two Central African pygmy populations (Biaka and Mbuti); and two archaic hominin populations (Neanderthal and Denisovan)." They found that the source of the "putative archaic" segments was not from any of these populations. (Here, again, the Khoisan, Jul'hoan, Biaka, and Mbuti populations were positioned conceptually as somewhat temporally older than the Yoruba or the Mende.)

In the end, the researchers' algorithms calculated an ancestor who appeared so old, possibly a million years or more, that it did not immediately find a clear home in the clades and trees that scientists diagram to reconstruct the relationships of organisms to give a picture of evolution. Without a referent, or knowledge about the extinct contributor to the African samples' genomes, they thus imagined a ghost species.

This ghost was raised not from the dead, but rather from the genes of the living. What is noteworthy here is that many in this field hope to use the ancient genomes mapped to date to learn something about how the living, us contemporary humans, have benefited from positive selection that may have been ramped up by these instances of interbreeding. At issue is how modern humans integrated this DNA. What have the effects of possible positive selection from this integration been? (In 2020, during the height of the pandemic, researchers found that Neanderthal DNA *may* make some people more susceptible to COVID, but this did not in any way account for COVID health disparities.)[20]

Another point of curiosity has centered on how ancient DNA might bear on the different ways that diverse people on the earth today express certain traits (albeit not necessarily in the ways people might imagine). At a more meta level, what is eminently clear concerns the circular dance that spins human DNA in a latched embrace with that of the ancient ghost. These scientists have used contemporary genomes to learn about a phantom ancestor, and they hope to use the phantom ancestor to learn about the people of today.

The ghost story made news around the world, but most prominently in the United States and Europe. The *New York Times* ran the headline "Ghost DNA Hints at Africa's Missing Humans"; the BBC, "'Ghost' Human Ancestor Discovered in West Africa"; France's *Liberation*, "Une espèce humaine «fantôme» hante l'ADN des Africains de l'Ouest"; while *Univision* rolled out a more scenic orientation for readers: "Vivieron en

África hace medio millón de años: Los científicos la llaman la especie 'fantasma.'" Many other publications picked up aspects of the discovery from shared news organization reprints.

FACING UP TO THE "MISSING" ANCESTORS IN THE AMERICAS

In the remainder of this chapter, I want to move beyond the world of simulations that result in clues of ghost ancestry to real-world instances where missing data represents more recent lives. I want to query the idea and the gesture of *the acknowledgment*, succinctly stated by Beerli in 2004, about missing data. Although Beerli himself focused on computational methods to infer genetic admixture in diverse organisms and animal species like Mediterranean frogs, fish, or parasites, many researchers working on human admixture have refined methods where they can forego the need for a missing "reference population"—that is, the actual ancestral people, or their descendants. On the one hand, the acknowledgments offered by geneticists who have attempted such studies are refreshing. On the other hand, when technical fixes amount to working around the nonexistence of a group who are actually missing for reasons of violence, protest, or colonial dispossession, we should pause to consider—and when possible redress—why they are absent. To proceed apace by using proxies, surrogates, or simulations in an effort to conveniently carry on business as usual presents its own set of ethical oversights.

During a 2010 interview I conducted with renowned geneticist David Reich, who has led many genetic admixture mapping efforts, including those on prostate cancer in African Americans, our interaction made me reflect on some of these issues. Reich and his team are not medical researchers per se; rather, they are known in their field for their methods that zero in on risk alleles that differ in frequency between different populations, while describing population history and time scales of human admixture in ancient DNA studies. As noted earlier in the chapter, the scientists at the Harvard-MIT Broad Institute, prominently Reich himself, were also part of the Neanderthal and Denisovan genome projects.

Reich, a careful, soft-spoken yet detailed narrator, was obviously steeped in the methodological intricacies of his field. He was quick to acknowledge some of the problems with the technologies he worked with and his belief in trying to accurately "model" human history. He presented the issues that his team were currently facing as sticky points of frustration, compared to their relatively easier string of successes in finding genetic associations in African Americans using their particular methods of admixture mapping.[21] He first described how the model that his team constructed was able to approximate African American ancestry by using putative ancestral populations of present-day Yoruba and Utah CEPH samples (of European-descended white Americans) to create admixture maps of Black Americans, which have yielded some of the results regarding prostate cancer genetic risk found on chromosome 8 (discussed in chapter 5).

Despite the majority of attempts that resulted in negative findings when he and his colleagues have tried admixture mapping methods to locate risk alleles, Reich listed off a number of "hits" concerning conditions that are now associated with some background African ancestry. These range from alleles involved in kidney failure to BMI to neutrophil count to multiple sclerosis. In our broader conversations about the field, he confided that there was little motivation or enough "time in life" to describe all of the failed tries when this method has *not* worked to find associations. As he poignantly put it: "We are not real epidemiologists in the sense that we report 'hits.' We do not report 'misses.'" After these hits were found mostly in Black Americans, Reich and his team struggled to strike success more broadly. In discussing African American–focused studies, he immediately jumped to the other "admixed" group that he assumed their technology should work for—that is, Latinos. At the time, the machinery was acting up in ways that raised questions about what aspect of their reconstructed population history was missing and throwing off their model.

In asking him specifically about what might be at issue, the door opened to a much more profound set of reflections. "So why do you think that is [why isn't it working]?"

"I think the problem is much more difficult in Latinos," Reich told me. "I mean, one of the problems we're struggling with in the laboratory is how to do these scans in Latinos. What happens is when you do the ancestry inference, we get systematic errors. So there are places in the genome where we know we must be messing up."

"Hmm," I nodded for him to continue.

"For example, there's a place on chromosome 6, on the HLA locus, which codes for an immune related system of genes. We keep estimating too much African ancestry, more than the average in the genome. And maybe that's really true, maybe that's really what you're seeing in Latinos. But I actually think that it's more likely to be the case that it's a complicated region of the genome and our ancestry inference machinery is messing up."

"Okay," I offered, sensing his frustration with his technology, but letting him know that his story was still interesting.

"And I think we're just making a systematic overcalling of African ancestry in this region." He went back to how well their computational inference programs work in African Americans before continuing. "And I think this is probably because we don't have [reference] populations for Latinos that are as close to the Native American ancestral populations of Latinos, as we do for African Americans. So, for African Americans, we have Yoruba and European Americans [the Utah CEPH samples]—who are not really close in terms of—it's not the same tribe, it's not the same exact populations that contributed genes to [African Americans], but West Africa and Europe are populations with big population sizes and so the frequencies [of the specific panel of markers] don't differ much amongst populations in West Africa and Europe. So using one as a surrogate for the other is not a bad idea. But in Native Americans, I think it's quite clear that the frequencies are dramatically different across populations. And we haven't found a Native American population that—well, that seems to be a good surrogate in terms of its frequencies for the *true* ancestral Native American populations of, say, Mexican Americans in Los Angeles, which is where most of the studies have been done."

"So, does—" I started but he interrupted to continue about the glitchy results.

"So we've had data like this for a while. We haven't published it because the technical analysis in the data shows that the analysis is not reliable. It's an important problem for us.

I wanted to understand the issue even more. "So you feel that you're overestimating African ancestry . . . ?"

"At that locus," he interjected quickly.

"At that locus," I repeated, slowing things down to review my mental notes. I started to put together for the first time that there was a larger issue that had not been explicitly named up until this point in our conversation. I wondered what the so-called technical issue, regarding sampling "true" Native American ancestral populations, actually stemmed from? I then offered: "What you've described in terms of not finding an appropriate proxy for Native Americans' parental contribution—it seems that's the sticking point."

"I think what's happening," he explained, "is that bits of true Native American ancestry are being erroneously called as 'African' . . ."

"Hmm."

"Because we don't have—" He interrupted himself. "You're looking for, the methods look for—you look at a panel of ancestral samples to try to find a bit of DNA that is similar to what you have in the population you're studying, in your admixed population. And it's not present in any of your samples. So it [the inference algorithm] just uses Africans, because Africans have so much diversity. So it just finds *something* kind of, a little bit, similar and it says, '*Ah, this must be it.*' And it sticks in an African haplotype."

Reich suddenly realized the strangeness of what he was describing. Faced with explaining this to someone not in his field, and not in the thought collective of his lab, he simply laughed at the sound of it and blushed. He wanted to make clear that the statistical ancestry inference program's actions of just sticking in a "diverse" African haplotype was not acceptable. "But—it's a mistake."

I was relieved that he laughed before regaining his prior composure, and continued apace to describe the problem—mostly as a technical one. In this, he made no explicit reference to what might be haunting his efforts.

"Because in fact what's probably happening," he explained, "is that the true Native American ancestral population of Mexican Americans, you just haven't sampled it. And it did have this variant that is more similar to what you see in Africans, and it really is a Native American ancestral segment."

As was becoming clear, the "population" that "just" had not been sampled, when spoken of in the past tense—as in, "it did have this variant"—was not just an instance of missing data but also a missed opportunity to discuss the historical violence and dispossession that allowed for such a statement to be made *and not made* at the same time. I decided to acknowledge the ghost in the room. "And so do you think that some of the Native populations who *were* here that contributed—the fact they are *no longer here* may be the issue?

His pace slowed a bit. "Yeah. I think so."

"In fact, we have lots of evidence of genocide basically—that many of those groups are just no longer around, right?" I teased out a bit further.

His response was quick and crisp: "I think that's a good hypothesis."

From that moment on, although he never explicitly used language of historical violence or genocide, Reich began to detach from simply thinking about data and the model working to produce correct results. He wanted me to know that he was committed to accurately representing history—not just difficult genetic regions. The actual people and what happened to them, however, still remained lodged in a distanced and somewhat safely sanitized narrative of the Spanish arrival and mestizo "admixture."

He continued: "I mean—our working hypothesis is exactly that. That the Native American ancestral populations that were there in 1500 when Cortez got to *um*—you know, when the Spanish got there—may no longer exist in an unadmixed form. The big agricultural ones in the central valley of Mexico, they may have all mixed with Spanish. And so, all you see are the mestizos, the Latino admixed populations. And the Native American populations you still see are kind of more isolated populations that didn't get admixed. So they may not be *typical* of the agricultural populations that gave rise to most of the ancestors of these populations. That's one working hypothesis."

I asked him about other research groups in my ethnography that were working on the same problem of trying to figure out how to use admixture models to infer Indigenous ancestry.

He responded more bluntly and unequivocally to hurdle past what was clearly some discomfort to find the usefulness in modeling efforts. "These models are clearly *terrible* models. And that's reflected in the fact that our ancestry inferences are biased. [. . .] You know, the saying in statistics is that '*All models are wrong, but some models are useful*.'"

"And you believe that?'

"And I believe that," he responded without missing a beat. "And I think it's *also* useful to have models that are inspired by some view of what history might have, kind of, approximately, been like."

HAUNTINGS FOR SCIENTISTS OF COLOR

In practice, the admixture idea offers today's DTC ancestry genetic testers possible results of not just mixed genes but also of mixed emotions. For people of color in the United States, these can arise as a kind of haunting around their heritage being linked to Europeans who brutally seized the Americas while summarily violating Indigenous and African people in the process. In the course of fieldwork, sometimes scientists of color relayed such unsettling feelings to me, sentiments they had when they underwent similar kinds of tests. For instance, in the Kittles lab, Wenndy, a researcher from Central America, shared her experience of testing herself for the markers. We were in the lab one morning as she was mentoring a younger intern who grew up in Nigeria. The two women were discussing a paper when I asked them about how they viewed Ancestry Informative Markers and if they themselves had taken the test.

The intern smiled and only slightly hesitated before she confidently answered: "Do I really need to do some kind of ancestry test? I'm *pretty sure* I'm going to come back 100 percent Nigerian." Her mentor agreed and told both of us that the technology is mostly for people in the "New World" before adding, "but it could be interesting though! You never know. There might be a surprise." The intern said that she didn't think so. After some more talk about the paper that they were concentrating on, the younger student left and Wenndy and I were alone at her work area. She suddenly went back to our earlier conversation and leaned in.

"When I got my results I was really happy that at least my mitochondrial DNA turned out to be Native American. Because my AIMs said that I was about 50 percent European, and only some 30 or so percent Native American."

"What about the other ancestries? Did anything else show up?"

"The African was around 12 percent, or something like that. I can't recall the Asian exactly."

"Oh okay, so pretty low."

"Honestly, I thought I would have much more Native American. I was really surprised to have so much European—*and this was not a good surprise*. The fact that my mtDNA is Native reflects what happened to Native women. It brings all that up, which we know. But I'm glad I at least have some Native, and that my maternal line is Indigenous."

Wenndy also used a lineage test that analyzes clonally inherited haplogroups to trace mutations in mitochondrial DNA mentioned in chapter 5. Still, she was troubled by the amount of European ancestry the autosomal genetic markers indicated. She pointed out the distress of feeling that a Native American woman in her line could not fend off the Europeans who forcefully took over Native bodies, lands—and now, much of her genome. She allowed this recognition to surface and quickly safeguarded her Indigenous line in a gesture of survival.

In a different context, when I was discussing AIMs with Ebony, she did away with the Old versus New World distinctions altogether. She had not submitted her DNA for an AIMs analysis but said that she had always felt a close affinity to Africa and Africans. In her telling, we see her resuscitating the ghost of her lost African language: "I don't know what it is, I love Africa, I've never been there. I don't know what it is. There is this silent rivalry with Africans. Even [a Black collaborator] said, 'I'm not no African.' This is brainwashing. This is part of the hate and separation that was brought upon us. I very much feel that I am from Africa [. . .] My mom went to Senegal after she was diagnosed with breast cancer. And she brought back these photos. People were running their own business— that kind of stuff you never hear from the media. I would want to know where I come from so maybe I could learn the language. I know I wasn't speaking *English* [back] then."

Ebony effortlessly transported us back to West Africa in the immediacy of placing herself in the past while speaking—or rather not speaking— English. She collapsed the time frames of pre- and post-New World in the recognition that enslavers cut Africans off from their mother tongues and cultures. The wish to speak an African language—as she abruptly jutted her English anachronistically into the past for a moment—was a temporal reflex, an attempt to reunite with her missing ancestors in some form.

THE ONGOING INJURY OF COLONIALISM

The project of genomic world building that, as we have seen, usually starts with European conquest, in 1492, is only one of several histories that must be examined. In the early 1990s to the early 2000s, five hundred years on, the building of actual genetic resource repositories, gene banks, and DNA libraries constituted efforts on the part of researchers and in some cases governments to build databases where scientists could access different populations' DNA for study. One such project was the formation of the Human Genome Diversity Project (HGDP) in 1991, which sought to genetically catalog five hundred populations, many of which were Indigenous groups in the Americas, Canadian First Nations, and aboriginal peoples throughout the world. In the words of the scientists at the helm of the project, populations were initially targeted for inclusion if they were seen to "constitute linguistic isolates" and "might be especially informative in identifying the genetic etiology of important diseases." They emphasized the humanitarian goal of the project, as they saw it, to focus on "populations that are in danger of losing their identity as recognizably separate."[22] For many of those targeted, this very framing would evoke the neglect, marginalization, dehumanization, and past harms that they suffered and continue to live with. Somehow, those at the helm of the project did not initially take the devastating and ongoing injury of colonialism into account.

In the introductory chapter to her book *The Human Genome Diversity Project*, Dutch anthropologist Amade M'charek describes this dynamic as the HDGP's "controversial character." That is, the very structure of the project appeared to be rooted in a "blunt 'science for the West, genes from the rest'" mind-set.[23] This stemmed from both the assumptions and the design of the massive effort. The original project organizers—Allan Wilson, a professor of biochemistry at the University of California, Berkeley, and Luca Cavalli-Sforza, a professor of population genetics at Stanford University—formed a largely American team linked to elite institutions, far removed from the issues and concerns of the people whose genes they were interested in "conserving." As sociologist of science Jenny Reardon has chronicled in her analysis of the HGDP's planning, the project scientists and those who came out against it had very different notions of what exactly the "political" pitfalls of the project were.[24]

The geneticists' focus on conserving so-called isolated people's biological data in Western-based repositories was shot through with long histories of colonial actors dogging people for their goods with little concern for their lives, while racializing their bodies, mental capacities, and deeper biologies to nefarious and genocidal ends. Many felt that the scientists were proceeding ahead in paternalistic ways without acknowledging past wrongs or seeking true permissions, which would have required respectful relations in the present.[25] At the same time, the scientists involved tried to address these concerns in the first few years of the project. In their 1993 summary report, they forecasted that "by leading to a greater understanding of the nature of differences between individuals and between human populations, the HGD Project will help to combat the widespread popular fear and ignorance of human genetics and will make a significant contribution to the elimination of racism."[26] In the end they viewed some of those coming out against the project as the ones with the obvious (i.e., overt) "politics," meaning the politics of protest.

Medical anthropologist Margaret Lock brings in a larger set of issues regarding how the planners were not sensitive enough to how those targeted would see the effort as exploitative in highly specific ways depending on the group in question in an era of new technologies. At issue were distinctly novel techniques, such as PCR, that would easily amplify genetic material, while another, the ability to immortalize people's cell lines, would allow indefinite storage and unknown future use. All of this was happening as new legal cases in the United States resulted in laws that would allow scientists to patent donors' biological material, all the while excluding them from profits made from the commodification of their body parts. Of course, people who were dispossessed of their lands, violently colonized, and perpetually socially and economically marginalized were keenly aware that it was these callous incursions that often led to their present conditions of "isolation."

As Lock pointed out, labeling "the San" in South Africa, "the Eta" (Baraku) in Japan, and "the Yuchi" in the United States "remote"—without reference to imperial histories, racism, new forms of capitalism, and the normalization of how people's marginal status came to be—was more than mere ethical oversight of the HGDPG.[27] This was evidence of scientists' privilege and power; the protests were thus remonstrations of a much longer history of colonial behaviors where those entitled displaced and disordered countless people's lives in the constant impulse of economic extraction. The HGDP's interest in preserving Indigenous genes as a kind of valuable human patrimony (now framed as an advantageous anthropological past vaulted in their genetic code) did not exactly translate to an exemplary mode of care concerned with their survivance, health, or actual futures.

THE GENETIC GAME OF RECOGNITION

I take anthropologist Audra Simpson's diagnosis of ongoing settler colonialisms as highly relevant for thinking about various genomic projects' attempts to "recognize" and "include" Indigenous people without a deeper reckoning of how and why people deemed "isolated" were construed as human valuables on the brink of becoming rare. Clearly such framings miss point after point. Simpson calls out "the game" that colonial projects continue to put forth.[28] These are usually legal, seemingly rational, contractual sets of formal givens, which are shot through with constraints for American Indians where they are boxed in to maneuver within the limitations of settler governance. This often results in "agreements" to which they would have never consented had they been able to live histories of fair conditions all along. Their "refusal" is one where Indigenous people cannot and will not blot from their consciousnesses the violent past that birthed their current encumbered political conditions.

Groups such as the Rural Advancement Foundation International (RAFI), The Third World Network, and the Indigenous People's Council on Biocolonialism (IPCB) all issued statements informing people featured on the HDGP's priority list about the perils of participating in the project. Debra Harry, a member of the Northern Paiute Nation in Canada, wrote the following in an information statement in the IPCB's newsletter in 1995: "The Human Genome Diversity (HGD) Project is an international consortium of scientists, universities, governments and other interests in North America and Europe organized to take blood, tissue samples (cheek scrapings or saliva), and hair roots from hundreds of so called 'endangered' indigenous communities around the world."[29] The statement went on to highlight several principal fears about being used in a way that underscored the maddening recurrence of a Western extractivist mentality and Native people's disposability. "On the assumption that indigenous peoples are inevitably going to disappear and some populations are facing extinction sooner than later, scientists are gathering DNA samples from the living peoples before they disappear. [. . . There is a] sense of urgency in collecting the DNA samples of indigenous peoples in order to 'avoid the irreversible loss of precious genetic information' due to the danger of physical extinction."

A few years earlier, the Third World Network posted a warning on Native-L, an aboriginal and First Peoples listserv. It read: "We find such initiatives emerging from the West totally unethical and a moral outrage. We call on all groups and individuals concerned with indigenous people's rights to mobilise public opinion against the case of human communities as material for scientific experimentation and patenting. Indigenous communities are not just 'isolates of historical interest.' They have a right to be recognised as fully human communities with full human rights which includes decisions about how other countries will relate to them."[30]

To varying degrees, each of the opposition statements expressed indignation as well as utter fatigue at the gross violence and mistreatment of Native peoples throughout the Americas and the world due to Western colonizing forces. In no uncertain terms these protests were refusals to pretend that people could simply move on from the accrued effects wrought by colonialism and its normalization of dispossessing people of their lands, their bodies, their cultures. In Simpson's view refusal is not simply resistance to the state or the dominant power that attempts to set the terms. Rather, "refusal is a symptom, a practice, *a possibility for doing things differently, for thinking beyond the recognition paradigm that is the agreed-upon 'antidote' for rendering justice in deeply unequal scenes.*"[31]

Eventually, many "priority" listed peoples refused participation in the HGDP. The project scaled down after this period of intense controversy whereby the organizers were charged with "racism, colonialism, and vampirism."[32] Today, the DNA collected before the HGDP's sampling era of Indigenous people ended are stored at the Centre d'Étude Polymorphisme Humain in France. The CEPH and the samples sit in an uncelebrated, immigrant, and working-class neighborhood in the tenth arrondissement of Paris on the grounds of the Louis Pasteur Hospital. But they also circulate the globe for use in population genetic studies and in the creation of ancestry testing technologies in ways likely unforeseen by many of the people who gave their DNA.[33]

In 2001 and 2002, when the US government through NIH established the International Haplotype Map project, some of these issues arose anew. As discussed in the introduction, the "HapMap" project's goal to collect DNA and create a map of human genetic variation in select global populations was a primary resource for scientists selecting markers that were classed as "African," "European," and, to a lesser degree "Asian," in admixture mapping studies. Native Americans' samples, however, were not included in the HapMap project.

The consortium's choice of populations to include—as well as *not* to include—was linked to Indigenous people's vocal critiques and still resonant moral concerns about large-scale DNA mapping efforts that started with the HGDP. As the first publications on the HapMap began to appear, members of the consortium published a "Perspectives" piece on the project's "ethics and science," in the journal *Nature*. Of primary importance were key instructions for those using the data to avoid making the HapMap samples stand-in for other groups, to never treat the samples as natural categories, and to eschew any racializing language when describing the particular HapMap populations as discussed in the introduction. The authors explained the imperfect and limited assemblage of their genetic data resource tool, limitations that centered on leaving DNA samples from Native Americans out of the project. "For [. . .] many American-Indian tribes and small isolated indigenous groups," they wrote, "population-history findings from genetic studies of members' samples could conflict with religious or cultural understandings about their origins, or legal or political claims that relate to land or items of cultural patrimony. For these groups, the decision to construct the HapMap with samples from named populations could signal renewed emphasis on a disfavoured aspect of genetic variation research."[34]

Although this rationale succinctly focuses on divergent ideas of how genetic data on population history could challenge people's beliefs and threaten their political claims, the details of potential land disputes, tribal recognition, and what exactly the potential conflicts might entail were not laid out. The authors simply nod to previous tensions that arose a few years earlier with the HGDP—gently worded here as the renewal of "a disfavored aspect of genetic variation research." Despite the many reasons that those who protested the HGDP cited in their resistance to the study, the HapMap consortium's summary of the tensions were limited to the reasons vaguely stated above. Its acknowledgment concludes: "Indeed, it is mainly for these reasons that no samples from members of American–Indian tribes are included among those being initially analyzed for the Project."[35]

THE "ILLUSION OF INCLUSION"

May 24, 2019

Dear Tribal Leader and Urban Indian Organization Leader:

In accordance with the HHS Tribal Consultation Policy, the National Institutes of Health (NIH) is announcing a series of **Tribal Consultation and Listening Sessions on the** *All of Us* **Research Program**.

The *All of Us* Research Program, part of NIH, was established to accelerate health research and medical breakthroughs to enable an era of precision medicine for all. Precision medicine is an approach to disease treatment and prevention that seeks to maximize effectiveness by taking into account individual variability in environment, lifestyle, and biological makeup. With the great promise this program represents for strategically addressing health challenges, it is our intent to respectfully engage with Tribal Nations to facilitate the inclusion of American Indian and Alaska Native (AI/AN) populations in this program.[36]

In 2015 the first African American president of the United States, Barack Obama, launched the Precision Medicine Initiative that would become the *All of Us* program. The effort, which is now under way, seeks to compile a database of one million Americans' genetic and broader health and lifestyle information with the hope of developing therapies for what is called personalized medicine. *All of Us* critically emphasizes "diversity," goals to reduce health disparities, and the inclusion of people who for various reasons have not been represented in research in respectful ways, as equals. The project has invested a campaign-like drive into establishing trust with prospective populations, getting feedback through "consultations" and "listening sessions," and especially in creating the ground for collaborative research protocols with Indigenous American tribes with the expectation that some will participate.

In a March 2021 report on summarizing the NIH Tribal Consultations, the health arm of the US government laid out its acknowledgment of past harms and its intention to do better by tribes: "Health equity is important to *All of Us*, and the program is committed to building a cohort that is demographically, geographically, and medically diverse, including groups that may not have had the opportunity to benefit from breakthroughs in biomedical research because they were left out of research studies in the past or did not feel welcome to participate. *All of Us* and all other research programs that seek to address underrepresentation must acknowledge the role of historical trauma, at times perpetuated through research, and actively avoid the mistakes of the past."

Currently, an *All of Us* Tribal Collaboration Working Group and the NIH Tribal Advisory Committee hope to convey that, if tribes participate, the NIH will respect a list of twelve commitments and promises. These include but are not limited to a respect for tribal sovereignty, collaborative outlines for terms of tribal approvals for recruitment protocols and collaborative efforts in defining research, the inclusion of scientists and scholars from tribes to play key roles in defining research questions and carrying out research from the data provided, and promises not to share data or reveal the tribal affiliation of participants without prior tribal consultation and approval. Several of the items emphasize respect, tribal expertise, and tribal health priorities.[37]

Before the publication of the report, as the listening sessions were taking place in 2020, Native Hawaiian geneticist Keolu Fox published his own conscientious call in the *New England Journal of Medicine* for the government to heed Native peoples' concerns. The poetically alliterative yet candid title zeroed in on what Audra Simpson laid out as the aforementioned "game," which Fox termed "The Illusion of Inclusion." Nearly twenty years after the halt of the HGDP collections, this time the attention was on the *All of Us* initiative. The hope was that NIH researchers would finally realize that structuring genetic projects in ways that data sharing could lead to the commercialization of people's DNA, and large profits for private companies, does not inspire trust or make Indigenous people want to get involved. For Fox, "commodification of data and policies permitting unrestricted access to them extend histories of marginalization and disempower Indigenous people from making decisions about how and under what circumstances their data can be used."[38]

Fox highlighted the embedded assumptions of genetic biomarker-based difference itself as commodifiable through a string of projects over the past few decades. He wrote: "Previous government-funded, large-scale human genome sequencing efforts, such as the Human Genome Diversity Project, the International HapMap Project, and the 1000 Genomes Project, provide examples of the ways in which open-source data have been commodified in the past. These initiatives, which promised unrestricted, open access to *data on population-specific biomarkers*, ultimately enabled the generation of nearly a billion dollars' worth of profits by pharmaceutical and ancestry-testing companies."[39] As a geneticist himself, Fox is not opposed to all genetic research, and is certainly not anti-science in his stance. According to his article, he's not even anti-capitalist, if commodification can be rethought. He ends his piece with several examples of benefits-sharing arrangements where profit-driven research would be owned by or have to include Indigenous groups in "collective-interest models (i.e., community trusts) or individual-interest models (i.e., fractional ownership of stock or a shareholder model)."[40]

NATIVE GENOMIC SOVEREIGNTY

Keolu Fox is only one of several early career geneticists whose research agendas grapple with how to "indigenize biomedical research."[41] As of 2019, he and other leading scientists and bioethicists, including Gutherie Decheneux (Lakota–Cheyenne River Sioux), Joseph Yracheta (Pūrepecha–Mexican Indigenous), Kristal Tsosie (Diné–Navajo Nation), and Matthew Anderson (Tsalagi–Eastern Cherokee Descent), have established a non-profit organization called the NativeBioData Consortium ("Nativebio" for short) that promises "research for Natives by Natives."[42] Their focus is on

Indigenous-led genomics concerning health issues that disproportionately affect Native peoples as well as on pharmacogenomics. But beyond the human body, they are also interested in the life forms connected to it. In addition to "omics" research and gene editing, they study "soil health and wealth" with an emphasis on the microbial and botanical species that constitute the land's ability to remain a resource for humans and animals as well as an agricultural resource for communities.

Most telling in terms of research infrastructure concerns Nativebio's message on the importance of building its own biobank. As Joseph Yracheta, the Nativebio Board president, put it: "If we give our data, and nothing on the other side is changed, there's no way that the benefits are going to come back to us and more likely it's going to be exploitative. [. . .] So just like we were exploited for land and crops, medicines and gold and oil, we will be exploited for the biological treasures that are in our genome."[43] The group has also been vocal about changing how the NIH and others recruit Indigenous Americans who are not living within tribal sovereign areas and who may not be aware that giving individual consent could implicate family and tribal members far beyond themselves. In a clear reproach of the NIH, they made sure to emphasize that the onus of care was on the *All of Us* program to take these issues seriously and to not exploit people's lack of knowledge about how easy re-identification in genomics can be.

> This consent model foregrounds the individual as the sole recipient of potential benefits and harms, which might not be appropriate for a member of a tribal community. Especially with small populations, there is a risk that an individual's genomic information can be used as a representative sample to make statistical inferences about his or her community. Indeed, this has been the source of many controversies occurring recently in genomics studies within tribal nations. While many researchers acknowledge that "de-identified" individuals can be bioinformatically re-identified, this is a greater issue for members of small tribes, despite data safeguards. If *All of Us* is, in fact, asking participants to indicate their tribal affiliation, then re-identification of tribes is a certainty. For Indigenous people, these risks implicate the communities as a whole, which thus transfers potential risks onto the group.[44]

Rather than rely on the US government to proceed with this degree of care and respect for their concerns about open data sharing, commodification,

and re-identification of tribal identities that could lead to a lack of protections and invasions of privacy, the move to build their own databases has
become the more secure and logical response.[45]

The Nativebio repository lives under these scientists' care in South
Dakota, where, they write, "the samples will be stored on sovereign Native
American land, of the Cheyenne River Sioux Tribe, which will provide
additional protections for the participants' samples from exploitation."[46]
In addition to Nativebio, other efforts, such as the Summer Institute for
INdigenous Peoples in Genomics (SING)—which started in the United
States as an endeavor to train more Indigenous scientists who would be
able to carry out, translate, and also advocate for genomic research to benefit tribes—now has three additional locally run nodes in Australia,
Canada, and New Zealand. At each semiautonomous but linked site, students and future scientists as well as professionals in the field discuss past
harms done to Native peoples and the latest genomic technologies that
will equip them to take research into their own hands and avoid simply
being studied and potentially exploited.[47]

In his concluding remarks to the conceptual architects at the helm of
All of Us, Fox emphasized the trust that remains to be built with tribes on
the part of scientists from the outside. "Given the fraught history of genetic
studies involving Indigenous peoples—including the example of
Havasupai v. Arizona State University, in which the tribe successfully
sued the university for improperly using its members' blood samples—
tribal communities continue to be wary about participating in the NIH's
newest endeavor." If the *All of Us* project hopes to appeal to Indigenous
Americans, the NIH outreach and goal of respectfully coming to research
agreements with tribes through the NIH Tribal Council is a crucial first
step. That said, these storylines—of governments granting information on
the terms that they would like Indians to agree to in the form of treatises,
forced removal, contracts for land allotments, and through extractive
labor practices pitched as economic opportunities that have made them
sick—have been heard many times before.[48]

One of the most egregious and painful instances of state neglect
and medical mistreatment that links tribal land, genetics and disease-
inducing labor took place on Navajo (Diné) lands rich with uranium ore
that was mined between the 1940s and the 1980s. As miners worked in

unsafe conditions, those who owned the mines grew rich from the extraction. Families who had members who toiled getting uranium out of the earth had astronomically high rates of lung cancer, 56 percent higher than the national average. Even worse were their rates of liver cancer, at 200 percent higher than most Americans.[49] In addition to these cancers, these families had elevated rates of other complex conditions, such as diabetes and kidney disease, that genetic studies like *All of Us* emphasize the power of genetics to solve.

Most unsettling, because of the way that researchers approached it, was the question of a rare disorder that came to be called "Navajo neuropathy." This relatively new syndrome, where people develop hooked fingers and toes as well as damaged eyes and livers, was first detected in 1959 as the mines began to be abandoned. In trying to assess its cause, government scientists overlooked many of the obvious environmental factors such as systematic and extensive radioactive toxins contaminating water sources in the area, and as journalist Judy Pasternak writes, they instead focused on the "familial pattern" among patients. They concluded the most likely cause of this now population-specific neuropathy was an "inborn error of metabolism."[50] Such was the language at the time to indicate a genetic cause.

Subsequently, researchers continued down this path and found a genetic mutation that was hypothesized to stem from a genetic "founder effect" that had only manifested in the population in 1959, fifteen years after the mining started in earnest. The founder effect idea—that in this case an Italian conquistador might have passed on the gene to the Navajo in the late 1500s, since a few rare cases have also been seen in Italy—have since been ruled out based on the fact that the mutations found in each set of families studied were different.[51] The fact that this fatal condition, where a specific mutated gene that leads to mitochondrial DNA depletion, is almost solely found in Navajos in the Southwest of the United States, and emerged only after the people who live there were massively exposed to uranium radiation, renders any genetic explanations that are stripped of environmental triggers highly suspect.

In 2020 the writer Oscar Schwartz published a profile piece of Navajo/ Diné geneticist Rene Begay in the *Washington Post* as the *All of Us* program rolled out its blue vans into Native communities. In this long-form article he included various instances of government abandonment, cases of research where there was no follow-up with the community, as well as damage caused by extractive economies that led to tribes' mistrust of outside scientists. He emphasized how the sudden appearance of "Navajo neuropathy" was geneticized from the outset by experts investigating the new phenomenon. The seemingly apolitical genetic diagnosis and discourse quietly diverted fault from the US government agencies and the mining companies that the tribe blamed for so much suffering—suffering that if acknowledged could have been redressed much sooner and possibly with less resistance. Schwartz retells the events in brief: "Some studies conducted since suggest that the neuropathy is, at least in part, genetic. But by offering this genetic explanation without also examining the particular environmental history, blame shifted from the mine owners, the Environmental Protection Agency and Congress—all of whom had failed to clean up the mess the uranium extraction left behind—and ended up in the DNA of the Navajo people. The memory of this injury, which is inscribed in the land and the bodies of the Navajo, lingers."[52]

How this particular "genetic" predisposition was understood to affect the Navajo people in the Southwest of the United States, as a population, should give us all pause. This is an instance where radioactive material contaminated water sources, land, air, and peoples' bodies for decades into the present. The studies that linked the poor health of Navajo to their genes were the subject of articles that started to appear in the medical literature in the 1990s just as the Human Genome Project and the HGDP were getting under way. As Judy Pasternak details in her book *Yellow Dirt: A Poisoned Land and the Betrayal of the Navajos*, it took decades of painstaking documentation that put the onus on those affected to prove the harm done to them.[53] It required families and advocates to travel to Capitol Hill, to consistently contest dismissive government narratives, and most importantly to suffer high rates of disease and death before the US government finally agreed to compensate miners who were exposed.

The first Radiation Exposure Compensation Act (RECA) for the families was proposed in 1990 and enacted with time limits set on claims. It required people to have meticulous workman's compensation records, which many lacked since the onset of cancers and other effects of radiation appear long after such claims might have been made. Over the years the tribe demanded amendments to RECA's hair-splitting restrictions, while they gathered more data on the devastation not just on miners but on people in surrounding areas as well as on the water supply and on the wrecked quality of the land.[54] The time frames for payment, which have never been sufficient, and in the early years were issued as IOUs, were once again set to expire in 2022. After more concerted effort on the part of Navajo advocates, the window was yet again extended by the House and Senate and approved by President Biden for another nineteen years in June of 2022.[55] Yet, of course, the uranium lives on in people's bodies and is passed onto future generations via the womb. In 2022, Navajo pregnant women were found to have uranium levels that were almost three times the level for women of childbearing age more broadly in the United States.[56]

In October of 2021 the podcast *The Experiment*, hosted by journalist Julia Longoria and affiliated with *The Atlantic* magazine, also featured Rene Begay and her stance toward the *All of Us* program titled, "What Does It Mean to Give Away Our DNA?" In the episode Begay recounts that her work in the field of genetics was never truly divorced from her broader tribal ethos of centering the importance of relationships over the importance of extracting DNA. I present this excerpt from the introductory segment of the program to show how one young scientist whose tribe, the Navajo Nation, has banned outside genetic research from taking place within its sovereign national boundaries, sees value in studying genetics for health reasons. That said, she would not want to implicate her tribal community in such endeavors if, as a collective, they did not share the sense that genetic research on health could take place without compromising or in some way taking advantage of Diné or other Native groups.

When Begay herself encountered the *All of Us* program, it was during one of its consultation and listening sessions in Denver not far from where she had relocated for work. She and other members of the public, including tribal leaders representing various Indigenous Americans, wanted more transparency and actions that demonstrated a fundamental sense of care on the part of US government scientists for Indigenous people and their concerns. I include the programming soundscape that animates the upbeat, persuasive tone of the NIH advertising PSAs and also contrasts the mood of Begay and others when confronted with the US government representatives. This introduction to the program ends with an inconclu-

sive musical soundscape "dwelling in the strings." The musical track signals the inconclusive ongoing dynamic (or is it a game?) of Indigenous
Americans being asked for their DNA for their own good. The scene
doesn't have a resolution or secure landing place. Instead, it offers an aural
ellipsis of time that evokes a flight above a wary ground that has been
tread lifetimes and lifetimes over again.

> *The Experiment* Podcast (excerpt)
>
> (Birds chirp over the sound of a synthesized whistle, moving in and
> out of harmonies and through branches, deeper into an electronic forest)

JULIA LONGORIA: I wanted to talk to Rene because being both Navajo and geneticist is a pretty rare combination. The Navajo Nation actually banned genetic research from taking place at all on tribal lands.

So Rene sits at the center of two worlds that have been at odds with each other for a long time.

But recently, the US government has tried to change that.

RENE BEGAY: They conducted these, um, engagement meetings in the city—and this was in Denver.

LONGORIA: In 2019, government officials invited Native Americans to a meeting in a federal building in downtown Denver to essentially make a pitch to them for a new, ambitious genetic research program. This was obviously right up Rene's alley. So she decided to go.

BEGAY: I knew there was tribal leaders there from different tribes.

LONGORIA: And she watched the scene unfold. Tribal leaders sat at a rectangular conference table facing each other as a group of white women, dressed in business casual, began their presentation.

NATIONAL INSTITUTES OF HEALTH (NIH) PSA VOICE-OVER 1: *(Over upbeat, inspirational music)* The *All of Us* Research Program is calling on 1 million people to join us as we try to change the future of health.

LONGORIA: The program is called *All of Us*.

NIH PSA VOICE-OVER 2: *(Over more inspirational music)* What lies inside of all of us is more than data. It's life.

LONGORIA: They've aired quite a few PSAs on TV. What they're trying to do is recruit people to take part in this huge study from the National Institutes of Health.

BEGAY: They said, the goal of *All of Us* is to create health care that is individualized for each person.

NIH PSA VOICE-OVER 3: *(Over light, bouncy strings)* Instead of one-size-fits-all! By gathering health data from one million people, our country's best researchers will be able to develop treatments that are as unique and complex as we are.

LONGORIA: To accomplish this, they want our family health history, our geographic location, and samples of our spit, our urine, and our *blood* to create a giant genetic database.

(Over more emotional, cinematic music)

NIH PSA VOICE-OVER 2: And what we find there will unlock mysteries, heal the sick, and eradicate disease.

BEGAY: I think it sounds like a really great idea, right? You know, really catering health care and medicine to an individual, because I know I'm not like anybody else.

NIH PSA VOICE-OVER 4: It is important for minorities to be a part of this, or we will again be left with medications that are created for—really—other populations.

NIH PSA VOICE-OVER 5: Why not? You know, why not participate? Why not be part of it?

BEGAY: They never really said, like, "This is a biobank. This is, like, a place where things are going to be stored, like bio-specimens and data." It was very superficial language. And I still don't think tribal leaders understood what they were doing.

(The dramatic music fades out.)

LONGORIA: Then the presentation was over.

BEGAY: Somebody did ask for more clarification. They basically said, "Well, it's on the website, and here's the link."

LONGORIA: This meeting was part of a big outreach campaign by the NIH. They've worked with many Native leaders and researchers to try to connect with their communities around the country in a meaningful way. But, according to Rene, tribal leaders at *this* particular session in Denver still did not seem like they were on board.

BEGAY: They kind of were like, "Okay." Like, "This program seems okay, but . . . I don't know. I just don't understand it fully." And they just kinda were like, "Okay, whatever. Do what you want, because the US does whatever they want anyway."

(A bed of strings play in a resonant, empty space. As bows cross the strings, the notes slide up and down, a metal sound vibrates in the emptiness gently.)

BEGAY: I think that's the wrong way to go about it.

(A long moment dwelling in the strings.)[57]

We could all learn from the clear and bold actions advanced by Indigenous scientists who are increasingly investing in establishing what might be thought of as firewalls of care—boundaried protections that will help to ensure that their data is responsibly governed, and not simply shared and exploited in ways that do not benefit them. The more options that people have to create their own research agendas, and to manage and protect their own data, the better. To this end, Stephanie Russo Carroll and her team at the Collaboratory for Indigenous Governance at the University of Arizona and others have meticulously mapped out the terms and the stakes of what they call "data sovereignty."[58] The principles include: "The articulation of Indigenous Peoples' rights and interests in data about their peoples, communities, cultures, and territories is part of reclaiming control of data, data ecosystems, data science, and data narratives in the context of open data and open science. Control, coupled with a focus on collective benefit and equity, repositions Indigenous Peoples, nations, and communities from being subjects of data that perpetuate unequal power distributions to self-determining users of data for development and wellbeing."[59]

To rejoin the various filaments that thread through this chapter, I want to speak again to the problem of accounting for who is missing and the importance of addressing the why of their absence head on. In this, I want to mark the importance of the gesture of acknowledgment. But like the land acknowledgments that are so prevalent today, and that are in danger of becoming routinized without real meaning, the words are simply not enough. Various forms of material and land restitution are required to redress the violence of America's foundational history. Acknowledgments should be followed with actions on the part of governments and private institutions who benefited from such brutality in Australia, Canada, the United States, and New Zealand, where Indigenous people are not vanishing but are still finding ways to survive despite imperial settler states' attempts to culturally and physical erase them at almost every level throughout our selectively told histories. What is required are actions that recognize the dehumanization of Indigenous people that continues to manifest as crippling poverty, high rates of preventable disease, water and land insecurity, and a more general persistent questioning of people's expressions of their sovereignty, and at times their refusals to continue to play settler societies' games.

Clearly as Indigenous geneticists Keolu Fox, Kristal Tsosie, Joseph Yracheta, Rene Begay and, with the support of efforts like SING, so many others are finding career homes in genomics, there are ways to build research collaborations, establish DNA biorepositories, and proceed with genetic research on questions that affect and may potentially benefit tribes. The equation should not be one of join the US government-run projects or be left behind—"left with medications that are created for— really—other populations," or more broadly, be left out of the genomic revolution. This mind-set of "inclusion or else" often blithely characterizes the hopeful "inclusion and diversity" discursive line aimed at increasing underrepresented groups in genetic studies. The point is not to say that people should not ever participate in research.

My hope is that people will do so not out of fear that they will miss out on care, precision medicine, and futuristic promises of gene-based solutions for broad bodily and social ailments, but rather will participate when enough safeguards, evidence of fairness, and assurances are in place. These would include forms of equity that level the obscenely unequal playing fields in most domains of social life. This future will be one where the ethos of research ceases to fuel ongoing histories of extraction. It will be informed by, but not scripted by, the past as a time and a concept that is not ever cut neatly into periods that are "behind us." It is a slightly *tweaked* future tense of our current social climate in the United States where Black and Brown people— who were historically targeted in colonists' entitled usurpations that defined imperial governance, slavery, and conquest—are validated, *and thus respected*, by governments, technologists, and scientists in their concerns that they and their families should not have to fear being tracked by their DNA for applications other than those to which they originally consented.

To return to the apparition in the midst, the ghost in this last instance might also evoke the future presence of that *would-be* comfortable and solaced subject who could trust that the research apparatus has their best interests as its ultimate goal. There will surely be discomfort that comes from facing American historical violence and ongoing tendencies of data captures, profit captures, and product captures that continue current modes of captive inclusion that so many genetics projects perpetuate in their calls for inclusion in a surface reading of diversity.[60] What would it mean to begin to redress the American past that has informed many Indigenous people's refusals? How might we learn to see the inherent value of various kinds of conscientious objections to genetic research, such as the Navajo Nation's moratorium that has been in place since 2002, or the Canadian Institutes for Health Research Guidelines for Research Involving Aboriginal People's recommendation for a "DNA on Loan" model where a person's DNA and the data derived from it would be given temporarily, under specific conditions, so that individuals and tribes could retain control of their information through ongoing negotiations and consent?[61]

As STS scholar Kim Tallbear reminds us, "indigenous peoples generally embrace a global definition of indigeneity that *facilitates survival and acknowledges the historical rupture of colonialism*."[62] Indigenous scientists' own rupture of an ongoing colonial-born capitalism actively addresses the rupture of colonialism, genocidal campaigns, and attempted violent erasures of Native peoples. Their actions hone a point of articulation that helps us ask: What does it mean for any of us to go beyond the commodifiable biogeographical surface reading of ancestry? What cultural work is required to change how reductive concepts of human difference influence geneticists' methods of optimizing models that attempt to homogenize, and that actively *socially reconstruct*, biomarker-based ancestry panels that resuscitate the ghost of racialized biology? It is this racialized biology shrouded in the language of "ancestry" and "population specificity" at a genetic level that continue to show up in efforts that target underrepresented groups' "inclusion" for biobanks like *All of Us*. How to decouple these precepts of continental and racialized difference from notions of inclusion—to expand the prospect of becoming involved in genetic research beyond a call to have one's people's genes represented, or else be left out of innovation?

To enlarge the conceptual terrain, Tallbear's parsing of genetic ancestry as distinct from Indigenous relatedness may be useful in thinking toward a truly more inclusive science. She writes: "Indigenous peoples' 'ancestry' is not simply genetic ancestry evidenced in 'populations' but biological, cultural, and political groupings constituted in dynamic, long-standing relationships with each other and with living landscapes that define their people-specific identities and, more broadly, their indigeneity."[63] These emphases are seen in projects like Nativebio, SING, and many of the research and ethics protocols put forth by US tribes, Canada's First Nations as well as by Māori and Australian aboriginal groups.[64]

And for their own part, those within the long-standing American institutions of genetic and biomedical sciences would do well to engage in sustained true "listening sessions." And to do so for some time. The actors in these cultural domains need to look up from the screen where the tabula raza resides and see the humans whose DNA is viewed as fungible and valuable even as they themselves remain locked in cycles of poverty and disease. Those belonging to mainstream cultural spheres of science would also do well to look beyond the white lights of the lab. To broach the dark, and possibly contend with their own ghosts—both past and present.

Conclusion

In the preface I offered two very different scenes where human tissue and DNA were extracted from people's bodies and transformed into scientific abstractions in the process. In the first instance, in the Kittles lab, descriptions and enumerations of people's DNA floated on a young researcher's computer screen in separate spreadsheet cells, neatly ordered in rows and sorted by columns. They appeared as anonymized aggregated genetic data, collected and tagged by race. The collection of them would serve as a larger resource to discover risks for prostate and other cancers in African-descended people, with a focus on Black Americans. As we saw in later chapters, this team was made up of people of color mostly from the African diaspora. The principal focus of the lab was to learn how genetic ancestry might shed light on prostate cancer risk, vitamin D level, and skin pigmentation in Black men.

The second instance of a person's biology being abstracted by a different young scientist over a half a century earlier centered on Mary Kubicek, the twenty-one-year-old research assistant to Dr. George Guy, who removed Henrietta Lacks's cervical cancer cells in 1950. Lacks, the young mother (and patient) who died at age thirty-one, became a valuable specimen for the physician-researchers at Johns Hopkins University who had

Kubicek extract Henrietta's tumor cells without her consent. Lacks's cells, now immortalized as HeLa, continue to reproduce and will likely outlive anyone reading these pages today. And yet, they were first grown from that long ago interaction between Kubicek and the "live woman," whom she had not "thought of that way" until she was confronted with Henrietta's very humanly painted toenails on the autopsy table.[1]

With that first immortalized cell line, the face of medicine would be forever changed. But what about the face of the person whose bodily matter becomes the abstracted material that can efface them almost entirely in the process? What about their human lives and concerns, and how to keep these in view while also carrying out the potentially important work of science?

Many scientists' intense focus on the therapeutic value and market power of bodily-extracted tissue imbues these with a veritable and sometimes virtual second life that is too often dramatically disconnected from their first lives—that is, those they lived. In both scenes above, the researchers were brought back to the race-based human relationships that allowed for the biological material used by their labs to be lifted from research subjects' bodies in the first place. In other words, with some degree of a jolt from an outside observer—someone interested in how scientists' social relationships to the life of tissue and DNA mirrors something about race relationships in the United States as a whole—these young professionals were brought back to a conscious vision of the sample as part of a person.

I want to end this book asking the following: Could other scientists who work at such distances, and who become somewhat emotionally removed from the people who constitute their data, regain the recognition of the human and generate a renewed sense of care demonstrated by the researchers in the preface? When we leave the domain of the medical research study, what might this look like? For instance, what modes of care might be afforded to people's DNA gleaned for medical or genealogical purposes but that has ended up in databases used in policing, where care as a precept is not exactly how the system is run?[2] In other words, how can biological data drawn from individuals' blood, tissues, cells, and DNA retain or, more specifically, regain some notion of the person, and forgo merely being rendered a useful and manipulable datapoint?

There are certain technical and ethical issues to consider concerning where we are as a society and our ability and willingness to carry this line of questioning forward. As we saw in chapter 7, part of the promise of so many genetic studies involves the guarantee of anonymity that is seen to be possible with the de-identification of samples. However, as data scientists like Yaniv Erlich and Latanya Sweeney have shown by re-identifying real-world people from their public yet anonymized genome sequences in the US government's 1000 Genomes data, and also by re-identifying participants in the PGP who wanted to keep their identities private, this promise no longer truly holds. At issue is the contemporary reality that reams of other personalized data are available in the public domain that can be matched with small descriptors about the genomic dataset in question.

If people can, and in some cases will, be re-identified, this prospect will affect some racial groups more than others as concerns corporate, police, and state surveillance. One may be fine with that, but I would like to leave the reader with the invitation to think about the following: When have you made the conscious decision to help build government, research, private commercial, and/or law enforcement databases? If you are drawing a blank, that does not mean that the answer is never. These entities rely almost wholly on the public—seen as "donors," "participants," "clients," or "suspects"—to amass the data that makes their machinery work for the disparate yet connected domains that now run on genetic information. These are DNA ancestry tracing, large-scale medical genomic studies, and policing where forensic sleuthing sometimes involves DNA garnered from medical and ancestry contexts to compare these to DNA left at a crime scene.

Technically, data scientists who work in the societal wide-reaching domains of health, law, and commercial ancestry fields can very easily determine who people are, what social groups they belong to, and what once private concerns they may hold (especially if this data is correlated with public information that users of social media post about themselves over the timeline of their lives). The researchers, and everyday citizens, who collect this data often know much more about people than the specific questionnaires or raw data of any sort show at face value. But as others have pointed out, "raw data" is an oxymoron.[3] Data have social contexts, will always be read in certain ways, are always historically dressed, and culturally layered both before and after analysis.[4]

With anonymity, some level of abstraction becomes written into the design of genetic scientific studies on people, at the same time that this is imagined as a clear norm for doing good science. In some cases, anonymity can mask certain social and power dynamics that will henceforth be buried from sight (as in the parts of this book that have been redacted, or identities anonymized, in an effort to protect certain people in my study who did not want to be named but whose humanness I still try to portray). As concerns anonymity to protect individuals in medical research studies and genetic databases, redaction of personal identifiers should not signal that the individual is no longer a consideration.

DNA donors should be able to choose what kinds of studies they want their genetic information to be used for. Researchers should collect information about their concerns. What if people do not politically agree with the uses to which IQ research has been put to racialize and naturalize the measured intelligence of Mexicans and African Americans during the American eugenics movement and, later, to track mostly poor students of color as well as validate students of privilege? Researchers should consider a respect for these potential recruits' concerns that could entail allowing them to check a box on a tiered consent form with the option to not be used in studies on the genetics of race and IQ, or IQ's close cousin called "educational attainment."

Similar consent structures should exist for protocols on forensics for people concerned with policing and bias. I would also advocate for the use of scenarios, or illustrated scenes, on consent forms that give real-world examples of how DNA can be used in these ways. This would also be especially helpful with regard to the possibility of reductive framings of health issues that can be narrowly described in racialized genetic terms. Here I'm thinking about research on asthma, cancers, and liver disease that overlook the risks and damages done to bodies because of colonial legacies of land dispossession, capitalist industrial mining, and radioactive polluting of the environments that people call home. These dynamics have brought on economic pressures that force many minorities to work jobs with high health risks. The local biology of diseases like "Navajo neuropathy" (discussed in chapter 8) require a broader approach, such as that advocated by medical anthropologists Margaret Lock and Vinh-Kim Nguyen.[5]

Consent conversations should also cover the possibility that the genetic research, or a database to which a member of the public gives their DNA, could aid in the development and the commercialization of pharmaceuticals that will be sold as "ethnic drugs." In these scenarios companies may profit from defining the market sector in narrow terms that result in higher costs for people sick with a condition marked by health disparities, while the pharmaceutical may in fact be useful for a broader swath of humanity. All of these points include open conversations about underrepresented groups' captive inclusion and the instantiation of the tabula raza as an economic structure as well as a political device to keep race and racialized bodies working overtime.[6]

I have spent less time in this book on the issues that arise when genetic and cellular material leaves the person's body and is slightly manipulated for a specific use or technological application that can be described as "novel." In such cases, courts in the United States have ruled that people no longer own their own biological material once it leaves the body—despite the fact that their biomatter's nucleotides contain more information about them than almost any other object they carry in their more standard, material lives. There are many possibilities that people's biosamples that have passed through the hands and labs of scientists discussed in this book have now been patented and profited from. This was a clear concern on the part of Indigenous groups and Native American geneticists (discussed in chapter 8), prompting them to protest certain government and private efforts, and to finally create their own databases in the name of what they have termed "data sovereignty."

There were, indeed, several instances when scientists that I studied were looking to commercialize or create market products from the basic science I witnessed them carrying out in their labs. These centered on some aspect of the research at several sites that was funded in part by pharmaceutical companies, private capital, philanthropies, the military, and various other arms of the US government (the Department of Defense, the Department of Energy, National Institutes of Health, and the National Institutes of Justice). Several of the PMT drug transporter studies in the Giacomini lab were on this trajectory. Some of the synthetic biology life forms as well as genotyping innovations in the Church lab were also envisioned to radically reshape the market with the new needs they could create. The courts' rulings in favor of commercial patents are one thing, and patents are publicly available.

A similar but less visible entity concerns products developed within the walls of for-profit companies that may not patent people's DNA or aspects of their tissue but that operate on proprietary trade secrets that utilize people's data. These constitute other economic spaces where scientists can benefit from extracting and abstracting parts of the human body. These are also sites where much of a company's internal activity remains out of the purview of independent assessment since these ventures are private, and often remain small. For better or worse, consumer protections in the United States do not include educating people on how valuable their DNA can be, especially when combined with thousands or millions of others' samples, which would require thinking beyond the atomized individual so favored in the American cultural psyche. But as more people realize the commercial value of their DNA, and the multiple uses to which it can be put, the public is gaining clarity on better genomic database governance.

A nationwide survey carried out by Forrest Briscoe and his team at Pennsylvania State's College of Business found that respondents were less willing to donate DNA if they knew that the party collecting it would share their data with others beyond the scope of the intended use. This included government sectors. Another key point of concern for most respondents was that they did not want their DNA data sold to pharmaceutical firms.[7] After 23andMe went public in 2021, it became clear to many that the business model for the company was to *lose* money on providing ancestry and some health information to consumers with the $99 "spit kits" they sell. This "loss" would be a small hit for a larger return on the more valuable access to people's DNA sequence information when aggregated as a multimillion-person dataset. When joined with the phenotypic and lifestyle information surveys, 23andMe has marketed clients' DNA to several pharmaceutical companies (Alnylam Pharmaceuticals Inc., Biogen, Genentech, Pfizer, P&G Beauty) to do all manner of studies.

In July of 2018, 23andMe gave GlaxoSmithKline, now called GSK, exclusive rights to mine their customers' data for drug targets for $300 million. In an interview with *Wired* magazine a week after the revelation, Kayte Spector-Bagdady, an ethicist and lawyer at the University of Michigan, spoke to science journalist Megan Molteni about the customer policies and disclosures around consent: "If you look at the documents carefully all of the information is there." They really do disclose it all, she explained. The crucible, however, for the lawyer: "The challenge is that people don't read it."[8] Backing up a bit, I would venture that the challenge is a bit more complicated. And that the lawyers, executives, and scientists at 23andMe just might have some sense of this. In fact, in a 2019 interview with *Forbes*, 23andMe's then chief scientist and head of therapeutics (an internal team that translates genetic data into discoveries to develop new drug therapies) offered a line that has stayed with me: "I thought it was genius actually, *that people were paying us* to build the database."[9]

As medical anthropologist Sandra Soo-Jin Lee has described, 23andMe embraces a culture of fun and recreation embedded in passive advertising that is explicitly spun through its products and website design. The company voices its desires in injunctions for the individual to "take control of their genetic information," strangely by giving it to 23andMe to analyze and profit from. Part of the deal, the literal face value commercial deal, is that the consumer will receive an ancestry report and some health "information." In the online interactive results page these combine to frame their genetic propensities and risks, and to link them to other contemporary humans around the globe or in their own household in new ways. In all of this, testers are invited to "re-create" some aspect of themselves that goes far beyond the recreational nature of fun and play, those design features that sell the product.[10]

I would wager that most of the now fourteen-million-plus people who have ordered these "ancestry" kits that simply match them to others whom I have called the "today people," which is to say contemporary humans living around the world, are not always aware of their role in helping the cheery start-up become a multibillion-dollar pharmaceutical enterprise.[11] Ancestry.com boasted an even larger database, of nineteen million, at the time that it sold itself to the private equity firm Blackstone, Inc. Shortly thereafter, at least one class-action suit was filed in an Illinois court for violating the Illinois Genetic Information Privacy Act (GIPA) for sharing clients' potential health and personal identifying information without their consent.[12] Yet, as the Michigan lawyer above pointed out, as concerns 23andMe, it was all there in the fine print. In this sense, 23andMe has proven to be a transparent, legally responsible company. For better or worse, much of the onus will be on individuals to become more educated about these issues, and also perhaps to be more demanding if they want profits shared, collectivized, combined, or otherwise distributed to the everyday people who create these companies' wealth.

Briscoe and colleagues' research showed that many participants feel that payment for the value of their genome is indeed warranted. Others, such as executives at the personal sequencing company Nebula Genomics, cofounded by George Church, want to invite users to exploit this data themselves (along with the company). In a bit of a bizarre stunt, Nebula created a nonfungible token (NFT) of Church's genome, which is currently up for auction at the time of this writing. The value of one person's DNA in this case comes from his role in advancing the field and being one of the first people to share their full genome sequence for study.[13] But that value, in market terms, has yet to be settled by a buyer. This could be because Church's genome is already available for free on the Web as part of the Personal Genomics Project.

Clearly the idea of people profiting from their own DNA is not a straightforward solution. I am not actually advocating for more commercialization and commodification of human bodies, or DNA. That said, I do think that people should know the myriad ways in which their information can and will be used. I invite publics to collectively ponder for themselves what kinds of conversations, arrangements, and plans they imagine will best contribute to human health and society. I also advocate for the onus to be on scientists to point out as many current applications and future scenarios that they can dream up for people's extracted DNA and other biological tissues stored in data vaults.[14] These professional experts are the ones designing and executing any such imminent capabilities. In this way, the PGP's consent form items regarding the possibility that someone could synthesize a donor's DNA and plant it at a crime scene, implicating the donor or someone related to them, is a good example of how scientists on the project have taken some aspects of that responsibility seriously.

Although many people would hope that science and business would be as transparent as possible, there is no way for the public, and little enticement for companies, to enforce this. The fact is that many direct-to-consumer ancestry testing companies are businesses with scientific aspects; that is, they commercialize products that are compelling because they glean data and information from scientific studies and public, government-funded genomic databases, such as the 1000 Genomes or the HapMap projects. Although the science is featured as what is on offer in packaged ancestry products, the larger business structure, with all of its lawyers and protections of trade secrets, its execution of NDAs to conceal proprietary knowledge, do not yield themselves to assessments of replicability or accuracy by independent review. In these often labyrinthine systems, products are born from scientific ingenuity and, quite often, from the DNA of donors from many walks of life—and death.

. . . . .

In 1991 the US General Services Administration (GSA) was set to begin construction on a federally funded thirty-four-story office tower at 290 Broadway in lower Manhattan. Like all federal works, the project planners were mandated to comply with section 106 of the National Historic Preservation Act of 1966, whereby a cultural survey and archeological digging led to "unearthing the 'Negroes Buriel Ground [sic].'" The six-acre graveyard contained upwards of fifteen thousand intact skeletal remains of enslaved and free Africans who lived and worked in colonial New York from the 1630s to 1795. It is the nation's largest and earliest African burial site to date.[15]

Rick Kittles (discussed in several chapters of this book) has spent his career trying to figure out why prostate cancer affects Black men more often, more aggressively, and earlier in life than other racialized groups. For all of the importance of this work, it was his role on the African Burial Ground project with a team led by anthropologist Michael Blakey at Howard University that thrust him into the public eye. Part of the research of the Howard team was to build a database, also funded by the GSA, to help identify aspects of these early African Americans' human remains and if possible to specify their origins. What could we know about lives, nutrition, and health of these enslaved Africans, and also their trajectories starting with the lands and peoples from whence they came by examining their bones and DNA? After less than a decade of building the African Burial Ground database and cataloging the samples with the larger team, Kittles began to make plans in early 2000 to offer contemporary African

Americans the same possibility: to be matched with people and locations in Africa through their DNA—for a fee.

Ever since Alex Haley's 1976 book *Roots: The Saga of an American Family* was published, followed one year later by the television series by the same name that aired in living rooms throughout the United States as well as the diaspora, Black Americans have been inspired to learn more beyond the depressing mainstream history offered in the United States that we basically originated with slavery.[16] Most high school history books did not bother to detail all of the ways that enslavers purposefully and strategically cut off the Africans they took as property from their languages, tribal customs, and longer rich histories.[17] The emotional and psychic toll of slavery continues to affect Black Americans beyond the legacy effects of massive inequities in wealth, education, health, and power in the United States and in the Americas more generally. Many want to know where they are from beyond the paper trail that often ends in archives that are almost always collections of documents where Black people and the ancestors figured in the wills and other legal documents that highlight the lives and humanness of white enslavers. Black lives were mostly listed as property or as subjugated citizens, given census or military records, within a state that oversaw their oppression.

Kittles wanted to offer Black people something more tangible and uplifting than dead paper trails that, at a maximal reach, cut off at points of slave ship debarkation and human cargo manifests. He had spent many of his graduate school years in reading groups to learn about and validate Black peoples' place in the world, and even took on the Smithsonian Museum for their representations of Africans, which resulted in some changes in their exhibits. So, in 2000 when a Washington-based reporter by the name of Sam Ford contacted Kittles to see if the scientist could test him for his genetic origins, while promising to air the story on his news program, Kittles agreed.[18] And as this possibility of finding one's roots hit the newswire, Kittles was slammed with requests to do the same as he did for Sam for countless other Black Americans, from celebrity academics to everyday working moms.

At the time Kittles was still at Howard, where the African Burial Ground Project was receiving federal funding. He came under fire for potentially capitalizing on the Howard team's humanistic archeological effort to understand more about what happened to a highly vulnerable past population, that of enslaved Africans. After a falling out with Blakey, Kittles would start his own for-profit venture with businesswoman Gina Paige a few years later. And thus the renowned African lineage genetic testing company African Ancestry was born. Today the company boasts thirty-three thousand African lineages across forty countries.

The excited fanfare and the critiques of African Ancestry have been high pitched from the start. Critics argued that Kittles shouldn't be capitalizing on people who were enslaved, especially while another project led by geneticists Bruce Jackson at Massachusetts Bay Community College and Burt Ely of the University of South Carolina aimed to give African Americans their genetic matches to Africa basically for free. Many, however, were so moved by the prospect of being connected through ancestral ties to Africa offered by Kittles's molecular biological reunifications that they nominated him for a Nobel Prize for "unlocking the Door of No Return."[19] Yet in all of the commotion, critiques, praise, and possibility that Kittles's database offered, few people, if anyone, asked who were these Africans whose DNA served as reference samples for the new technology to place Black Americans' origins? And how did African peoples' DNA end up in a database that came to be so potentially helpful for African Americans who understandably wanted a connection back to the family and heritage lines that American white supremacy and slavery took from them?

When discussing the issues of DNA sharing, database building, and how norms of consent had changed in the 2010s after the Havasupai tribe of Arizona sued the University of Arizona on the grounds that researchers there used the tribe's DNA in ways that they did not agree to, Kittles relayed his own story to me about how some of the African Ancestry database came to be. Certain samples came from his biomedical collaborations on the continent. The DNA taken from men with prostate cancer and their female relatives helped him begin to match Black people in the United States with lineages of living Africans. He told me that data sharing and future reuse was common in the late 1990s and early 2000s. Back then, many people had DNA in their freezers from colleagues who had collected it for other studies. It all seemed legitimate as long as the DNA was anonymized at the level of the individual person. I heard other murmurings to this effect at scientific conferences on ancestry DNA. With the

Havasupai case reemerging more broadly in 2010, scholars rallied to create clearer ethical guidelines. Bioethicists Amy McGuire and Laura Beskow conducted a 2010 review of consent standards for genetic data collection, with the prefacing comment: "How specimens and data are collected and stored, for how long, by whom, and for what purposes varies tremendously."[20]

.   .   .   .   .

In 2019 the *New York Times* covered a story that involved Yale geneticist Kenneth Kidd contributing genetic data to an abusive Chinese state mechanism that violated the human rights of China's ethnic and religious minority Uighurs.[21] In amassing DNA from all over the world, from colleagues and mentees who collected samples over decades, Kidd reported that he had perhaps unwittingly collaborated on studies that included DNA that was used in genetic surveillance technologies that forcibly took samples from the persecuted Muslim ethnic minority group.[22] Among geneticists, norms and behaviors around data sharing have changed over the decades. Still, there is a lack of transparency about not just what ethnic, racial, or geographic labels represent people in DNA databases, but an equally important question that needs some detail concerns who these people actually are. Are they patients giving a trusted local doctor blood in the course of medical care? Are they classes of ethnic minorities with restricted rights like Uighurs in China? Are they vulnerable due to poverty and lack of education, like many of the world's very poor, who, like the vast majority of educated Americans, may not fully understand the untold uses to which DNA can be put? All of this still spurs deeper questions about entrepreneurial research practices.

Like his competitors in the ancestry genetic genealogy market who wanted details about which ethnic groups Kittles had data on by name, government forensic scientists also pressured him multiple times to make his proprietary database public for their use. He always found ways to politely, or not so politely if the requests kept coming, say no. But there was one instance when Kittles's African diasporic and African American donors' samples may have ended up in a forensic database. When he found out about the potential breach, he stopped sharing his data with a scientist who, up until that point, had been a close collaborator and colleague. I interviewed Kittles about the tangled set of events. He recounted how the scientist and the CEO of the company that the man "consulted for" flew Kittles down to talk in person to see if he couldn't be persuaded. Kittles told me that he informed the two men that he would not help them "put more Black people in jail." He cited for them America's glaring disproportionate rates of incarcerating Black men and asked them to consider the harms that targeted policing inflicts on Black communities. In a word, Kittles told me that he let them know—in the particular tone he would call up in serious moments—"*I'm not gon' let you come into my house wit' ya dogs!*"

.   .   .   .   .

In the spring of 2004, before I had ever worked with Kittles, I flew in a very small plane with lots of turbulence to the high country of Pennsylvania to observe the lab of a scientist who shall remain unnamed. What I will say about the experience was that I decided not to stay in the lab past the short visit of a little over a week, mostly because he asked for my DNA for a project that, as I was coming to understand, had some "issues." He also wanted a photograph. When I learned more about the technology this scientist was building, I sensed that he was especially interested in me because I myself am biracial, or "admixed" in the lab's terminology.

In return for giving him my genes, he offered me my genetic biogeographical ancestry results "for free." I later realized, however, that I would have also ended up in a database where DNA like mine, coupled with a facial photo, would be used to generate algorithms to assess criminal suspects' ancestry percentages and facial features for police work. The data would go to a company that the scientist was consulting for, and that peddled in two seemingly different products that utilized the same clients' samples for both. These were called AncestrybyDNA for genetic heritage testing and DNAwitness for forensics. By using the same ancestry informative markers on DNA evidence left at crime scenes, the company claimed that the markers, when named and marketed as DNAwitness, could be used to "infer elements of physical appearance from crime scene DNA and allow forensic investigators to 'paint' molecular portraits of a suspect.'"[23]

When that company, DNAPrint Genomics, folded in 2009, I very well could have ended up in another forensic private venture's database that the scientist also consulted for—one that eventually developed what have been dubbed "digital mugshots" in 2014. These mugshots were alternatively named "DNA phenotyping" portraits. The second company is called Parabon NanoLabs. Their specific research to develop forensic DNA facial phenotyping was funded with a grant from the US National Institutes of Justice to *initially* optimize the technology in one particular group: Black people. The project was to reconstruct Black faces from DNA and facial morphology data (measuring things like eye distance, lip breath, nose width, and head size from photos like the one I was asked to submit) taken from people in the United States, Brazil, and Cape Verde. People in the latter two countries were thought to approximate ancestry mixes associated with Black Americans. Some of the first applications, developed by both companies, were used in police work to track criminal suspects in unsolved crimes who were (in the language of probability statistics) *most likely* Black. These were the databases that Kittles refused to help build with his data obtained from people in the African diaspora—data that at one time in his career he had shared with a scientist colleague for other research purposes on ancestry, skin pigmentation, and disease risk.[24]

To this day, Parabon NanoLabs produces eerie humanoid-looking portraits of criminal suspects whose would-be likenesses are drawn from an amalgamation of features gleaned from students and other "donors" who have given of their DNA, accompanied by a photograph.[25] The company has now expanded the technology to other populations, and for other uses beyond the re-identification of criminals, such as missing persons. They hired genetic genealogy expert CeCe Moore to conduct another type of forensic work where she searches large ancestry databases for relatives of people who may have committed violent crimes. The company surely realized the efficacy of this approach after investigators found and arrested the Golden State Killer in 2018 using the then public crowdsourced data hosted by GEDmatch. This is a different method than that used for DNA phenotyping, one that has more legitimate science behind it.

Shortly thereafter, with its new fame and clear utility, GEDmatch sold itself to Verogen, Inc., a sequencing company devoted exclusively to forensics. In May of 2019 the National Institutes of Justice released the first ever policy guidelines stating when police could use ancestry DNA databases in attempts to solve crimes. The guidelines limit this to violent crimes and prevent law enforcement from attempting to correlate medical or psychological traits with a suspect's DNA. GEDmatch, for its part, met with client pushback on issues of privacy after the case of the Golden State Killer genealogical match made headlines nationally. The company responded in good faith by giving its users the choice to "opt in" for forensic uses. That move reduced the number of people who were willing to be used in such searches by 90 percent, greatly decreasing the "value" of the database.[26]

Back when I was visiting the Pennsylvania site, I met a postdoc in someone else's lab on the same floor as the scientist who asked for my photo and DNA. They told me that he had been using a very popular lecture course in sociology, called "Race and Ethnic Relations," to give away "free" ancestry tests as part of a pedagogical exercise on the "social construction of race"—and in the process was obtaining students' DNA. The class regularly enrolled five hundred to seven hundred people at the time. I reflected on how we were witnessing a strange "alchemy of race and rights," to borrow Patricia Williams's poetic phrase, where one could simultaneously racialize students' DNA for the designs of policing technologies, in a system of law enforcement bias against racial minorities, and offer the social constructionist promise to liberate young college students in a diverse classroom from ideas of fixed racial thinking.[27]

Similar yet different moves have been orchestrated by plaintiffs in Supreme Court cases on affirmative action—with new ones brought against Harvard and the University of North Carolina in 2022. Conservative lawyers and legal advocates argued to end affirmative action by quoting sources such as the American Anthropological Association's very own statement on race. Legal scholar Mary Ziegler provided one of the first analyses about conservatives embracing the "social construction of race" idea to argue against university admission policies. Concerning amicus briefs and arguments in *Fisher v. the University of Texas II*, she writes: "Fisher did not transform the law of affirmative action, but the case did mark the appearance of important new arguments against race-conscious remedies."[28]

Two briefs written on behalf of plaintiffs in 2018 penned by the socially conservative American Law Center for Justice (ACLJ) and Judicial Watch,

Inc. (JWI) were the first to incorporate what was heretofore usually seen as a liberal line of reasoning. Ziegler writes: "JWI extensively quoted a statement made by the American Anthropological Association that 'race evolved as a worldview, a group of prejudgments that distort ideas about human differences and group behavior.' If racial categories had no scientific validity, and if race was a sociopolitical construct, then race-based preferences had to be inherently incoherent and wrong. . . . For JWI, the idea of race as a social construct first meant that racial categories are 'inherently ambiguous.' Racial definitions could be cultural, personal, or genetic, the brief argues, and individual racial groups are maddeningly hard to define."[29] She continued: "The ACLJ echoed these claims, arguing that race-conscious affirmative action is 'ultimately incoherent, as racial categories are both incoherent and porous.'"[30]

Some aspects of this argument resembled those made by socially conscious geneticists like Dr. Joseph Graves, who in his genome map era book, *The Emperor's New Clothes*, painstakingly detailed how biogenetic notions of race are not scientific.[31] The new conservative uptake of the "social construct" idea has gained traction in arguments subsequent to *Fisher II*.[32] That said, there is little discussion beyond the pat surface statement that if race is a construct, and simply a social figment, then it shouldn't determine policy. Yet there is a cavernous blind spot regarding what these policies were intended to redress: the lack of educational and other opportunities that years of slavery, Jim Crow, Termination, dislocation, dispossession, and discrimination have etched into our economic and social fabric for Black, Brown, and Indigenous groups.

Regrettably, the primary reasoning that informed Lyndon B. Johnson signing into law Executive Order 11246, stating that employers and contractors receiving federal funds had to take "affirmative action" to redress the legacy of America's racist past, got thrown out with the first case challenging the issue with regards to college admissions policies (*University of California v. Bakke* in 1978). That was when the concept of "diversity" entered the equation. Since then, we have been left with a much weaker foundation for considering race in admissions, mainly because "diversity" was meant to signal the value of different opinions on campuses, which is laudable, but it strips the initial policy of its judiciously particular intentions.[33] Johnson explicitly wanted to address the damage of slavery in his vision of a new "Great Society." In a commencement address he delivered to students, their families, and the faculty at Howard University in 1965, the then president of the United States described the necessary political work from within the government that might begin to change the racist culture of the United States. He spoke explicitly of how Black people in America were still captive, or "trapped." He enumerated many causes for this condition, settling on the "more deeply grounded," "more desperate" ones—those being "the devastating heritage of long years of slavery; and a century of oppression, hatred, and injustice."

Johnson's policies and programs (around labor, health care, housing, and education) created more attention and earmarked resources to begin redressing the historical violence, gross assaults, pervasive denigration, and systemic exclusions of Black, Brown, and Indigenous people since the founding of the United States. It had been illegal to teach enslaved African Americans to read and write in many jurisdictions for centuries.[34] Mexican young women and men were sterilized in the United States, including juveniles who did poorly on IQ tests, like the Stanford-Binet, in no small part because they spoke Spanish and were often classed as "moronic."[35] Indigenous children all over the country were sent to traumatizing boarding schools that most often prevented them from maintaining a clear sense of self, much less learning.[36] The marks of these abuses and continued discrimination do not set up America's underrepresented groups to pass on legacies of wealth, education, and social stability to their descendants, even if some have managed to thrive despite ongoing systemic bias and inequities.

.   .   .   .   .

In this book I have explored scientists' political stances, identity politics, rights discourses, social justice hopes, sensibilities of minority uplift, as well as how specific stories about conquest and slavery have all entered into how geneticists collect and analyze DNA. These political narratives have also primed scientists' views about which populations were to be selected as "important." Their visions often influenced how they designed their studies, wrote their algorithms, and built their models to infer composites of genetic continental origin.

My initial concern when I started this research, back in 2003, was less about recreational or personal genetics. I was curious about how ancestry concepts were being used and touted in health disparities research, and how ancestry marker panels served as tools for more effective genetic "admixture" mapping to localize candidate genes for disease. What I found was that these professionals often enrolled DNA markers, as points of biological difference, into a biogenetic, statistical, yet always already social, new species of the construction of race. When collected, classified, analyzed, circulated, and published on, DNA variation from bodies living in specific points on the globe were called up to serve racial categories in ways that were seen as beneficial. *Who could be against improving minority health?*

It was necessary that the people doing this work were racially diverse researchers themselves and that they were outspoken, out on the front lines, about the importance of using race in the field of genetics. If it had only been white men advancing such research, then the specter of the racial science of old might have been too threatening for many of the people who, as patients or publics, were being asked to contribute this valuable part of themselves to medicine. The DNA that many of these scientists collected paved the way for a racializing of genetic molecules.

What I quickly learned was that the basic science of genetic ancestry marker technology (AIMs), and the uses to which it can be put, didn't exactly "spill over" into other domains, as some have argued. Instead, the broader spheres of law, family history, personal identity, medicine, pharmacy, and state surveillance are often mutually characterized by normalized beliefs about distinct humans who are imagined to possess innate racial differences—differences that are still often understood to be genetic at the group level. I have seen aspects of these societal interests make themselves present alongside DNA in the various labs where I conducted fieldwork. Forensics, police surveillance, and personal ancestry were linked points of departure in the Pennsylvania lab where I observed daily ongoings briefly. In two other labs, in San Francisco and Chicago, health disparities in asthma and prostate cancer, respectively, were the foci. In yet another, racial differences in reactions to pharmaceuticals seeded their efforts. Finally, in the Boston-Cambridge area, producing individual sequences known as personal genomes, it was hoped, would carry us beyond using race as a proxy for disease risk or drug susceptibility, while detailing everything there is to know about a person (eventually). But what I also learned in working with so many different scientists was how personal their commitments were, how unique and storied their own journeys, how compelling their many reasons for trying to make sense of genes and human difference.

.   .   .   .   .

In March of 2015 I was invited by the Personal Genomics Education Project (PG-ED), led by Ting Wu of Harvard Medical School, to be part of a panel to brief Congress on law enforcement and civil rights issues that might arise from personal genetic ancestry testing.[37] Earlier that winter, media outlets from *Fox News* to the *New York Times* had run stories heralding the fact that geneticists could now construct the likeness of a perpetrator's face from DNA that he or she had left at a crime scene. The tool being touted was DNA phenotyping. In light of then President Obama's Precision Medicine Initiative, now called *All of Us* (discussed in chapter 8), certain legislators wanted to learn more about how constituents' DNA might be used and hear potential civil liberties concerns pertaining to consent for forensic applications.

At the end of the briefing, I asked George Church, who was in the audience and mingling with the crowd, if the PGP leadership was open to having participants' data used for forensic purposes. He told me: "Some years ago we thought that we could use the photographs to do the kind of work you see in DNA phenotyping." (Reader, please recall the adhesive ruler tape placed on the foreheads of the PGP-10 for their frontal and profile pictures, figure 5 in chapter 7.) Church continued: "It was in line with what we were trying to show about how the data could be used—and if people would want to put all of that data online with clear knowledge about how it could possibly be used by police, or others." I brought up the issue of PGP's open consent as opposed to broad consent, which usually gives researchers permissions to use data down the line for a wider array of studies. I wondered about how to start more explicit discussions on the sharing of DNA gleaned for "ancestry" so that people could decide if a forensic application was something that they would want to support.

"I think we're actually *addicted* to ancestry now," he responded. "It's like one of those things like fat or sugar. We're now living in excess of it and it's possibly hurting us. In the past you needed all of these ancestry connections, and it really mattered in terms of who you were, and sometimes for survival. Just like excess fat and sugar today, we have an excess of an obsession with ancestry that we now don't need in the same way we did in the past."

What exactly could an addiction to ancestry mean? Addicted to the products and their seemingly precise and quick results? Addicted to direct-to-consumer social Web interfaces, like those offered by 23andMe and ancestry.com, that mediate notions of new family and a sense that DNA reconciled relationships might tell us something about the invisible humans we contain within? Or was it addiction in the sense of lacking control? But then, there is no control over a past that both promises and simultaneously has taken so much. And so, we might comprehend the "excess," especially when it amounts to an emotional need for histories that were cut off for so many. We can understand the desire to resolve an impossible to resolve New World conquest that ruined the lives of millions.

We are now living in newly troubled and violent times in America. Racism and dismissals of the other are elemental to our atmosphere and fog our ability to imagine a collective future. At the same time, people from all walks of life are now invoking the ancestors in everyday speak all around me in ways that feel increasingly normal. At social justice protests, at BIPOC meditation retreats, at writers' workshops. Amid the alienation, loneliness, and frustration of America's stuck place in the fire of race, I understand this refuge to the past, to summon spirits and guardians to protect us in the present. I also want to invite us into deeper "promises of affiliation" with the many lives that, as scholar Saidiya Hartman reminds us, are still "at peril" in the present.[38] There are people for whom we could be better cousins or kin if we only called them up more, or were willing to channel our support, via our human technology of care while we are still alive on this heartbroken planet.

As concerns the genetic data used to broach the societal wounds that open wider with health disparities and denials of racial repair, the people from whom this DNA has been taken are not literally ancestors. They are contemporary humans in the here and now living all around us in a world that remains unequally shared. They are not ghosts, even though we often don't see them. And for reasons absorbed in the rationales of study protocols, we never will. Their names are lost, their faces gone, their common or uncommon plights—their lives—are overwritten in Sharpie-penned labels that grace vials of their valuable biomatter in subzero freezers. We merely see readouts, correlations, data, and in the commercial realm, matches with some small portion of their DNA.

Yet when called "ancestors," they are rendered "past," rendered old, rendered historical, rendered family cast as real yet distant, both temporally and spatially. In this, they can be phantasmal and doubly othered. The narrative irony is that we take them to be immediately and intimately *who we are* in the present via the DNA that captures and arranges them in a static still-life framed in the tabula. And through this frame, we too are captured. We are captured in a system of an aspiring, racialized "inclusive" science—a New World run through with seeming contradictions that begs to be better understood.

Acknowledgments

This book would not have been possible without the many people who have helped me think through ideas and logistics, and provided an ear for the many small and large plots and subplots of what eventually became this ethnography over two decades. I owe much to the many scientists included in these pages who were open to an anthropologist's study about the ins and outs of their work. It was not always easy navigating the dynamics of laboratory life. But most people were patient and wanted to aid the project in ways that only they truly could, given that, with time and space to reflect, scientists are uniquely positioned to help the public see the myriad ways the technologies they are bringing into the world may be used.

I thank my friends and mentors who offered so much valuable advice and support throughout the early years, especially the always intellectually bold late Paul Rabinow, the ever brilliant Rayna Rapp, and the keenly intuitive Emily Martin. Troy Duster has been a cherished friend, fellow traveler, and guide for nearly three decades. His long vision, humanity, and poetic curiosity have always inspired me to keep going and to find connections of broad social relevance. Too numerous to name, I thank the group of thoughtful colleagues on the RACEGEN listserv who have likewise provided a needed community for nearly twenty years. I am especially thankful to have been part of the early gang of us that Troy convened at New York University in the fall of 2003, specifically Alondra Nelson, Anne Morning, and Aaron Panofsky. Friends and colleagues in France who have shared their *amitié*, lightheartedness, and helpful outside eye about American visions of biological difference over the years include Luc Berlivet, Mélanie Gour-arier, the late Bruno Latour, and Pap Ndiaye.

During my time at Harvard various comrades and colleagues provided sources of support and always compelled me to think deeper—notably Arthur Kleinman, Tamara Kay, Nancy Krieger, Terence Keel, and Jason Silverstein. Evelynn Hammonds convened a serious group of New England scholars who met periodically, and we did some great thinking together. A special thanks goes to that group, to Evelynn, Lundy Braun, Anne Fausto-Sterling, Jennifer Hamilton, Alondra Nelson, and Susan Reverby.

When I moved to Stanford, I was fortunate to find myself in a sea of uniquely brilliant folks. I am especially appreciative for the various forms of support and kindness I have received over the years from Al Camarillo, Jen Brody, and Matt Snipp in CSRE as well as Paulla Ebron, Jim Ferguson, Angela Garcia, Thomas Blom Hansen, Matthew Kohrman, Liisa Malkki, Sharika Thiranagama, and Sylvia Yanagisako in the Department of Anthropology. I thank the many Stanford undergraduates and graduate students in my courses—"Race and Power," "Genes and Identity," and "Technology and Inequality"—for all of your dynamism and clarifying questions, and ultimately for your undying hope for the future. We need it.

Since 2021, the University of California Press Atelier Series writing workshop participants have helped me get this version of the book completed with more of an ear to language, communicability, and accessible prose. Series editor Kevin O'Neil's optimism and openness to experimentation in writing has been eternally refreshing. Anthropology editor Kate Marshall acquired the book, gave me useful suggestions on several sticking points, and was always encouraging about the writing itself. Assistant editor Chad Attenborough was always gracious, insightful, and dependable at the final stages of tying all the loose ends together. Thank you all. Atelier authors who gave me close readings and feedback include Alessandro Angelini, Sarah Besky, Tracy Canada, Laurie Denyer-Willis, Kate Mariner, and Marina Welker. Kevin O'Neil and Thomas Blom Hansen graciously provided institutional funds for a book workshop in early 2023. A special note of gratitude goes to the participants who read and commented on the full manuscript: Charles Briggs, Emily Martin, Elizabeth Povinelli, Noah Tamarkin, and Patricia Williams. Karla Brundage, Shilpa Kamat, Shawna Sherman, and Darina Sikmashvili read and edited parts in the final stretch. I thank the three anonymous peer reviewers for their time and judicious feedback.

Other attentive readers who engaged key sections of the book at critical moments were those of my beloved community of thinkers—Chris Loperena, Mariana Mora, Keenan Norris, and Meredith Palmer. Your intuition, creativity, and care remain unparalleled and always leave me humbled. Anya Bourg, Aaron Correa, Junot Diaz, Pape Daouda Ndoye, Jared Swanson, Mayowa Tomori, Erika Van Buren, and Eric Wise have listened, held me, and helped me get back on track when necessary. I thank my parents, Marietta and Harvey, as well as my siblings, especially my brother Perry, for their abiding encouragement. A special acknowledgment goes to my partner, Keenan, whose brilliance, insight, and energy of mind reflect back to me a spirit that is forever curious—and who believes that so much is possible.

Aspects of this research were funded by a National Science Foundation (NSF) postdoctoral fellowship, award #0208100, the Robert Wood Johnson Heath and Societies Scholar program at the Harvard School of Public Health, and an NSF Scholars Award, #0849109. The final stages of writing were made possible by the Enhanced Sabbatical Fellowship offered by the School for Humanities & Sciences at Stanford University in 2022-2023.

Notes

PREFACE: SKIN AND CODE

1. Most scientists in this book are high-profile figures, or involved in recognizable research labs. Participants who did not consent to attribution statements, as well as some junior researchers in labs, are given pseudonyms or not named at all. Culturally, to give or possess a name humanizes. But one could argue that the function of codes to conceal identity for protection, which most scientists do (and which provoked my gesture here), could also be done without losing sight of the person, without converting aspects of their lives into fungible commodifiable parts, if done attentively.

2. See Rebecca Skloot, *The Immortal Life of Henrietta Lacks* (New York: Random House, 2010), 90–91.

3. See Skloot, *Immortal Life*; and Evan Starkman, "What Are HeLa Cells?," *WebMD*, January 22, 2022, www.webmd.com/cancer/cervical-cancer/hela-cells-cervical-cancer.

4. Simon Tripp and Martin Grueber, "The Economic Impact and Functional Applications of Human Genetics and Genomics," TEConomy Partners LLC, May 2021, www.ashg.org/wp-content/uploads/2021/05/ASHG-TEConomy-Impact-Report-Final.pdf, 2.

5. For an overview, see Michael J. Joyner, Nigel Paneth, and John P. A. Ioannidis, "What Happens When Underperforming Big Ideas in Research Become Entrenched?," *JAMA* 316, no. 13 (2016): 1355–1356, https://doi.org/10.1001/jama.2016.11076.

6. See "African Ancestry—Trace Your DNA. Find Your Roots. Today," African Ancestry, https://africanancestry.com/ (accessed June 13, 2021).

7. See Keith J. Winstein, "Harvard's Gates Refines Genetic-Ancestry Searches for Blacks: Scholar Founds a Firm after DNA Tracer Put Forebear in Wrong Place," *Wall Street Journal*, November 15, 2007, www.wsj.com/articles /SB119509026198193566.

8. For more on the connection between Kittles's work on the lower Manhattan African Burial Ground and his company, see Alondra Nelson, *The Social Life of DNA* (New York: Basic Books, 2016).

9. Kristen V. Brown, "All those 23andMe Spit Tests Were Part of a Bigger Plan," *Bloomberg*, November 4, 2021, www.bloomberg.com/news/features /2021-11-04/23andme-to-use-dna-tests-to-make-cancer-drugs.

10. Emphasis added. Collins's longer explanation reads: "In Africa, because the population has been around longer, the [genetic variation] neighborhoods are smaller, and therefore the ability to shine a bright light on what the functional variant is that's actually responsible for that diabetes risk or that high blood risk is substantially better, and that is a resource that will help in tracking down these ancient variations that are probably present across the world but in Africa will be more easily delimited to a more precise interval." See Transcript, "Human Heredity and Health in Africa Announced in London," National Human Genome Research Institute, www.genome.gov/27539880/human-heredity-and -health-in-africa-announced-in-london (accessed February 20, 2012).

11. For an in-depth analysis of these dynamics, see Richard Ford, *Racial Culture: A Critique* (Princeton, NJ: Princeton University Press, 2004). Also see Ruha Benjamin, *Race after Technology: Abolitionist Tools for the New Jim Code* (Cambridge, UK: Polity Press, 2019), on her concept of "technological benevolence," chapter 4.

12. See the 1994 commemorative edition, Ralph Ellison, "Introduction," *Invisible Man* (New York: The Modern Library, November 10, 1981), xxv.

13. For other ethnographic studies on scientists who use these or similar technologies, see Nadia Abu El-Haj, *The Genealogical Science: The Search for Jewish Origins and the Politics of Epistemology* (Chicago: Chicago University Press, 2012); Catherine Bliss, *Race Decoded: The Genomic Fight for Social Justice* (Stanford, CA: Stanford University Press, 2012); Rajagopalan Ramya and Joan H. Fujimura, "Variations on a Chip: Technologies of Difference in Human Genetics Research," *Journal of the History of Biology* 51, no. 4 (2018): 841–873, https://doi.org/10.1007/s10739-018-9543-x; Duana Fullwiley, "The Biologistical Construction of Race: 'Admixture' Technology and the New Genetic Medicine," *Social Studies of Science* 38, no. 5 (October 2008): 695–735, https://doi .org/10.1177/0306312708090796; Michael Montoya, *Making the Mexican Diabetic Race, Science, and the Genetics of Inequality* (Berkeley: University of California Press, 2011); and Kim Tallbear, *Native American DNA: Tribal Belonging*

and the False Promise of Genetic Science (Minneapolis: University of Minnesota Press, 2013).

14. Duana Fullwiley, "DNA and Our 21st-Century Ancestors," *Boston Review*, February 4, 2021, www.bostonreview.net/articles/duana-fullwiley-dna-and -our-twenty-first-century-ancestors/.

15. If Max Weber's "iron cage" emphasized the idea of "a shell as hard as steel," in contradistinction to the "light cloak that could [be] thrown aside at any moment" as a feature of the spirit of capitalism, I want to call the bind of racial repair through racialized genetics the literal *communal* spirit of capitalism. It is based on the many bodies exploited, past and present, in the initial damage of extraction and exploitation as well as in the lure of political inclusion since emancipation. Here, as in Weber's world, the confinement of the shell performs like a cell. It subjugates. But steel, rather than the notion of *iron* deployed by Talcott Parsons in his translation, is man-made and malleable, suggesting that a new kind of being is also possible with the emergent and transformative natures of capitalism. See Peter Baehr, "The 'Iron Cage' and the 'Shell as Hard as Steel': Parsons, Weber, and the Stahlhartes Gehäuse Metaphor in the Protestant Ethic and the Spirit of Capitalism," *History and Theory* 40, no. 2 (2001): 153–154; cf. Max Weber, *The Protestant Ethic and the Spirit of Capitalism* (Oxford, UK: Oxford University Press, 2010), 123. It is important to keep in mind whether there exists a "light cloak" that can be donned or cast off with regard to the governmentality of race, in a mode akin to Michel Foucault's notion of "refusal." See Michel Foucault, "The Subject and Power," *Critical Inquiry* 8, no. 4 (1982): 785 (emphasis added), www.jstor.org/stable/1343197:

> Maybe the target nowadays is *not to discover what we are but to refuse what we are.* We have to imagine and to build up what we could be to get rid of this kind of political "double bind," which is the simultaneous individualization and totalization of modern power structures. The conclusion would be that the political, ethical, social, philosophical problem of our days is not to try to liberate the individual from the state and from the state's institutions but to liberate us both from the state and from the type of individualization which is linked to the state. *We have to promote new forms of subjectivity* through the *refusal* of *this kind* of individuality which has been imposed on us for several centuries.

16. Ocean Vuong, *On Earth We're Briefly Gorgeous* (New York: Penguin, 2019), 13.

17. Spillers writes: "[When] the procedures adopted for the captive flesh demarcate a total objectification, the entire captive community becomes a living laboratory." She continues: "Even though the captive flesh/body has been 'liberated,' and no one need pretend that even the quotation marks do not *matter*, dominant symbolic activity, the ruling episteme that releases the dynamics of naming and valuation, remains grounded in the originating metaphors of captivity. . . ." See Hortense Spillers, *Black, White and Color: Essays on American*

Literature and Culture (Chicago: University of Chicago Press, 2003), 208 (emphasis in the original).

INTRODUCTION: AMERICA AND THE TABULA RAZA

1. See Amerigo Vespucci, *Mundus Novus: Letter to Lorenzo Pietro di Medici*, trans. George Tyler Northrup (Princeton, NJ: Princeton University Press, 1916).

2. In Aristotle's major 350 BC treatise called *De Anima* in Latin, he wrote: "Thought is in a potential way identical with thinkable objects, though in an actualized way with none of them until it thinks. It must be present in it the same way as on a tablet on which there is nothing written [beforehand] in an actualized way." See Aristotle, *On the Soul and Other Psychological Works*, trans. Fred D. Miller Jr. (Oxford: Oxford University Press, 2018), 57. As for Locke, "the senses at first let in particular ideas, and furnish the *yet empty cabinet*; and the mind by degrees growing familiar with some of them, they are lodged in the memory, and names got to them. Afterwards the mind, proceeding farther, abstracts them, and by degrees learns the use of general names. In this manner the mind comes to be furnished with ideas and language, the materials about which to exercise its discursive faculty: and the use of reason becomes daily more visible, as these materials, that give it employment, increase." From John Locke, *The Works, Vol. 1 An Essay Concerning Human Understanding Part 1* (1689; London: Rivington, 1824), §15 (emphasis mine), https://oll.libertyfund.org /title/locke-the-works-vol-1-an-essay-concerning-human-understanding-part -1#lf0128-01_label_294.

3. From the beginning, the Spanish viceroys emphasized New World subjects' human "mixes" in terms of fractions and racializing labels. Obsessions with fifteenth-century Catholic "blood purity" (*limpieza de sangre*) in Spain during the Inquisition gave way to a classification system that tabulated people as one-half, one-quarter, one-eighth Christian and so on. In the New World, logics of fractionation influenced the establishment of a messy and vast caste system (*casta*), under which a serious government accounting of people's phenotypic coloring, social status, and genealogical ancestry determined their political and life chances in everyday dealings. As historian Ben Vinson III argues, *casta* later loosened for some types of "mixed" humans into a broad notion of *mestizaje* in many regions. Some social planners even came to eventually laud what geneticists today might call "admixed" subjects as constituting the "cosmic race" of Latin America (*la raza cosmica*). Thus the New World subject as mixed—yet meticulously tracked and marked by the details of mixture that were based in, yet went beyond, African, European, and Indigenous terms—has a long history. See Ben Vinson III, *Before Mestizaje: The Frontiers of Race and Caste in Colonial Mexico* (New York: Cambridge University Press, 2018), 29–34; 41–45.

Nonetheless, the category of the *mestizo*, along with notions of *mestizaje*, were often fraught with racializing and eugenic practices that pervaded medicine, reproduction, and the territorial planning of states over time. See Marisol de la Cadena, *Indigenous Mestizos: The Politics of Race and Culture in Cuzco, Peru, 1919-1991* (Durham, NC: Duke University Press, 2000); Fabiola López-Durán, *Eugenics in the Garden: Transatlantic Architecture and the Crafting of Modernity* (Austin: University of Texas Press, 2018); Ann Twinam, *Purchasing Whiteness: Pardos, Mulattos, and the Quest for Social Mobility in the Spanish Indies* (Stanford, CA: Stanford University Press; 2015); and Nancy Leys Stepan, *The Hour of Eugenics: Race, Gender, and Nation in Latin America* (Ithaca, NY: Cornell University Press, 1996). As will become clear in later chapters, people from Spain (as a key population representing "Europe") served as the one of primary reference datasets for scientists who wanted to understand genetic diversity, in terms of "admixture," in present-day Latinos. Also, see note 9 of this introduction for how other social scientists have approached similar issues of race and genetics in Latin America.

4. There were some exceptions, however. For example, explorers such as Jean de Léry, who penned a book-length monograph published in 1578, complicated this vision and even exalted certain Indigenous ways of life in contrast to that of his own "French culture"; see Jean de Léry, *History of a Voyage to the Land of Brazil, Otherwise Called America* (Berkeley: University of California Press, 1578). Brazil was the first territory that French explorers reached in the Americas, where they competed with the Portuguese until the 1600s. Two of their noted colonies were La France Antarctique (1555-1560), founded by Nicolas Durand de Villegagnon, and La France Equinoxiale (1612-1615).

5. See geographer Meredith Palmer's excellent treatment of Anglo Saxon legal and technical systems to survey, measure, and rationalize units and tools, such as the square mile, for allotment as a form of natural transaction that both racialized land and made it into property. These processes largely omitted Native peoples from negotiations, within which they nonetheless served as figments. Meredith Palmer, "Rendering Settler Landscapes: Race and Property in the Empire State," *Environment and Planning D: Society and Space* 38, no. 5 (2020): 793-810.

6. For a deeper understanding of how such visual orderings of the world in Europe and North America were shaped by Christian thinking, see Terence Keel, *Divine Variations: How Christian Thought Became Racial Science* (Stanford, CA: Stanford University Press, 2018), 69-70. For a wonderful discussion of taxonomies of *casta* in Latin America detailed by historian Ben Vinson III in their dizzying number of divisions that were bolstered by religious concepts, humoral ideas, astral positioning, and climate as these were thought to favor certain types of human variation, see Vinson, *Before Mestizaje*, xvii, 44-47. One of the most referenced of the *casta* paintings resides in the Museo Nacional del Virreinato in

Mexico. For the explanatory text that accompanies it and that contains the wording cited here, which translates to "[t]he natural state of human differences among the inhabitants of the Americas," see Museo Nacional del Virreinato, "Cuadro de Castas," accessed May 30, 2023, https://lugares.inah.gob.mx/es /museos-inah/museo/museo-piezas/8409-8409-10-241348-cuadro-de-castas .html?lugar_id=475.

7. See Karen Barad, *Meeting the Universe Halfway: Quantum Physics and the Entanglement of Matter and Meaning* (Durham, NC: Duke University Press, 2007), 136–137.

8. Povinelli writes: "Ancestral catastrophes are past and present; they keep arriving out of the ground of colonialism and racism rather than emerging over the horizon of liberal progress." See Elizabeth Povinelli, *Between Gaia and Ground: Four Axioms of Existence and the Ancestral Catastrophe of Late Liberalism* (Durham, NC: Duke University Press, 2021), 3.

9. Genetic ancestry "admixture" tools that are similarly structured to those in the United States have been taken up by researchers in Latin America, mostly Brazil and Mexico, who often collaborate with American-based scientists. Scholars working in these areas consistently try to make sense of how continental ancestry and ideas of pure racial types—African, European, Indigenous—can coexist with an outward discourse on racial mixing and *mestizaje*, respectively. See Sahra Gibbon, "Translating Population Difference: The Use and Re-Use of Genetic Ancestry in Brazilian Cancer Genetics," *Medical Anthropology* 35, no. 1 (2016): 58–72, https://doi.org/10.1080/01459740.2015.1091818; and Peter Wade, *Mestizo Genomics: Race Mixture, Nation, and Science in Latin America* (Durham, NC: Duke University Press, 2014). For a review of these technologies in the United States, see Troy Duster, "A Post-Genomic Surprise," *British Journal of Sociology* 66, no. 1 (2015): 1–27, https://doi.org/10.1111/1468-4446.12118.

10. See Amy Harmon, "Why White Supremacists Are Chugging Milk (and Why Geneticists Are Alarmed)," *New York Times*, October 17, 2018, www .nytimes.com/2018/10/17/us/white-supremacists-science-dna.html.

11. This applies to a singular notion of whiteness itself as well as to the fact that many African and other peoples have long been present in what is now Europe. On the first count, Patricia Williams writes: "Whiteness is a kind of neural toboggan run, encouraging good people to slide, to *slush* at high speed right on through the realm of reason. Whiteness is a kind of sociological clubhouse, a weird compression of tribal and ethnic animosities, some dating back to the Roman invasions, all realigned to make new enemies, all compromised to make new friends. . . . The notion of whiteness as any kind of racial purity is a cognitive blind spot blocking out the pain . . . of such histories as the Thirty Years' War, massacres in Scotland, and tyranny in Transylvania." From Patricia Williams, *Seeing a Color-Blind Future: The Paradox of Race* (New York: Farrar, Straus and Giroux, 1997), 52–53. For a good review of the fluid historical bound-

aries between what are now called "Africa" and "Europe," as well as the long presence of African peoples living in territories that are often only associated with early Europeans, see Olivette Otele, *African Europeans: An Untold History* (New York: Basic Books, 2012), chapters 1–4.

12. See Richard Cooper et al., "Race and Genomics," *New England Journal of Medicine* 348, no. 12 (March 2003): 1166–1170.

13. See Neil Risch et al., "Categorization of Humans in Biomedical Research: Genes, Race and Disease," *Genome Biology* 3, no. 7 (2002): 1–12, https://doi .org/10.1186/gb-2002-3-7-comment2007.

14. Luisa N. Borrell et al., "Race and Genetic Ancestry in Medicine: A Time for Reckoning with Racism," *New England Journal of Medicine* 384, no. 5 (2021): 474–480, https://doi.org/10.1056/NEJMms2029562; Akinyemi Oni-Orisan et al., "Embracing Genetic Diversity to Improve Black Health," *New England Journal of Medicine* 384, no. 12 (February 2021): 1163–1167; and Jessica P. Cerdeña, Vanessa Grubbs, and Amy L. Non, "Genomic Supremacy: The Harm of Conflating Genetic Ancestry and Race," *Human Genomics* 16, no. 18 (May 19, 2022): 1–5, https://doi.org/10.1186/s40246-022-00391-2.

15. For a detailed account of the history of the "admixture" concept, and its racialist connotations of pure groups in the genomic era, see Kostas Kampourakis and Erik L. Peterson, "The Racist Origins, Racialist Connotations, and Purity Assumptions of the Concept of 'Admixture' in Human Evolutionary Genetics," *Genetics* 223, no. 3 (March 2023): 3–4, https://doi.org/10.1093 /genetics/iyad002.

16. After this point in the book I drop the quotations around these terms, but they should still be questioned as catchall labels.

17. Contrary to what I observed in the labs studied herein, in 2020 an international team of scientists tried to take broader approaches to understanding global diversity and also looked at different kinds of mutations and their patterns to talk about genetic variation somewhat differently in 929 genomes across 54 populations. Based on their findings, they refuted the notion of continental specific markers taken in absolute terms to focus on the relative frequencies of alleles. They write: "We find no such private variants that are fixed in a given continent or major region." See Anders Bergström et al., "Insights into Human Genetic Variation and Population History From 929 Diverse Genomes," *Science* 367, no. 6484 (2020): eaay5012, doi:10.1126/science.aay5012, p. 5.

18. Fullwiley, "Biologistical Construction of Race," 700–705; and Duana Fullwiley, "The 'Contemporary Synthesis' When Politically Inclusive Genomic Science Relies on Biological Notions of Race," *ISIS* 105, no. 4 (2014): 806–808.

19. See Fullwiley, "Biologistical Construction of Race," 702. For an emphasis on what two STS scholars call "genome geography"—which they term "a thread between population, race, and 'genetic ancestry' that renders the three concepts difficult to untangle"—see Joan Fujimura and Ramya Rajagopalan, "Different

Differences: The Use of 'Genetic Ancestry' versus Race in Biomedical Human Genetic Research," *Social Studies of Science* 41, no. 1 (2011): 7. Also see Catherine Nash, *Genetic Geographies: The Trouble with Ancestry* (Minneapolis: University of Minnesota Press, 2015).

20. See Barad, *Meeting the Universe*, 176 and 182. She writes that "boundaries are interested instances of power, specific constructions with real material consequences. There are not only different stakes in drawing different distinctions, there *are different ontological implications*" (emphasis added).

21. Graham Coop, "Genetic Similarity and Genetic Ancestry Groups," July 23, 2022, doi:10.48550/arXiv.2207.11595, preprint. Also see Iain Mathieson and Aylwyn Scally, "What Is Ancestry?" *PLoS Genetics* 16, no. 3 (2020): 1–2, doi:10.1371/journal.pgen.1008624. For a detailed discussion about how genetic and genealogical ancestry differ, see Kostas Kampourakis, *Ancestry Re-imagined: Dismantling the Myth of Genetic Ethnicities* (New York: Oxford University Press, 2023), chapter 4.

22. Tracy Jan, Jena McGregor, and Meghan Hoyer, "Corporate America's $50 Billion Promise," *Washington Post*, August 23, 2021, www.washingtonpost .com/business/interactive/2021/george-floyd-corporate-america-racial-justice /?tid=usw_passupdatepg.

23. The American Society of Human Genetics, "Facing Our History—Building an Equitable Future," January 2023, www.ashg.org/wp-content /uploads/2023/01/Facing_Our_History-Building_an_Equitable_Future_Final _Report_January_2023.pdf.

24. As per the report, the "National Academies convened a committee of 17 members representing diverse expertise areas including human genetics; clinical genetics; population genetics; statistical and computational genetics and genomics; historical, ethical, legal, and social implications research; sociology and anthropology; and demography and population statistics." See National Academies of Sciences, Engineering, and Medicine, *Using Population Descriptors in Genetics and Genomics Research: A New Framework for an Evolving Field* (Washington, DC: National Academies Press, 2023), 44, 46, https://doi .org/10.17226/26902.

25. Barad, *Meeting the Universe*, 146. Barad writes that "apparatuses produce differences that matter—they are boundary-making practices that are formative of matter and meaning, productive of, and part of, the phenomenon produced."

26. Two round tables with ethnographers at the American Anthropological Association meetings in 2020 and 2021 dealt with these issues and displayed a range of projects and approaches to writing ethnographies on science and medicine for the future, see Duana Fullwiley, "Futures of Writing in Medical Anthropology: Voice and Practice (year 2)," panel, American Anthropological Association Annual Meeting, Baltimore, MD, Zoom, November 18, 2021; and Charles

Briggs, "Futures of Writing in Medical Anthropology," panel, "Raising our Voices," American Anthropological Association Annual Meeting, livestream, November 14, 2020. Also, on boundary jumping and "play" to reinvigorate how scholars might approach ideas, see Charles Briggs, *Unlearning: Rethinking Poetics, Pandemics, and the Politics of Knowledge* (Logan: Utah State University Press, 2021), 64.

27. Sociologist Ruha Benjamin fictionalizes speculative scenarios, which she calls "speculative methods." Her essay on the topic imagines futuristic possibilities of repair that are based on both racism and biotechnologies in the here and now. See Ruha Benjamin, "Racial Fictions, Biological Facts: Expanding the Sociological Imagination through Speculative Methods," *Catalyst: Feminism, Theory, Technoscience* 2, no. 2 (2016): 1–28. The difference with what I am doing here lies in my emphasis on the present, *nonfictive* realities that I have witnessed scientists grapple with as they bring them into being. I explore this further in chapters 1 and 7.

28. Throughout the book I use ellipses in conversations and reported dialogue in two ways. These are distinguished as regular ellipses for vocal delivery of various kinds—trailing off, halting speech, etc.—and bracketed ellipses for the elision of spoken material where light editing of conversations proved necessary for brevity or clarity. Regarding Chakravarti's comments, see "Committee on Use of Race, Ethnicity, and Ancestry as Population Descriptors in Genomics Research (Meeting 2 and Public Workshop)," National Academies, April 4, 2022, www .nationalacademies.org/event/04-04-2022/committee-on-use-of-race-ethnicity-and -ancestry-as-population-descriptors-in-genomics-research-meeting-2-and-public -workshop. To view the full meeting on YouTube, see NASEM Health and Medicine, "Welcome&Session 1," uploaded on April 14, 2022, YouTube video, 2:01:35, www.youtube.com/watch?v=ObtrFNlydSg&list=PLGTMA6QkejfjFG77TzlbT _-ieI3E7BwPH&index=1.

29. See Eric D. Green et al., "Strategic Vision for Improving Human Health at the Forefront of Genomics," *Nature* 586, no. 7831 (2020): 683–692, 690, https:// doi.org/10.1038/s41586-020-2817-4. See Box 5 item 4, which reads: "Research in human genomics will have moved beyond population descriptors based on historic social constructs such as race."

30. Fullerton quoted within "Committee on Use of Race, Ethnicity, and Ancestry as Population Descriptors."

31. International HapMap Consortium, "A Haplotype Map of the Human Genome," *Nature* 437, no. 27 (2005): 1299–1320 (emphasis added), https://doi .org/10.1038/nature04226.

32. One key hope of white nationalists, which is often espoused by David Duke as we will see in his open letter to scientists working on admixture in chapter 1, is to live apart. When Derek Black, son of a highly visible white nationalist

who started the website stormfront.org, defected from the hateful movement that he had been raised in since birth, he gave an interview to "The Daily," of the *New York Times*. In listing some of the beliefs he once held, he included ideas that Black people were intellectually inferior, the plea for land to live separately from minorities, and a belief that genetics could bear out white nationalists' ideas of their own superiority. See Derek Black, "Transcript: Interview with Former White Nationalist Derek Black," interview by Michael Barbaro, "The Daily" podcast, *New York Times*, August 22, 2017, 34:49, www.nytimes.com/2017/08/22 /podcasts/the-daily-transcript-derek-black.html. For coverage of the Buffalo shooter's use of population genetic findings on "European" "educational attainment," colloquially thought of as IQ, see Megan Moltini, "Buffalo Shooting Ignites a Debate over the Role of Genetics Researchers in White Supremacist Ideology," *Stat News*, May 23, 2022, www.statnews.com/2022/05 /23/buffalo-shooting-ignites-debate-genetics-researchers-in-white-supremacist -ideology/.

33. Scholars have traced the extensive and often surprising twists and turns of the concepts of the "Yoruba" and also the idea of the "Caucasian," which is often used for the "European" samples of white people from the state of Utah. Regarding the political and historical dynamics that have solidified a contemporary notion of "The Yoruba," see Akinwumi Ogundudiran, *The Yorùbà: A New History* (Bloomington: Indiana University Press, 2020). He writes: "Let me affirm that the Yorùbà whom I am referring to here did not have their origin in the mid-nineteenth century when the label was first used to define an ethnic identity for all the speakers of the dialect continuum whom we now know by that name. The project of that ethnic identity making responded to and was necessitated by the European racialization of the globe and European empires' insistence on classifying the world populations into 'nations' through the process of colonialism and industrial capitalism during the nineteenth century" (8). He goes onto define the Yorùbà as an ongoing "community of practice" (10–12).

Also see the wonderfully detailed account of Afro-descended English professors who went back and forth between Brazil and Nigeria and were integral to solidifying the contemporary ethnic and linguistic notion of Yorùbà by J. Lorand Matory, "The English Professors of Brazil: On the Diasporic Roots of the Yoruba Nation," *Comparative Studies in Society and History* 41, no. 1 (1999): 72–103, www.jstor.org/stable/179249. For an equally detailed tour de force on the origins of the concept "Caucasian" for Europeans, see Sara Figal, "The Caucasian Slave Race: Beautiful Circassians and the Hybrid Origin of European Identity," in *Reproduction, Race, and Gender in Philosophy and the Early Life Sciences*, ed. Susanne Lettow, 163–186 (Albany: SUNY Press, 2014).

34. See Wen-Wei Liao, Mobin Asri, Jana Ebler, et al., "A Draft Human Pangenome Reference," *Nature* 617 (2023): 314, https://doi.org/10.1038 /s41586-023-05896-x.

CHAPTER 1. GENOMIC WORLD BUILDING

1. In reality, the full map of the human genome was reported to finally be complete in March of 2022. The 2003 map that was lauded as a then complete map was only 92 percent realized. Problematic areas close to chromosomal ends (telomeres) and centers (centromeres) would take decades to detail. The NGHRI press release in 2022 (www.genome.gov/news/news-release/researchers -generate-the-first-complete-gapless-sequence-of-a-human-genome) included references to a special issue of *Science* that featured the new reports including Sergey Nurk et al., "The Complete Sequence of a Human Genome," *Science* 376, no. 6588 (2022): 44–53, https://doi.org/10.1126/science.abj6987. Still, there was a remaining missing piece of the Y-chromosome that was not completed until April 2022.

2. Richard C. Lewontin, "The Apportionment of Human Diversity," in *Evolutionary Biology*, ed. T. Dobzhansky, M. K. Hecht, and W. C. Steere, chapter 14, 381–397 (New York: Springer, 1972), https://doi.org/10.1007/978-1-4684 -9063-3_14. In the introduction to his piece Lewontin sets up the study with a series of social observations that can bias the science, one of which is the visual habit and organizing structure of racial thinking. He writes: "The erection of racial classification in man based upon certain manifest morphological traits gives tremendous emphasis to those characters to which human perceptions are most finely tuned (nose, lip and eye shapes, skin color, hair form and quantity), precisely because they are the characters that men ordinarily use to distinguish individuals" (382).

3. In a yearslong lively set of exchanges that started with British statistician, geneticist, and evolutionary biologist Anthony William Fairbank Edwards's paper provocatively titled "Human Genetic Diversity: Lewontin's Fallacy," where he argued that humans could be grouped into taxonomies if many loci and allelic variants were analyzed together, others weighed in. Some were in support of Lewontin's general idea about the lack of DNA variants to allow scientists to reliably predict a coherent notion of racial groups, while others pointed out that the critics and Lewontin were focused on different aspects of statistics and genetic diversity. See A. W. F. Edwards, "Human Genetic Diversity: Lewontin's Fallacy," *BioEssays* 25, no. 8 (July 2003): 798–801, https://doi.org/10.1002/bies.10315; Jonathan M. Marks, "Ten Facts about Human Variation," in *Human Evolutionary Biology*, ed. M. P. Muehlenbein (Cambridge, UK: Cambridge University Press, 2010), 270; and R. G. Winther, "The Genetic Reification of 'Race'? A Story of Two Mathematical Methods," in *Phylogenetic Inference, Selection Theory, and History of Science: Selected Papers of AWF Edwards with Commentaries*, ed. R. G. Winther (Cambridge: Cambridge University Press, 2018), 489.

4. For a detailed overview of the uses and limitations of AIMs, as well as other ancestry genetic markers (such as mitochondrial DNA and Y-chromosome

markers), see John Relethford and Deborah Bolnick's *Reflections on Our Past: How Human History Is Revealed in Our Genes*, 2nd edition (New York: Routledge, 2018), 187–188. Concerning AIMs, they write:

> This method assumes that we can obtain accurate estimates of allele frequencies in the ancestral populations. These estimates would ideally come from studies of ancient DNA extracted from skeletal remains, which provide a direct window onto gene pools that existed in ancient times. However, because ancient DNA is difficult and expensive to study researchers often sample present-day populations instead and treat them as proxies. . . . In other words, we must assume that there has been no significant evolutionary change in any of these populations. . . . [This] is often problematic due to changes stemming from gene flow, genetic drift, natural selection, and other factors.

5. See Jada Benn Torres, "Race, Rare Genetic Variants, and the Science of Human Difference in the Post-Genomic Age," *Transforming Anthropology* 27 (2019): 38, https://doi-org.stanford.idm.oclc.org/10.1111/traa.12144. For an overview of the assumptions and limitations that go into models, and the need to emphasize ideas of inference and "estimates," rather than exact notions of ancestral inheritance as many people understand it, see Jada Benn Torres and Gabriel A. Torres Colón, *Genetic Ancestry: Our Stories, Our Pasts* (London: Routledge, 2021), 14–22.

6. I use pseudonyms for this lab and for all researchers at this site.

7. Reluctance on the part of potential recruits might stem from the ways that health institutions have treated them in the past. Or it might come from their many daily interactions with power structures, ranging from schools to employers to immigration officials that leave them less than trusting to hand over such highly identifying information.

8. Norah L. A. Gharala, *Taxing Blackness: Free Afromexican Tribute in Bourbon New Spain* (Tuscaloosa: University of Alabama Press, 2019), 32–33.

9. Gharala, *Taxing Blackness*, 34, 86–89.

10. Gharala, *Taxing Blackness*, 89–90.

11. For updated clarifications in 2017, see "NIH Policy and Guidelines on the Inclusion of Women and Minorities as Subjects in Clinical Research," NIH Grants and Funding, https://grants.nih.gov/policy/inclusion/women-and-minorities/guidelines.htm (accessed June 10, 2022).

12. Steven Epstein, "Inclusion, Diversity, and Biomedical Knowledge Making: The Multiple Politics of Representation," in *How Users Matter: The Co-Construction of Users and Technologies*, ed. N. Oudshoorn and T. Pinch, 173–190 (Cambridge, MA: MIT Press, 2003); Steven Epstein, "Bodily Differences and Collective Identities: The Politics of Gender and Race in Biomedical Research in the United States," *Body & Society* 10, no. 2–3 (2004), https://doi.org/10.1177/1357034X04042942; and Jonathan Kahn, "Patenting Race," *Nature Biotechnology* 24 (2006): 1351, https://doi.org/10.1038/nbt1106-1349.

13. Epstein, "Inclusion, Diversity, and Biomedical Knowledge Making," 183.

14. See Executive Office of the President, "Standards for Maintaining, Collecting, and Presenting Federal Data on Race and Ethnicity," Federal Register, Office of Management and Budget (OMB) 81, no. 190, notices (September 30, 2016): 67401, www.federalregister.gov/documents/2016/09/30/2016-23672 /standards-for-maintaining-collecting-and-presenting-federal-data-on-race-and -ethnicity.

15. "Anthropological" here refers to biological anthropology or what is also called physical anthropology. See Executive Office of the President, "Standards for Maintaining, Collecting, and Presenting Federal Data on Race and Ethnicity."

16. Karl Marx, *The German Ideology*, in *The Marx-Engels Reader*, ed. Robert C. Tucker (New York: Norton, 1978), 165 (emphasis added).

17. Audrey Smedley, *Race in North America: Origin and Evolution of a Worldview*, 4th ed. (Boulder, CO: Routledge, 2011), chapter 9, "The Rise of Science and Scientific Racism," https://searchebscohostcom.stanford.idm.oclc.org/login.aspx ?direct=true&db=nlebk&AN=421158&site=ehost-live. Also, anthropologist Alan Goodman writes: "There are two foundational myths in genetic ancestry work that lead to simplification and the sense that complexities do not exist. The first is that individuals did not move around much before 1492. If anything, the opposite is true. The lessons of archaeology, read first through trade goods and now through chemical analyses of the humans themselves, are that individuals moved, and moved, and moved. They moved in predictable fashions and stochastically and in small ways." See Alan Goodman, "Toward Genetics in an Era of Anthropology," *American Ethnologist* 34 (2007): 228, https://doi-org.stanford .idm.oclc.org/10.1525/ae.2007.34.2.227.

18. During her in-depth study of experimental psychology labs, anthropologist Emily Martin also grappled with the subtle, captive power of everyday tables. She writes: "If the table can be thought of as a kind of trap, to capture and contain a subject, it is a disarming one—it looks so placid and innocent for something that has the potential to be a powerful constraint." See Emily Martin, *Experiments of the Mind: From the Cognitive Psychology Lab to the World of Facebook and Twitter* (Princeton, NJ: Princeton University Press, 2021), 157.

19. On re-identification of samples in large databases, see Yaniv Erlich and Arvind Narayanan, "Routes for Breaching and Protecting Genetic Privacy," *Nature Reviews Genetics* 15, no. 6 (2014): 409–421, https://doi.org/10.1038 /nrg3723; Melissa Gymrek et al., "Identifying Personal Genomes by Surname Inference," *Science* 339, no. 6117 (2013), https://doi.org/10.1126 /science.1229566; and "Case: Big Data & Genetic Privacy: Re-identification of Anonymized Data," Online Ethics Center for Engineering and Science, https:// onlineethics.org/cases/big-data-life-sciences-collection/case-big-data-genetic -privacy-re-identification-anonymized (accessed November 29, 2022). On the

use of ancestry testers' DNA to identify suspects who are related to them for forensic purposes, see Rafil Kroll-Zaidi, "Your DNA Test Could Send a Relative to Jail," *New York Times*, December 27, 2021, www.nytimes.com/2021/12/27/magazine/dna-test-crime-identification-genome.html.

20. See Melinda C. Mills and Charles Rahal, "A Scientometric Review of Genome-Wide Association Studies," *Communications Biology* 2, art. no. 9 (2019): 1–11, 3–4, https://doi.org/10.1038/s42003-018-0261-x. The authors write: "We show that 71.80% of participants are recruited from only three countries; the US, UK, and Iceland. Although participants from the United States are most frequently the basis for the largest number of studies (41.01% of all studies), the United Kingdom dominates in terms of the number of participants (40.50% of all participants) analyzed. Conversely, although 1.13% of recorded studies involve Icelandic participants, the small Icelandic population (around 334,000) represents 11.52% of all participants contributed to GWAS research." Beyond country they explain, "Our results . . . concur with existing estimates, showing on aggregate, ancestry in genetic discovery has been highly unequal and dominated by participants of European ancestry (86.03% discovery, 76.69% replication, 83.19% combined)." For earlier commentaries about the need for diverse genomics, see Sarah K. Tate and David B Goldstein, "Will Tomorrow's Medicines Work for Everyone?," *Nature Genetics* 36, suppl. 11 (2004): S34-S42, https://doi.org/10.1038/ng1437. Also see Carlos D. Bustamante, Francisco M. De La Vega, and Esteban G. Burchard, "Genomics for the World," *Nature* 475 (2011): 163–153, https://doi.org/10.1038/475163a.

21. See Melinda C. Mills and Charles Rahal, "The GWAS Diversity Monitor Tracks Diversity by Disease in Real Time, *Nature Genetics* 52 (2020): 242–243, https://doi.org/10.1038/s41588-020-0580-y. On the actual tracker site the following information can be found: "Total GWAS participants diversity Version 1.0.0. Last check for data: 2022-12-03 00:16:20." On December 3, 2022, at the time of this note revision, the site shows figures for groups studied listed by name and percentage of GWAS for all time: 95.41% "European"; 3.00% "Asian"; 0.14% "African"; 0.43% "African American or Afro-Caribbean"; 0.31% "Hispanic or Latin American"; and 0.71% "Other/Mixed," see "Total GWAS participants diversity," GWAS Diversity Monitor, https://gwasdiversitymonitor.com (accessed December 3, 2022).

22. Marx, *German Ideology*, 161.

23. To see how ethnographers chronicle how this happens, see Nadia Abu El-Haj, *The Genealogical Science: The Search for Jewish Origins and the Politics of Epistemology* (Chicago: Chicago University Press, 2012); Duana Fullwiley, *The Enculturated Gene: Sickle Cell Health Politics and Biological Difference in West Africa* (Princeton, NJ: Princeton University Press, 2011); Margaret Lock, "The Epigenome and Nature/Nurture Reunification: A Challenge for Anthropology," *Medical Anthropology* 32, no. 4 (2013): 291–308, doi:10.1080/01459740.2012

.746973; Michael Montoya, *Making the Mexican Diabetic: Race, Science, and the Genetics of Inequality* (Berkeley: University of California Press, 2011); and Noah Tamarkin, *Genetic Afterlives: Black Jewish Indigeneity in South Africa* (Durham, NC: Duke University Press, 2020).

24. See Travis A. Hoppe et al., "Topic Choice Contributes to the Lower Rate of NIH Awards to African-American/Black Scientists," *Science Advances* 5, no. 10 (October 9, 2019): 1–12, https://doi.org10.1126/sciadv.aaw7238. The authors focus on a 1.7-fold increase. They write: "The funding rate for WH scientists remains approximately 1.7-fold higher than for AA/B scientists—16.1% AA/B versus 29.3% WH in fiscal year (FY) 2000–2006 and 10.7% AA/B versus 17.7% WH in FY 2011–2015," 1.

25. Hoppe et al., "Topic Choice Contributes to the Lower Rate of NIH Awards," 1.

26. See Emily Vaughn, "What's behind the Research Funding Gap for Black Scientists?," *NPR*, www.npr.org/sections/health-shots/2019/10/18/768690216 /whats-behind-the-research-funding-gap-for-black-scientists (accessed May 30, 2022).

27. See note 32 of the introduction.

28. See Nancy Krieger, Jaquelyn L. Jahn, and Pamela D. Waterman, "Jim Crow and Estrogen-Receptor-Negative Breast Cancer: US-Born Black and White Non-Hispanic Women, 1992–2012," *Cancer Causes & Control* 28, no. 1 (2017): 49–59, https://doi.org/10.1007/s10552-016-0834-2; and Eric Bock, "Housing Segregation a Central Cause of Racial Health Inequities," NIH Record, https://nihrecord.nih.gov/2021/03/05/housing-segregation-central-cause-racial -health-inequities (accessed November 15, 2022).

29. Bill Picture, "First: Do No Harm?," *AsianWeek.com*, http://news .asianweek.com/news/view_article.html?article_id=1352b889890ad56828d85 2d98d0d50e4 (accessed August 1, 2008).

30. "Profile: David Duke," Anti-Defamation League, www.adl.org/resources /profiles/david-duke (accessed June 10, 2022).

31. David Duke, "Race and Medicine: A Reply," March 10, 2007, https:// davidduke.com/race-and-medicine-a-reply-from-david-duke-to-a-quote-of-dr -esteban-burchard/.

32. See Elhawary Borrell et al., "Race and Genetic Ancestry in Medicine: A Time for Reckoning with Racism," *New England Journal of Medicine* 384, no. 5 (2021): 474–480, https://doi.org/10.1056/NEJMms2029562, for an example of how geneticists using genetic admixture and ancestry analyses include racism and race as an important way to track epidemiology yet still fundamentally deploy racialized ancestry as derived from purified continental populations in the same age-old terms African, Asian, European, and Native American.

33. Borrell et al., "Race and Genetic Ancestry in Medicine." The researchers focus on *APOL1* genetic variants leading to a variety of kidney diseases in people

of West African descent, variants that have been hypothesized to have evolved in areas with high incidence of sleeping sickness; genetic variants in the 6q25 locus that are protective for breast cancer in Latinas and associated with Indigenous ancestry; variants in the 8q24 region associated with prostate cancer and increased burden of PCA in Black men; and genetic variants in CYP219 that affect metabolism of a cardiovascular drug called Plavix (clopidogrel) that may affect up to 75 percent of people classed as Asian Pacific Islanders in a limited study in Hawaii. Yet like so much of the focus in this field, these variants affect a small subset of the "populations" in question rather than the majority of people classed as belonging to those groups.

34. Borrell et al., "Race and Genetic Ancestry in Medicine," 477. They cite ideas of "inclusion" and the numbers: "Globally diverse populations must be studied because genetic variation and genome architecture vary among populations. More than 80% of participants in existing genome-wide association studies are of European background; Black and Latino people, who account for more than 30% of the U.S. population, are dramatically underrepresented (about 2% and < 0.5%, respectively)." As we see, this vexing tension remains. There is a simplicity of such category usage for genetics. Researchers are currently debating the possible futility of an overinvestment in what are usually probabilistic genetic risk scores that explain very little of the variance when it comes to diseases, while social determinants of health have indeed been established. See David Williams and Chiquita Collins, "Racial Residential Segregation: A Fundamental Cause of Racial Disparities in Health," *Public Health Reports* 116, no. 5 (2001), https://doi.org10.1093/phr/116.5.404. Also see Nancy Krieger, "Structural Racism, Health Inequities, and the Two-Edged Sword of Data: Structural Problems Require Structural Solutions," *Frontiers in Public Health* 9, art. no. 655447 (April 2021): 1–10, sec. Life-Course Epidemiology and Social Inequalities in Health, https://doi.org/10.3389/fpubh.2021.655447.

CHAPTER 2. FROM *MUNDUS* TO MODEL TO *MUNDUS* AGAIN

1. In general, scientists in this book were focused on US underrepresented minorities and health disparities that affected Black and Brown people when they used AIMs. For this reason they only used three "populations"—Africans, Europeans, and Native Americans. Those grouped as Asian were sometimes added for other purposes, mostly commercial, or when pharmacogenetic studies were done more broadly.

2. Carl D. Langefeld et al., "Genome-wide Association Studies Suggest That APOL1-Environment Interactions More Likely Trigger Kidney Disease in African Americans with Nondiabetic Nephropathy Than Strong APOL1-Second

Gene Interactions," *Kidney International* 94, no. 3 (2018): 599–607, https://doi
.org/10.1016/j.kint.2018.03.017.

3. See Leslie A. Bruggeman, John F. O'Toole, and John R. Sedor, "Identifying
the Intracellular Function of APOL1," *Journal of the American Society of Nephrology* 28 (2017): 1008, https://doi.org/10.1681/ASN.2016111262

4. Giulio Genovese et al., "Association of Trypanolytic ApoL1 Variants with
Kidney Disease in African Americans," *Science* 329, no. 5993 (2010): 845,
https://doi.org/10.1126/science.1193032.

5. Genovese et al., "Association of Trypanolytic ApoL1 Variants," 845.

6. In a 2021 review article, just over a decade on from their key paper in *Science*, David Friedman and Martin Pollak write: "Not all individuals with the
APOL1 high-risk genotype develop kidney disease. Although there are likely to
be some genetic modifiers, data to date suggest that environmental influences
may play a larger role. . . . Recent reports have linked APOL1 nephropathy to
infections with Parvovirus B19 in native kidneys and with cytomegalovirus and
BK virus in the transplant setting. Provocative, indirect evidence raises the possibility of a complex relationship between JC viruses, APOL1 genotype, and kidney disease, although the precise nature of this interaction is not yet fully understood." See David J. Friedman and Martin R. Pollak, "APOL1 Nephropathy:
From Genetics to Clinical Applications," *Clinical Journal of the American Society of Nephrology* 16, no. 2 (2021): 299, https://doi.org/10.2215/CJN.15161219.

7. In an editorial for a special issue on APOL1 cellular mechanisms in the
Journal of the American Society of Nephrology, Leslie Bruggeman and colleagues
write: "Although APOL1 kidney disease–associated variants are common (13%
of United States blacks carry the risk genotype), only a subset of these individuals develops advanced CKD, which suggests that a second hit or environmental
stressor is required to initiate disease. HIV-associated nephropathy, the CKD
most strongly associated with APOL1 risk variants, is the clearest example of
this environmental second hit theory." See Bruggeman, O'Toole, and Sedor,
"Identifying the Intracellular Function of APOL1," 1008.

8. See Bruggeman, O'Toole, and Sedor, "Identifying the Intracellular Function of APOL1."

9. See Friedman and Pollak, "APOL1 Nephropathy," 294–295.

10. See Pollak and Friedman, "APOL1 Nephropathy," table 1. They write:
"APOL1 risk variants increase the risk of hypertension-attributed ESKD
(H-ESKD; 7- to 11-fold), FSGS (17-fold), HIV-associated nephropathy (HIVAN;
29- to 89-fold), nondiabetic CKD(2- to 4-fold), and other kidney disease manifestations [Table 1]," 297.

11. For one in a recent series of articles on APOL1 and some of the issues discussed here, see Gina Kolata, "Targeting the Uneven Burden of Kidney Disease
on Black Americans," *New York Times*, May 17, 2022, www.nytimes
.com/2022/05/17/health/kidney-disease-black-americans.html.

12. Pollak's team writes: "The G2 mutation prevents the SRA virulence factor produced by *T. b. rhodesiense* from binding to and inactivating ApoL1. Even 10,000-fold dilutions of plasma containing these mutations (particularly G2) are active against the parasite. This raises the possibility that transfusion of small volumes of plasma, ApoL1-containing HDL particles, or recombinant protein might be effective treatment for trypanosomiasis caused by *T. b. rhodesiense*" (Genovese et al., "Association of Trypanolytic ApoL1 Variants," 845).

13. Mario Apata et al., "Human Adaptation to Arsenic in Andean Populations of the Atacama Desert," *American Journal of Physical Anthropology* 163, no. 1 (2017): 193, https://doi.org/10.1002/ajpa.23193.

14. Apata et al., "Human Adaptation to Arsenic in Andean Populations," 193.

15. Apata et al., "Human Adaptation to Arsenic in Andean Populations," 196.

16. For a longer discussion about the early debate on grid sampling, see Jenny Reardon, *Race to the Finish* (Princeton, NJ: Princeton University Press, 2005), 77.

17. See "Whose DNA Was sequenced?," The Genome Project, National Human Genome Research Institute, www.genome.gov/human-genome-project/Completion-FAQ (accessed April 25, 2022), where the actual group identities are not revealed beyond the following: "Candidates were recruited from a diverse population. . . . About 5 to 10 times as many volunteers donated blood as were eventually used, so that not even the volunteers would know whether their sample was used. All labels were removed before the actual samples were chosen." It must be said, however, that the initial published draft of the human genome was a competitive effort between two teams: one was the government project that the NIH/NHGRI completed map has evolved from; the other was a private venture funded by Celera Genomics, whose CEO, J. Craig Venter, hoped to accelerate the speed of the mapping effort, prompting a competition that indeed resulted in the map being completed ahead of schedule.

18. James Shreeve, *The Genome War: How Craig Venter Tried to Capture the Code of Life and Save the World* (New York: Knopf, 2004), 220.

19. With regard to African ancestry, see Chao Tian et al., "A Genomewide Single-Nucleotide-Polymorphism Panel with High Ancestry Information for African American Admixture Mapping," *American Journal of Human Genetics* 79, no. 4 (2006): 642, https://doi.org/10.1086/507954. Under the section heading "Validation and Exclusion Methods," readers are given a detailed explanation of how the researchers in question made populations from the continent of Africa "homogenous." They treated populations from the continent of Europe similarly. As part of this process, researchers in the paper discuss how they screened out "heterogeneity" between groups from the same continent to create the homogeneity they so desired in order to construct their African ancestry panel.

20. See Joshua M. Galanter, Juan Carlos Fernandez-Lopez, Christopher R. Gignoux, et al., "Development of a Panel of Genome-Wide Ancestry Informative

Markers to Study Admixture throughout the Americas," *PLOS Genetics* 8, no. 3 (2012): 11, https://doi.org/10.1371/journal.pgen.1002554. The researchers hailed from the United States (the lead author and lab), Bolivia, Chile, Columbia, Mexico, Peru, Spain, and Venezuela.

21. See Galanter et al., "Development of a Panel of Genome-Wide Ancestry Informative Markers."

22. See Galanter et al., "Development of a Panel of Genome-Wide Ancestry Informative Markers," 10.

23. For a detailed history on the use of race in pulmonary medicine going back to the Civil War, as well as the ways that it has been embedded in spirometry measurements, see Lundy Braun, *Breathing Race into the Machine: The Surprising Career of the Spirometer from Plantation to Genetics* (Minneapolis: University of Minnesota Press, 2014).

24. See PBS, "Faces of America | A Piece of the Pie," *Faces of America*, video, 04:18, https://ca.pbslearningmedia.org/resource/foa10.sci.living.gen.piecepie/faces-of-america-a-piece-of-the-pie/.

25. See "White Supremacists, You Won't Like Your DNA Results," *The Late Show with Stephen Colbert*, October 17, 2017, YouTube video, 05:27, https://youtu.be/yvS2gjMMXBQ.

26. Kenneth M. Weiss and Jeffrey C. Long, "Non-Darwinian Estimation: My Ancestors, My Genes' Ancestors," *Genome Research* 19 (2009): 706, https://doi.org/10.1101/gr.076539.108.

27. Karen Knorr Cetina, *Epistemic Cultures* (Cambridge: Harvard University Press, 1999), 3.

28. Esteban González Burchard, Luisa N. Borrell, Shweta Choudhry, Mariam Naqvi, et al. "Latino Populations: A Unique Opportunity for the Study of Race, Genetics, and Social Environment in Epidemiological Research," *American Journal of Public Health* 95, no. 12 (2005): 2161, https://doi.org/10.2105/AJPH.2005.068668.

29. González Burchard et al., "Latino Populations," 2161.

30. The term "Moor" was a disparaging term that Christians used for Muslims. María Rosa Menocal, *Ornament of the World: How Muslims, Jews and Christians Created a Culture of Tolerance in Medieval Spain* (New York: Back Bay Books, Little Brown and Co., 2002), 10.

31. Chris Lowney, *A Vanished World: Muslims, Christians and Jews in Medieval Spain* (Oxford: Oxford University Press, 2005), 63; and Richard W. Bulliet, *Conversion to Islam In the Medieval Period: An Essay in Quantitative History* (Cambridge, MA: Harvard University Press, 1979), www-fulcrum-org.stanford.idm.oclc.org/concern/monographs/hq37vn72q, chapter 5.

32. Menocal, *Ornament of the World*, 32–34.

33. The vast market in enslaved people was far-reaching, from West to East, as well as from East to West. Within al-Andalus trading sites, the enslaved were

said to be "Slav, Greek, Frank and Lombard" (from Lombardy in Italy). Others were captives of Vikings who posed a threat to al-Andalus and attacked Seville in 844 and Algeciras in 859. They both "traded and raided" and brought with them enslaved "Irish and Anglo-Saxons." Richard Fletcher, *Moorish Spain* (Berkeley: University of California Press, 2006), 42.

34. Despite some descriptions of skin color and physique in early historical writings going back to ancient Greece, most scholars agree that race as a concept to categorize people by biophysiological traits was not one that people in these historical periods used. Anthropologist Audrey Smedley writes: "What seems strange to us today is that the biological variations among human groups were not given significant social meaning. Only occasionally do ancient writers ever even remark on the physical characteristics of a given person or people. Herodotus, in discussing the habits, customs, and origins of different groups and noting variations in skin color, specifically tells us that this hardly matters. The Colchians are of Egyptian origin, he wrote, because they have black skins and wooly hair 'which amounts to but little, since several other nations are so too.'" See Audrey Smedley, "'Race' and the Construction of Human Identity," *American Anthropologist* 100, no. 3 (1998): 691–693, www.jstor.org/stable /682047; also see Alan Goodman and Joseph Graves, *Racism, Not Race: Answers to Frequently Asked Questions* (New York: Columbia University Press, 2021), 23–27.

35. In these latter regions, the enslaved were also Africans from the Maghreb, especially Muslim captives called *morisco/as*—a term deployed after the Albaicín rebellion in Grenada in 1501. Spain enslaved people in the North African coastal regions following the seizures of Oran in 1509 and Tunis in 1535, while large numbers of captives were rounded up following the later uprising in Alpurarras, Grenada, or the War of Grenada (1568–1571). See Eloy Martín Corrales, "The Spain That Enslaves and Expels: Moriscos and Muslim Captives" (1492 to 1767–1791)," in *Muslims in Spain, 1492–1814* (Leiden, Netherlands: Brill, 2020), 76.

36. Leo J. Garofalo, "The Shape of a Diaspora: The Movement of Afro-Iberians to Colonial Spanish America," in *Africans to Spanish America: Expanding the Diaspora*, ed. Sherwin K. Bryant and Rachel Sarah O'Toole (Champaign: University of Illinois Press, 2012), 28, https://doi.org/10.5406/illinois /9780252036637.003.0001. Also see his extensive note 6.

37. Garofalo, "Shape of a Diaspora," 29, also see his note 8.

38. Fletcher, *Moorish Spain*, 53.

39. For a summary see Darío Fernández-Morera, *The Myth of the Andalusian Paradise* (Delaware: Intercollegiate Studies Institute, 2016), note 78 in chapter 7.

40. Menocal, *Ornament of the World*; Fletcher, *Moorish Spain*; Lowney, *A Vanished World*; and Fernández-Morera, *Myth of the Andalusian Paradise*.

41. See Raymond Kabo, "Les esclaves Africains face à l'inquisition espanole: Les procès de sorcellerie et de magie," thesis, Université de Montpellier, 1984, www.sudoc.fr/041209532; and Christine Fournié-Martinez, "Contribution à l'étude de l'esclavage en Espange au Siècle d'or: Les esclaves devant l'inquisition," thesis, Ecole Nationale des Chartes, Paris, 1988, 75–81, https://bibnum.chartes .psl.eu/s/thenca/item/58599.

42. See interviews with historians Abigail Balbale, Frank Peters, and María Rosa Menocal in the PBS film *The Ornament of the World*, accessed December 17, 2019, video 1:55:31, www.pbs.org/show/ornament-world/.

43. The massive number of converts to Islam, which have been methodically traced through historical and genealogical studies, as well as through the vast amount of cultural interchange that scholars like Menocal present (as well as archeological evidence), shows how the archbishop's narrative was overblown.

44. Lowney, *Vanished World*, 254.

45. Lowney, *Vanished World*, 251.

46. Lowney, *Vanished World*, 254.

47. Carol Delaney, "Columbus's Ultimate Goal: Jerusalem," *Comparative Studies in Society and History* 48, no. 2 (2006): 261, https://doi.org/10.1017 /S0010417506000119.

48. Lowney, *Vanished World*, 255.

49. María Elena Martinez, *Genealogical Fictions: Limpieza de Sangre, Religion and Gender in Colonial Mexico* (Stanford, CA: Stanford University Press, 2008).

50. Martinez, *Genealogical Fictions*, 11–13. Martinez makes it clear that race as a concept does not happen in a unified way with one history. Therefore, it is not anachronistic to talk about racial thinking before the enlightenment as some historians argue. The differentiation and exclusion practices that sort and demean populations based on cultural, physical aspects that are naturalized and essentialized are part of racial thinking even if the term is not applied or if other terms are used to describe such demeaned groups.

51. Nancy Krieger, "Stormy Weather: Race, Gene Expression, and the Science of Health Disparities," *American Journal of Public Health* 95 (2005): 2155–2160, https://doi.org/10.2105/AJPH.2005.067108. Krieger writes: "As emphasized in recent scholarship, choice of time scale—often shaped by unconscious belief as well as conscious design—can exert profound effects on scientific analysis. This is because the framing of scientific questions depends heavily on assumptions, usually more implicit than explicit, regarding the appropriate time frame, level and scale of analysis" (2157). For assumptions about the ancestral population, see González Burchard et al. "Latino Populations."

52. See Fullwiley, "DNA and Our Twenty-first-century Ancestors."

53. Carrie L. Pfaff et al., "Information on Ancestry from Genetic Markers," *Genetic Epidemiology* 26 (2004): 310–311, https://doi.org/10.1002/gepi.10319.

54. Rick A. Kittles and Kenneth M. Weiss, "Race, Ancestry, and Genes: Implications for Defining Disease Risk," *Annual Review of Genomics and Human Genetics* 4, no. 1 (2003): 48, https://doi.org/10.1146/annurev.genom.4.070802.110356.

55. David M. Eberhard, Gary F. Simons, and Charles D. Fennig, eds., *Ethnologue: Languages of the World*. Twenty-sixth edition (Dallas, TX: SIL International, 2023), www.ethnologue.com.

CHAPTER 3. MAKING RACE

1. See Aihwa Ong, *Fungible Life: Experiment in the Asian City of Life* (Durham, NC: Duke University Press, 2016), 13. In her work on the pan Asian SNP consortium and genomics in Singapore, Ong writes: "By coding and valorizing ethnic and racial variability, Biopolis scientists are better able to be competitive in the international arena of bioscience research and pharmaceutical investments while also continually re-emphasizing a reinvestment in the diverse 'races' that compose Singapore and Asia at large."

2. Sunder Rajan emphasizes market capitalism's creation of this large class of unemployed and underemployed surplus labor. The large population of people in this position comes to comprise what Sunder Rajan refers to as a "melting pot for clinical trials." See Kaushik Sunder Rajan, *Biocapital: The Constitution of Postgenomic Life* (Durham, NC: Duke University Press, 2006), 95.

3. Sunder Rajan, *Biocapital*, 95 (emphasis in original).

4. Katherine A. Mason's *Infectious Change: Reinventing Chinese Public Health after an Epidemic* (Stanford, CA: Stanford University Press, 2016), 113–114.

5. The market rationale of tailored medicine has been questioned from various perspectives. One would think blockbuster generalized drugs would be more marketable, yet there are also examples of when "ethnic drugs" can create value by virtue of being sold to specific segments of the population who suffer more from certain health disparities, see Dorothy Roberts's excellent discussion of these dynamics in chapters 7–8 of *Fatal Invention: How Science, Politics and Big Business Re-create Race in the Twenty-first Century* (New York: New Press, 2012). Also see Jonathan Kahn's initial analyses of the first such racial pill called BiDil. Jonathan Kahn, "Getting the Numbers Right: Statistical Mischief and Racial Profiling in Heart Failure Research," *Perspectives in Biology and Medicine* 46, no. 4 (2003): 473–483, https://doi.org10.1353/pbm.2003.0087. The rise of race-based patents has also incentivized this trend, as Khan points out in later work; see Jonathan Khan, "Mandating Race: How the USPTO Is Forcing Race into Biotech Patents," *Nature Biotechnology* 29, no. 5 (2011): 401–403, 401, https://doi.org/10.1038/nbt.1864. Kahn writes:

Biotech patents have increasingly included race-specific claims since the completion of the Human Genome Project in 2003. The question of how and why this phenomenon is occurring was first addressed in these pages in November 2006. . . . At that time there appeared to be two basic forces at work: first, some inventors were using race defensively, to buttress broader claims not specific to race; second, others were using race affirmatively to capture race-specific markets. A more recent review of select patent prosecutions before the United States Patent and Trademark Office (USPTO) indicates a third and potentially more troubling dynamic at work: USPTO examiners are requiring applicants to include racial categories in the claims sections of some of their biotech patent submissions.

Regarding the first race-based patent in the United States, concerning the heart disease medication BiDil, a team of social scientists and geneticists carried out a survey among cardiologists and found that "most participants (59.2%) perceived race as defining biologically distinct individuals. Respondents prescribed BiDil more often to African American patients than non-African American patients. However, they prescribed the generic components that makeup BiDil to African Americans and non-African American patients similarly." See Shawneequa L. Callier et al., "Cardiologists' Perspectives on BiDil and the Use of Race in Drug Prescribing," *Journal of Racial and Ethnic Health Disparities* 9, no. 6 (2022): 2146, https://doi.org/10.1007/s40615-021-01153-x.

6. Elsewhere I describe several instances where researchers working on these projects experienced a marked cognitive dissonance when faced with the prospect of sorting themselves into the categories they deployed to sort their study subjects' DNA. They described their confusion first and foremost as a personal inability to make sense of American racial categories. See Duana Fullwiley, "The Molecularization of Race: Institutionalizing Human Difference in Pharmacogenetics Practice," *Science as Culture* 16, no. 1 (2007): 18–19, https://doi .org/10.1080/09505430601180847.

7. Fullwiley, "Molecularization of Race."

8. And this of course includes the pharmaceutical industry with regard to precision medicine. As Kahn observed the rise of race-based patents for drugs over a ten year period, he writes:

From 1998 to 2005 there were twelve uses of racial and ethnic categories in granted patents. Between 2006 and 2016 that number grew to 63. Similarly, with respect to the use of racial and ethnic categories in patent applications filed, the number of uses rose from 65 in the years between 2001 and 2005 to 384 in the years between 2006 and 2016. Far from abating with new genomic discoveries, the use of racial categories in biomedical patenting has increased aggressively. What we see happening here is the normalization and routinization of inserting racial and ethnic categories into biomedical patents over the past decade.

See Jonathan Kahn, "Revisiting Racial Patents in an Era of Precision Medicine," *Case Western Reserve Law Review* 67, no. 4 (June 2017): 11–56, https:// ssrn.com/abstract=2982539.

9. "China Omeprazole Market Report, 2018–2022: Sales, Competition, Manufacturers, Prices & Market Prospects," PR Newswire, Research and Markets, www.prnewswire.com/news-releases/china-omeprazole-market-report-2018-2022 -sales-competition-manufacturers-prices—market-prospects-300743001.html (accessed June 21, 2022).

10. "Study Of PHarmacogenetics in Ethnically Diverse Populations (SOPHIE Study) (SOPHIE)," ClinicalTrials.gov, US National Library of Medicine, https:// clinicaltrials.gov/ct2/show/NCT00187668 (accessed June 21, 2022)

11. Fullwiley, "Molecularization of Race," 15–16. The postdoc recounted the situation as "weird" and in imagining an explanation, veered toward the possibility that people were "actually" mixed rather than considering the pattern of genetic diversity might be shared between people whom she was being conditioned to see as "opposites" as she recounted: "Something's weird in OCT 2. Some of the haplotype frequencies have completely changed to the opposite race. Ones that were African American are now Caucasian, and ones that were mostly Caucasian are now African American. We've already seen discrepancies between Coriell and SOPHIE for OCT 1. Maybe this is the same thing. Maybe some of those Coriell people were not really African Americans, or, Caucasians. Maybe they were mixed, because Coriell was just based on self-report." She eventually concluded that the samples could be "contaminated."

12. This was the only lab or set of scientists that I encountered who used this language. There are references in the scientific literature that look at genetic diversity of "cosmopolitan populations," or cosmopolitan animal and plant life, but nowhere else did I come across "cosmopolitan" used for variants found in what were taken to be a diverse set of humans who might broadly stand in for larger continental groups. There are scientists who use the idea of cosmopolitanism, however, as a means to include people from "Africa, Asia, Europe" in large-scale reference panels. See Bryan Howie, Jonathan Marchini, and Matthew Stephens, "Genotype Imputation with Thousands of Genomes," *G3 Genes|Genomes|Genetics* 1, no. 6 (2011): 457–470, https://doi.org/10.1534 /g3.111.001198.

13. Anthropologist and STS scholar S. Lochlann Jain chronicles how deep these dynamics run in his analysis of the oral polio vaccine in the Belgian Congo and over to what was then Ruanda-Urundi. In an exploration of thinking deeper about the possibility of "fluid bonding" (the "non-barrier intimate relations through which fluid sharing occurs"), he writes: "By opening bodies to each other, biomedical and technological infrastructures choreographed zones that fundamentally altered and expanded viral dynamics and exchanges. The biomedical infrastructure undergirding vaccine research and production spurred justification of these risks; assurances of the positive potential and intent of biomedicine became the foil and norm against which real or possible hazards have been adjudicated, understudied, and sometimes silenced." See S. Lochlann Jain,

"The WetNet: What the Oral Polio Vaccine Hypothesis Exposes about Globalized Interspecies Fluid Bonds," *Medical Anthropology Quarterly* 34 (2020): 506, https://doi.org/10.1111/maq.12587.

14. According to the FDA, "Many drugs are broken down (metabolized) with the help of a vital enzyme called CYP3A4 in the small intestine. Grapefruit juice can block the action of intestinal CYP3A4, so instead of being metabolized, more of the drug enters the blood and stays in the body longer. The result: too much drug in your body." See "Grapefruit Juice and Some Drugs Don't Mix," Consumer Updates, US Food and Drug Administration,www.fda.gov/consumers /consumer-updates/grapefruit-juice-and-some-drugs-dont-mix (accessed June 21, 2022). Saint John's Wort interacts in various ways with a number of drugs and affects the activity of multiple drug metabolizing enzymes. See "Saint John's Wort," The Mayo Clinic, www.mayoclinic.org/drugs-supplements-st -johns-wort/art-20362212 (accessed June 21, 2022).

15. In his book *Making the Mexican Diabetic*, anthropologist Michael Montoya observes a similar dynamic while ethnographically studying scientists working on the genetics of diabetes on the US/Mexico border. He writes:

> Researchers report that comparing the expected haplotype frequencies with those found within the sampled group establishes that the data set is representative of the group as a whole. Homogeneity, then, refers to the degree to which Mexicana/o DNA data sets are homologous to and thus representative of Mexicanas/os in Sun County writ large. Interestingly, Finns and Germans are never described as admixed: Europeans, for the purposes of the genetic epidemiology of type 2 diabetes, are a homogeneous (not admixed) population. . . . In this way, Mexicana DNA is thus evaluated for its homogeneity, only in this instance, it is the homogeneity of Mexicana admixture that is of concern for genetic researchers. In other words, this haplotype model is founded upon a notion of *a standardized admixed Mexican body: simply put, a pure Mexican.*

See Michael Montoya, *Making the Mexican Diabetic: Race, Science, and the Genetics of Inequality,* 1st ed. (Berkeley: University of California Press, 2011), 93 (emphasis added).

CHAPTER 4. FOR THE LOVE OF BLACKNESS

Epigraphs: Amiri Baraka, "Ka Ba," in *I Am the Darker Brother: An Anthology of Modern Poems by African Americans,* ed. Arnold Adoff (New York: Simon and Schuster, 1997), 38, 39; and "Prostate Cancer," The Mayo Clinic, www.mayoclinic.org/diseases-conditions/prostate-cancer/symptoms-causes/syc -20353087 (accessed July 2, 2022).

1. Nina G. Jablonski and George Chaplin, "Hemispheric Difference in Human Skin Color," *American Journal of Physical Anthropology* 107, no. 2 (December

1998): 221–223, discussion 223–224, https://doi.org/10.1002/(SICI)1096-8644 (199810)107:2221::AID-AJPA83.0.CO;2-X.

2. Although Kittles is careful about trying not to collapse ancestry and race, he wondered if 8q24 might mean that the social descriptor of race indeed overlaps with a "deep" biology that traces back to Africa. This is revisited in chapter 5.

CHAPTER 5. LOOK, A BLACK GUY!

1. Frantz Fanon, *Black Skin, White Masks* (London: Pluto Press, 2008), 84.

2. James Baldwin, "A Stranger in the Village," in *Notes of a Native Son* (1955; reprint, Boston: Beacon Press, 1984), 168.

3. Baldwin, "Stranger in the Village," 173.

4. This cohort is explained more fully later in the chapter.

5. "American Cancer Society: Cancer Facts and Figures 2022," American Cancer Society, www.cancer.org/content/dam/cancer-org/research/cancer -facts-and-statistics/annual-cancer-facts-and-figures/2022/2022-cancer-facts -and-figures.pdf (accessed December 4, 2022). The report reads: "In 2022, an estimated 268,490 new cases of prostate cancer will be diagnosed in the US and 34,500 men will die from the disease. The incidence of prostate cancer is 73% higher in non-Hispanic Black men than in non-Hispanic White men for reasons that remain unclear." Also see Prashanth Rawla, "Epidemiology of Prostate Can- cer," *World Journal of Oncology* 10, no. 2 (2019): 63–89, https://doi.org/10.14740 /wjon1191.

6. Anuradha Jagadeesan, Ellen D. Gunnarsdóttir, Sunna Ebenesersdóttir, Valdis B. Guðmundsdóttire et al., "Reconstructing an African Haploid Genome from the 18th Century," *Nature Genetics* 50, no. 2 (2018): 199–205, https://doi .org/10.1038/s41588-017-0031-6.

7. For more on the life and eventual court case of Hans Jónatan, see Gísli Páls- son, *The Man Who Stole Himself: The Slave Odyssey of Hans Jonatan* (Chicago: University of Chicago Press, 2016).

8. See Nicholas Wade, "A Genomic Treasure Hunt May be Striking Gold," *New York Times*, June 8, 2002, www.nytimes.com/2002/06/18/science/a-genomic -treasure-hunt-may-be-striking-gold.html; and David Ewing Duncan, "An Interview with Kari Steffanson, M.D.," *The Believer*, July 1, 2003, www .thebeliever.net/an-interview-with-kari-steffanson-m-d/. For more on the gene- alogical stories of the sagas, see Ari Þorgilsson, *Islendingabok, Kristnisaga: The Book of the Icelanders, the Story of the Conversion*, trans. Siân Grønlie (Exeter: Viking Society for Northern Research, 2006).

9. For an overview of the critiques, see Martin Enserink, "Physicians Wary of Scheme to Pool Icelanders' Genetic Data." *Science* vol 281, no. 5379 (1998): 890-

891, https://doi.org/10.1126/science.281.5379.890; and Michael Fortun, *Promising Genomics: Iceland and DeCODE Genetics in a World of Speculation* (Berkeley: University of California Press, 2008).

10. Matthew L. Freedman et al., "Admixture Mapping Identifies 8q24 As a Prostate Cancer Risk Locus in African-American Men," *Proceedings of the National Academy of Sciences of the United States of America* 103, no. 38 (2006): 10471, https://doi.org/10.1073/pnas.0605832103.

11. See "Composition of the Cohort," The Multiethnic Cohort Study, University of Hawaii Cancer Center, www.uhcancercenter.org/for-researchers /mec-cohort-composition (accessed December 2, 2021).

12. From this work, two areas or "regions" within the larger region of 8q24 were established as region 1 and region 2. A third, or region 3, would be discovered by a third group, and then a fourth, region 4, by Kittles's group and later a second "region 4" by the scientist's team who qualified their finding as "region 4 for whites" (discussed earlier in this chapter).

13. Freedman et al., "Admixture Mapping Identifies 8q24," 14071.

14. Freedman et al., "Admixture Mapping Identifies 8q24."

15. They write: "Our data add to the continually growing body of data supporting the presence of prostate cancer risk loci at 8q24 and provide justification for further genetic investigations in this region to facilitate the identification of causal alleles. As prostate cancer disproportionately affects African Americans, the discovery of true risk alleles could have important implications for early detection of prostate cancer in this high-risk population." See Christiane Robbins, Jada Benn Torres, Stanley Hooker, Carolina Bonilla, et al., "Confirmation Study of Prostate Cancer Risk Variants at 8q24 in African Americans Identifies a Novel Risk Locus," *Genome Research* 17, no. 12 (2007): 1720–1721, https://doi .org/10.1101/gr.6782707.

16. They write: "For DG8S737, the population attributable risk (PAR) of 16% in African Americans was considerably higher than the PAR for the European populations studied (5%–11%) and suggests that this allele may partially account for the significant difference in incidence seen among African American men." See Robbins et al., "Confirmation Study of Prostate Cancer Risk Variants," 1717.

17. They write: "Mean West African ancestry was significantly higher among the PCa cases than controls for both local 8q individual ancestry (LIA) and global individual ancestry (GIA). We determined that the slight associations for rs2124036 and rs780321 were likely due to local admixture. Two [other] SNPs remained significantly associated with prostate cancer." See Robbins et al., "Confirmation Study of Prostate Cancer Risk Variants," 1720–1721.

18. Freedman et al., "Admixture Mapping Identifies 8q24." Also see Julius Gudmundsson et al., "Genome-wide Association Study Identifies a Second Prostate Cancer Susceptibility Variant at 8q24," *Nature Genetics* 39, no. 5 (2007): 631–637, https://doi.org/10.1038/ng1999; and Fredrick R. Schumacher et al., "A

Common 8q24 Variant in Prostate and Breast Cancer from a Large Nested Case-Control Study," *Cancer Research* 67, no. 7 (2007): 2951–2956, https://doi .org/10.1158/0008-5472.

19. There is a typo in the publication on the SNP rs1001979; it should be rs16901979 (A allele). There is no SNP with rs1001979 in any of the tables and diagrams that compare all the SNPs discussed in the paper. There is only the aforementioned rs16901979 that is consistent with the rest of the text as well. This is not the Kittles team's prize discovery in this paper; however, for that one, rs7008482 appears correctly. Robbins et al., "Confirmation Study of Prostate Cancer Risk Variants," 1720.

20. See Laurent Berlant, *Cruel Optimism* (Durham, NC: Duke University Press, 2011), 25. We can think of how political conservatives are now citing the "social construction of race" to aid in arguments to dismantle affirmative action. I come back to these issues in the conclusion.

21. See Donna Haraway, "Situated Knowledges: The Science Question in Feminism and the Privilege of Partial Perspective," *Feminist Studies* 14, no. 3 (1988): 595, https://doi.org/10.2307/3178066 (emphasis added).

CHAPTER 6. A FAMILY AFFAIR

1. Elijah Muhammad, *Message to the Blackman in America* (Phoenix, AZ: Secretarius MEMPS, 1973), 58–59.

CHAPTER 7. SCI NON-FI

1. For an overview of the project as described by the scientists involved, see Jeantine E. Lunshof et al., "Personal Genomes in Progress: From the Human Genome Project to the Personal Genome Project," *Dialogues in Clinical Neuroscience* 12, no. 1 (2010): 47–60, https://doi.org/10.31887/DCNS.2010.12.1 /jlunshof.

2. For the anthropologists the concept was more focused on disease groups lobbying for specific funds and attention to rare disease entities to be included in government research priorities. Their activism also included normalizing and destigmatizing their conditions, see Deborah Heath, Rayna Rapp, and Karen-Sue Taussig, "Genetic Citizenship," in *A Companion to the Anthropology of Politics*, ed. David Nugent and Joan Vincent, chapter 10 (Hoboken, NJ: Blackwell Publishing Ltd., 2007), https://doi.org/10.1002/9780470693681.ch10.

3. Jeantine E. Lunshof et al., "From Genetic Privacy to Open Consent," *Nature Reviews Genetics* 9, no. 5 (2008): 408, https://doi.org/10.1038/nrg2360.

4. Lunshof et al., "From Genetic Privacy to Open Consent."

5. Lunshof et al., "From Genetic Privacy to Open Consent."

6. Also see Madeline Price Ball et al., "Harvard Personal Genome Project: Lessons from Participatory Public Research," *Genome Medicine* 6, no. 10 (2014), https://doi.org/10.1186/gm527.

7. Lunshof et al., "From Genetic Privacy to Open Consent," 409; also see Kareem Ayoz, Erman Ayday, and A. Ercument Cicek, "Genome Reconstruction Attacks Against Genomic Data-Sharing Beacons," *Privacy Enhancing Technologies Symposium* 3 (March 2021): 28–48, https://doi.org/10.2478/popets-2021-0036.

8. On Church's Harvard website he transparently lists over one hundred companies, funders, nonprofits, and other affiliations. See "George M. Church's Tech Transfer, Advisory Roles, and Funding Sources," https://arep.med.harvard.edu/gmc/tech.html (accessed December 5, 2022).

9. Many large databases now have controlled access protocols, including those of the NIH and many DTC companies, within this the NIH and others have called for large-scale responsible data sharing with as much protections as possible; see Kendall Powell, "The Broken Promise That Undermines Human Genome Research," *Nature* 590, no. 7845 (2021): 198–201, https://doi.org/10.1038/d41586-021-00331-5.

10. Nils Homer et al., "Resolving Individuals Contributing Trace Amounts of DNA to Highly Complex Mixtures Using High-Density SNP Genotyping Microarrays," *PLoS Genetics* 4, no. 8 (August 2008), https://doi.org/10.1371/journal.pgen.1000167.

11. See Ericka Check Hayden, "Privacy Protections: The Genome Hacker," *Nature* 497 (May 2013): 172–174, https://doi.org/10.1038/497172a.

12. Latanya Sweeney, Akua Abu, and Julia Winn, "Identifying Participants in the Personal Genome Project by Name," Harvard University, 2013, https://privacytools.seas.harvard.edu/files/privacytools/files/1021-1.pdf; also see Adam Tanner, "Harvard Professor Re-Identifies Anonymous Volunteers In DNA Study," *Forbes*, April 25, 2013, www.forbes.com/sites/adamtanner/2013/04/25/harvard-professor-re-identifies-anonymous-volunteers-in-dna-study/?sh=383d932092c. Tanner writes: "Sweeney's latest findings build on a 1997 study she did that showed she could identify up to 87% of the U.S. population with just zip code, birthdate and gender. She was also able to identify then Massachusetts Gov. William Weld from anonymous hospital discharge records." This latter bit was often discussed by the PGP participants as a point of reference when discussing the futility of many promises of privacy.

13. One might also analyze this in Lauren Berlant's terms of "cruel optimism" (Berlant, *Cruel Optimism*, 24–25).

14. When it actually came to putting their information on the Web, once they began getting their data, not all of the participants chose to link their names with their genomic and trait information, although they knew anyone with the

intention to do that could do so quite easily; see Susannah F. Lock, "Meet my Genome: 10 People Release Their DNA on the Web," *Scientific American*, October 21, 2008, https://blogs.scientificamerican.com/news-blog/meet-my -genome-10-people-release-th-2008-10-21/. Also see Emily Singer, "The Genome Pioneers: Early Adopters of Personal Genome Sequencing Gather To Reflect on What We Need To Do To Move the Field Forward," *MIT Technology Review*, April 30, 2010, www.technologyreview.com/2010/04/30/204008/the -genome-pioneers/ (accessed October 15, 2010).

15. "Participant Profiles," Personal Genome Project, https://my.pgp-hms.org /users (accessed November 23, 2022).

16. Reginald Wilson, "Barriers to Minorities Success in College Science, Mathematics and Engineering Programs," in *Access Denied: Race, Ethnicity and the Scientific Enterprise*, ed. George Campbell Jr., Ronni Denes, and Catherine Morrison (Oxford: Oxford University Press, 2006), 195; Beatriz Chu Clewell and Jomills Henry Braddock II, "Influences on Minority Participation in Mathematics, Science and Engineering," in *Access Denied*, ed. Campbell, Denes, and Morrison, 94–95; 97; and Steven Epstein, *Inclusion: The Politics of Difference in Medical Research* (Chicago: Chicago University Press, 2007), 193–195.

17. Between 2006 and 2010 the project was featured in *Scientific American*, the *Wall Street Journal*, *Bloomberg News*, *NPR*, *U.S. News and World Report*, *Fox News*, the *Washington Post*, *Nature*, the *Boston Globe*, the *New Scientist*, the *New York Times*, PBS, and *Wired Magazine*, among a slew of smaller venues such as the *MIT Technology Review* and the Harvard news sources. Most of the press hyped the project's "ingenuity" and forward-thinking. Only as an aside did some journalists include a discussion about the PGP's "controversial" nature regarding the ethical hurdles of "open" genetics.

18. "Public Profile—hu604D39," Personal Genome Project, https://my.pgp -hms.org/profile_public?hex=hu604D39 (accessed August 21, 2020).

19. See George Church, "The Personal Genome Project," *Molecular Systems Biology* 1, no. 1 (2005): 1–3, https://doi.org/10.1038/msb4100040.

20. Sherley pointed out the many ways that institutional and personal racism were to blame for his tenure denial at MIT in a speech he delivered at his alma mater, Johns Hopkins University, in 2007. The talk later appeared as a book chapter. In it he implores his audience to recognize various levels of resistance and refusal to truly integrate Black faculty and scientists in US academia; see James L. Sherley, "Including the Excluded: Concepts for Successful Integration of the U.S. Academy," in *Racism: Global Perspectives, Coping Strategies and Social Implications*, ed. Tracey Lowell, chapter 3, 37–46 (Nova Science Publishers), https://novapublishers.com/wp-content/uploads/2019/04/Concepts-for -Successful-Integration-of-the-US-Academy.pdf.

21. June Q. Wu, "Harvard Prof's Personal Genome Project Reveals DNA Secrets: 'PGP-10' Volunteers Release Their Medical Records and Personal

Genome Sequences Online," *Harvard Crimson*, October 23, 2008, www
.thecrimson.com/article/2008/10/23/harvard-profs-personal-genome-project
-reveals/.

22. Martin Bashir, "Confessions of a Sperm Donor," ABC-News, August 31,
2006, https://abcnews.go.com/Nightline/Health/story?id=1982328&page=1.

23. Documentary filmmaker Marilyn Ness has documented the initial PGP-
10 participants, which captured some of these dynamics; see Marilyn Ness,
"GENOME: The Future Is Now WEBISODE 1," uploaded on July 29, 2009,
YouTube video, 07:21, www.youtube.com/watch?v=mVZI7NBgcWM; Marilyn
Ness, "GENOME: The Future Is Now WEBISODE 2," uploaded on August 6,
2009, YouTube video, 06:06, www.youtube.com/watch?v=2r9DpthvNKM; and
Marilyn Ness, "GENOME: The Future Is Now WEBISODE 3," uploaded on
August 13, 2009, YouTube video, 06:53, www.youtube.com/watch?v=mgXAO8pv
-X4. I was given a full transcript of the meeting by the PGP director of commu-
nity, Jason Bobe.

24. Indeed, one can order his cells through Coriell, but it does not list names.
Numerous bios and write-ups on Sherley call him PGP-10, which is also how
his cell line is referred to on the Coriell site for purchase. See "GM21846 LCL
from B-Lymphocyte," Coriell Institute for Medical Research, www.coriell.org/0
/Sections/Search/Sample_Detail.aspx?Ref=GM21846&Product=CC (accessed
November 23, 2022). For an example of publicly available information about him
that can easily link him to his genome, microbiome, and cell line data, see, for
instance, "About Us, James L. Sherley, M.D., Ph.D. Associate Scholar," The Charlotte Lozier Institute, https://lozierinstitute.org/team-member/james-l-sherley
(accessed November 23, 2022).

25. National Archives and Records, "Executive Order 13505—Removing Barriers to Responsible Scientific Research Involving Human Stem Cells, Memorandum of March 9, 2009," *The Federal Register* 74, no. 46 (2009): 10667–
10668, www.federalregister.gov/documents/2009/03/11/E9-5441/removing
-barriers-to-responsible-scientific-research-involving-human-stem-cells.

26. The memorandum that allowed the case to go forward and argued that
the plaintiffs did have standing read: "Congress has spoken to the precise question at issue—whether federal funds may be used for research in which an
embryo is destroyed. The Dickey-Wicker Amendment provides that no federal
funds shall be used for 'research in which a human embryo or embryos are
destroyed, discarded, or knowingly subjected to risk of injury or death greater
than that allowed for research on fetuses in utero under 45 C.F.R. § 46.204(b)
and section 498(b) of the Public Health Service Act (42 U.S.C. 289g(b)).' Pub. L.
No. 111-8, § 509(a)(2). Thus, as demonstrated by the plain language of the stat-
ute, the unambiguous intent of Congress is to prohibit the expenditure of federal
funds on 'research in which a human embryo or embryos are destroyed.'" See
Royce C. Lamberth, "Memorandum Opinion." Civ. No. 1:09-cv-1575 (RCL), US

District Court for the District of Columbia, August 23, 2010, https://ecf.dcd
.uscourts.gov/cgi-bin/show_public_doc?2009cv1575-44.

27. "Induced Pluripotent Stem Cells (iPS)," Broad Cell Stem Research Center,
https://stemcell.ucla.edu/induced-pluripotent-stem-cells (accessed November
24, 2022).

28. Meredith Wadman, "High Court Ensures Continued US Funding of
Human Embryonic-Stem-Cell Research," *Nature* (2013), https://doi.org/10
.1038/nature.2013.12171.

29. See Misha Angrist, *Here Is a Human Being: The Dawn of Personal Genomics* (New York: Harper Perennial, 2011).

30. Jenny Reardon, *The Postgenomic Condition: Ethics, Justice and Knowledge after the Genome* (Chicago: Chicago University Press, 2017).

31. See Steven Pinker, "My Genome, My Self," *New York Times*, January 7,
2009, www.nytimes.com/2009/01/11/magazine/11Genome-t.html. One of the
PGP-10, Pinker writes: "For all the narcissistic pleasure that comes from poring
over clues to my inner makeup, I soon realized that I was using my knowledge of
myself to make sense of the genetic readout, not the other way around. . . . Individual genes are just not very informative. Call it Geno's Paradox. We know from
classic medical and behavioral genetics that many physical and psychological
traits are substantially heritable. But when scientists use the latest methods to
fish for the responsible genes, the catch is paltry." Also see Goetz, "The Gene Collector"; and Church, "Personal Genome Project."

32. This phrase comes from Richard Lewontin's *Biology as Ideology* (New
York: Harper Perennial, 1993).

33. "What We Do," PG-ED: Personal Genomics Education Project, www
.transvection.org/pged (accessed November 21, 2022).

34. The PGP planned to eventually biopsy one hundred thousand people, but
some dreamed of the project reaching one million at some point.

35. In many domains of experimental science where researchers are
creating life in vitro, sentience and the development of a human brain
constitute a significant threshold, one that is rife with ethical barriers for moral
trespass.

36. "Open Source Next Generation Sequencing Technology," Church Lab,
Harvard Medical School, https://arep.med.harvard.edu/Polonator/ (accessed
December 5, 2022).

37. See Mark Schoofs, "Advance in Quest for HIV Vaccine," *Wall Street Journal*, July 9, 2010, www.wsj.com/articles/SB10001424052748703609004575353
5072271264394; and Stephen R. Walsh and Michael Seaman, "Broadly Neutralizing Antibodies for HIV-1 Prevention," *Frontiers in Immunology* 12 (2021),
Sec. "Vaccines and Molecular Therapeutics," https://doi.org/10.3389
/fimmu.2021.712122.

CHAPTER 8. SEEING GHOSTS

1. For the UCLA study, see Arun Durvasula and Sriram Sankararaman, "Recovering Signals of Ghost Archaic Introgression in African Populations," *Science Advances* (February 12, 2020): 1–9. Regarding the sequences see, "1000 Genomes Project," Coriell Institute for Medical Research, www.coriell.org/0/Sections/Collections/NHGRI/1000genome.aspx?PgId=664&coll=HG (accessed October 22, 2022). This number is from phase 3 of the 1000 Genomes Project. The first phase "lightly sequenced" 179 samples. This resource database also offers cell lines and DNA itself. For example, researchers can buy these materials via mail order. For the Mende population, the description of the product reads: "This panel contains 2 micrograms of DNA, normalized to a concentration of 100ng/µl, from 97 unique and unrelated Mende in Sierra Leone. Of these, 47 are males and 50 are females. Cell lines for these individuals were established from blood samples collected from these unrelated individuals and DNA was isolated from the established cell cultures. Samples for these individuals are also available as individual DNA aliquots (50µg of DNA) and as lymphoblastoid cell lines." See "MPG00021 Human Variation Panel," Coriell Institute for Medical Research, www.coriell.org/0/Sections/Search/Panel_Detail.aspx?PgId=761&Ref=MGP00021 (accessed October 22, 2022).

2. See Peter Beerli, "Effect of Unsampled Populations on the Estimation of Population Sizes and Migration Rates between Sampled Populations," *Molecular Ecology* 13, no. 4 (2004): 828, https://doi.org/10.1111/j.1365-294x.2004.02101.x. For other uses of the term "ghost ancestry," see Kamporakis, *Ancestry Reimagined*, 78–84.

3. Richard E. Green et al., "A Draft Sequence of the Neandertal Genome," *Science* 328, no. 5979 (2010): 710 (emphasis added), https://doi.org/10.1126/science.1188021.

4. Green et al., "Draft Sequence of the Neandertal Genome."

5. For more on the porous relationship between religion and science, see Keel, *Divine Variations*. Regarding further analysis of white supremacists' embracing studies of population genetic difference, see Aaron Panofsky and Joan Donovan, "Genetic Ancestry Testing among White Nationalists: From Identity Repair To Citizen Science," *Social Studies of Science* 49, no. 5 (2019): 653–681, https://doi.org/10.1177/0306312719861434. For an insightful read on the recuperation of Neanderthal heritage, see James Doucet-Battle, "Ennobling the Neanderthal: Racialized Texts and Genomic Admixture," *Kalfou: A Journal of Comparative and Relational Ethnic Studies* 5, no. 1 (2018): 61-67.

6. "Celebrate Your Ancient DNA with a New Neanderthal Report," *The 23andMe Blog*, https://blog.23andme.com/ancestry-reports/new-neanderthal-report/ (accessed October 22, 2022).

7. Bonham as quoted in "Complete Neanderthal Genome Sequenced DNA Signatures Found in Present-Day Europeans and Asians, but Not in Africans," National Human Genome Research Institute, www.genome.gov/27539119/2010 -release-complete-neanderthal-genome-sequenced (accessed October 23, 2022).

8. See the Max Planck press release for the finding: "Neandertal Mother, Denisovan Father!," August 22, 2018, www.mpg.de/12208106/neandertals -denisovans-daughter; and the published paper: Viviane Slon et al., "The Genome of the Offspring of a Neanderthal Mother and a Denisovan Father," *Nature* 561, no. 7721 (2018): 113–116, https://doi.org/10.1038/s41586-018 -0455-x. The Denisova cave site has been home to many species of animals as well as ancient and modern humans. The last human inhabitant was an eighteenth-century "Old Believer" hermit who wanted to maintain the rituals and rites of the Russian Orthodox Church as the Greek Orthodox Church was being introduced. He was called Dionisiy or Dionisij (modernized as Denis), which led to the site being called Denisova as a memorialization of its last long-term human inhabitant. See *Wikipedia*, s.v. "Denisova cave," https://en.wikipedia.org/wiki /Denisova_Cave (accessed October 22, 2022). According to a "wonders of the world" site, throughout many thousands of years the "Denisova Cave has served as a shelter both for people and animals. In total there have been found 20–22 cultural layers. . . . Finds include remnants of 27 species of large and middle-sized mammals, 39 species of small mammals and remnants of other vertebrates. . . . These include a tusk of mammoth, ostrich eggshell, remnants of small cave bear (*Ursus rossicus*), cave hyaena (*Crocuta spelaea*), cave lion (*Panthera spelaea*), wooly rhinoceros (*Coelodonta antiquitatis*) and many others"; see "Denisova Cave: The Only Find of Denisovan Humans," Wonders of the World," *Wondermondo*, www.wondermondo.com/denisova-cave/ (accessed October 19, 2022).

9. See Matthias Meyer et al., "A High-Coverage Genome Sequence from an Archaic Denisovan Individual," *Science* 338, no. 6104 (August 2012): 222–226, https://doi.org/10.1126/science.1224344.

10. Meyer et al., "High-Coverage Genome Sequence."

11. Lu Chen et al., "Identifying and Interpreting Apparent Neanderthal Ancestry in African Individuals," *Cell* 180, no. 4 (January 2020), https:// doi.org/10.1016/j.cell.2020.01.012.

12. See Michael F. Hammer et al., "Genetic Evidence for Archaic Admixture in Africa," *Proceedings of the National Academy of Sciences of the United States of America*, 108, no. 37 (2011): 15123–15128. doi:10.1073/pnas.1109300108. Also see Joseph Lachance et al., "Evolutionary History and Adaptation from High-Coverage Whole-Genome Sequences of Diverse African Hunter-gatherers," *Cell* 150, no. 3 (2012): 457–469, doi:10.1016/j.cell.2012.07.009.

13. Ping Hsun Hsieh et al., "Model-based Analyses of Whole-Genome Data Reveal a Complex Evolutionary History Involving Archaic Introgression in Cen-

tral African Pygmies," *Genome Research* 26, no. 3 (2016): 291–300, doi:10.1101/gr.196634.115.

14. See Durvasula and Sankararaman, "Recovering Signals." Also see University of California Television, "Sriram Sankararaman: Recovering Signals of Ghost Archaic Introgression in African Populations," recorded February 21, 2020, www.youtube.com/watch?v=vXOyVG-LNoA.

15. Hammer et al., "Genetic Evidence for Archaic Admixture in Africa," 406.

16. Laura T. Buck, "Homo heidelbergensis," in *Encyclopedia of Animal Cognition and Behavior*, ed. Jennifer Vonk and Tad Shackelford, 3187–3192 (Switzerland: Springer Nature, 2020), https://doi.org/10.1007/978-3-319-47829-6_1151-1.

17. Early on a 2011 publication in *PNAS* describes how researchers sequenced sixty-one autosomal noncoding regions in samples of three sub-Saharan populations, "Mandenka [*sic*], Biaka and San" for a similar query. The researchers did simulations that allowed them "to infer that contemporary African populations contain a small proportion of genetic material ($\approx$2%) that introgressed $\approx$35 kya from an archaic population that split from the ancestors of anatomically modern humans $\approx$700 kya." See Hammer et al., "Genetic Evidence for Archaic Admixture in Africa."

18. Wikipedia, s.v., "Introgression," https://en.wikipedia.org/wiki/Introgression (accessed October 22, 2022).

19. Durvasula and Sankararaman, "Recovering Signals," 3 (emphasis added). The subsequent quotations in this discussion are from text in figure 3 and page 4 of this same article.

20. See Hugo Zeberg and Svante Pääbo, "The Major Genetic Risk Factor for Severe COVID-19 Is Inherited from Neanderthals," *Nature* 587, no. 7835 (2020), https://doi.org/10.1038/s41586-020-2818-3. For a thoughtful review of this and other findings that a segment on chromosome three contains a haplotype that could put people at a higher risk for COVID-19, see John Hawks, "Neanderthals and Covid-19, Beyond the Hype: What a Gene Discovery Means to Understanding the Biology of This Pandemic," *Medium*, July 5, 2020, https://medium.johnhawks.net/neanderthals-and-covid-19-beyond-the-hype-c258dc8bc2c9. Hawks brings his insights on the science to a position of caution about the hype that may detract a real focus from how to prevent COVID health disparities. He writes:

> The virus is raging in the United States today, and people's Neanderthal ancestry is not making any visible difference to that. The highest incidence is among Blacks and Latinos, a function of the higher rate of spread in urban areas and—over the last few weeks—in Texas, Florida, and Arizona. The Navajo Nation and some other tribes have also experienced very high incidence of the virus. Epidemiologists point to many factors that underlie the disparity of Covid-19 spread in these groups, from multigenerational households and reliance on public transportation, to obesity and other long-term health stresses, to a greater proportion of people working in

front-line professions. None of those factors are genetic. Behavioral and cultural factors are leading this pandemic, many of them reinforced by a history of poverty and racism.

21. Several of the methods and software this lab uses are explained as (1) "The EIGENSTRAT method uses principal components analysis to explicitly model ancestry differences between cases and controls along continuous axes of variation; the resulting correction is specific to a candidate marker's variation in frequency across ancestral populations, minimizing spurious associations while maximizing power to detect true associations. The EIGENSOFT package has a built-in plotting script and supports multiple file formats and quantitative phenotypes"; (2) "ANCESTRYMAP 2.0, ANCESTRYMAP finds skews in ancestry that are potentially associated with disease genes in recently mixed populations like African Americans. It can be downloaded for either UNIX or Linux or MAC"; (3) "ADMIXTOOLS (Patterson et al. 2012) is a software package that supports formal tests of whether admixture occurred, and makes it possible to infer admixture proportions and dates. Please contact Nick Patterson if you have any questions about the software and for scientific questions. * The latest ADMIXTOOLS release is available at GitHub." There are still other methods and supplementary tools on offer as well. See "Software," David Reich Lab, Harvard University, https://reich.hms.harvard.edu/software (August 8, 2022).

22. Amade M'charek, *The Human Genome Diversity Project: An Ethnography of Scientific Practice* (Cambridge, UK: Cambridge University Press, 2005), 8.

23. M'charek, *Human Genome Diversity Project*, 14.

24. Jenny Reardon, *Race to the Finish: Identity and Governance in an Age of Genomics* (Princeton, NJ: Princeton University Press, 2005), 161–162.

25. It is important to distinguish between genocidal killing off of Native lifeways, as well as massacres, targeted land grabs, and displacements that led to starvation, sickness, and death of Native peoples throughout the Americas, and the reality that these acts did not in fact eliminate all Native peoples. Indigenous Americans and First Nations peoples in Canada have had to contend with the contradictions of mainstream cultural narratives that they are "vanishing" while at the same time calling attention to the ways that colonial settler states have tried to make them disappear. For a review of these issues and some of the complexities they raise regarding acknowledging colonial violence, while emphasizing survivance in the present, Theodore Fontaine, a residential "school" survivor, writes: "The term genocide was not applied to the deliberate, Parliament sanctioned action to 'kill the Indian in the child,' and the Indian residential schools' policy was portrayed as a well-intentioned, though misguided, brief event in history. . . . Some individuals say that the term genocide is appropriate only to convey the complete destruction of a race. Ironically, given that we Indigenous Nations have inconveniently survived by the threads of our own spirituality, resilience, and courage, these same individuals use our survival to deny the

truths of our history, whether or not it is called genocide." Theodore Fontaine, foreword to *Colonial Genocide in Indigenous America*, ed. Andrew Woolford, Jeff Benvenuto, and Alexander Hinton (Durham, NC: Duke University Press, 2014), ix.

26. HUGO, *The Human Genome Diversity (HGD) Project*, summary document, December 31, 1993, https://digital.library.unt.edu/ark:/67531 /metadc692124/, 1.

27. Lock gives detailed examples of the simplistic ways that San in South Africa and Baraku in Japan were taken at face value when anthropologists have described the violent and highly discriminatory ways that both groups were made marginal, cut off from larger societal forms of care and inclusion, and indeed relegated to "isolated" social spheres in these processes. Margaret Lock, "The Alienation of Body Tissue and the Biopolitics of Immortalized Cell Lines," *Body & Society* 7, no. 2–3 (2001): 80, https://doi.org/10.1177/1357034X0100700204.

28. See Audra Simpson, "The Ruse of Consent and the Anatomy of 'Refusal': Cases from Indigenous North America and Australia," *Postcolonial Studies* 20, no. 1 (2017): 21, https://doi.org/10.1080/13688790.2017.1334283. In her much larger discussion about Mohawk/Kahnawà:ke efforts to assert true sovereignty, she writes:

> If there was a "structure" of settler colonialism that was discernible through time, there was also a structure of refusal. These refusals were symptomatic of that structure and manifested in the games of settler colonial governance, and in particular, the play that would signal consent. Paramount among these refusals is citizenship itself, the aforementioned actions that would signal consent and belonging within a settler political system and would move Mohawks out of their own sovereignty into an ambit of "consent" and with that, settler citizenship and the promise of whiteness. All of this pointed analytically to the deeply unequal scene of articulation that people were thrown into and were remaking through the quotidian and the grand.

29. Debra Harry, "The Human Genome Diversity Project and Its Implications for Indigenous Peoples," Indigenous People's Council on Biocolonialism, www .ipcb.org/publications/briefing_papers/files/hgdp.html (accessed October 17, 2022).

30. Third World Network's warning as cited in M'charek, *Human Genome Diversity Project*, 12.

31. Simpson, *The Ruse of Consent*, 21 (emphasis added).

32. Reardon, *Race to the Finish*, 158.

33. Krystal S. Tsosie, Keolu Fox, and Joseph M. Yracheta, "Genomics Data: The Broken Promise Is to Indigenous People," *Nature* 591 (2021): 529, https:// doi.org/10.1038/d41586-021-00758-w. See also Kelou Fox, "The Illusion of Inclusion—The 'All of Us" Research Program and Indigenous Peoples' DNA." *New England Journal of Medicine* 383, no. 5 (2020): 411, https://doi .org/10.1056/NEJMp1915987.

34. International HapMap Consortium, "Integrating Ethics and Science in the International HapMap Project," *Nature Reviews Genetics* 5, no. 6 (June 2004): 470, https://doi.org/10.1038/nrg1351.

35. International HapMap Consortium, "Integrating Ethics and Science in the International HapMap Project."

36. "*All of Us* Tribal Leader and Urban Indian Organization Letter," NIH *All of Us* Research Program, May 24, 2019, https://allofus.nih.gov/about/tribal -engagement/all-us-tribal-leader-and-urban-indian-organization-letter.

37. "*All of Us* Research Program Tribal Consultation Final Report," NIH *All of Us* Research Program, March 2021, https://allofus.nih.gov/all-us-research -program-tribal-consultation-final-report.

38. Fox, "Illusion of Inclusion," 411–412.

39. Fox, "Illusion of Inclusion," 412 (emphasis added).

40. Fox, "Illusion of Inclusion."

41. See "Keolu Fox," Department of Anthropology, UC San Diego, https://anthropology.ucsd.edu/people/faculty/faculty-profiles/Keolu%20Fox.html (accessed October 18, 2022).

42. See "Research for Natives, by Natives," NativeBioData, https://nativebio .org (accessed October 18, 2022).

43. Yracheta as quoted in Melissa Gismondi, "How Indigenous Scientists Are Using Biomedical Research To Seek 'Genomic Justice,'" *CBC Radio*, September 28, 2021, www.cbc.ca/radio/ideas/how-indigenous-scientists-are-using-biomedical -research-to-seek-genomic-justice-1.6190855.

44. Krystal Tsosie, Joe Yracheta, and Donna Dickenson, "Overvaluing Individual Consent Ignores Risks to Tribal Participants," *Nature Reviews Genetics* 20, no. 9 (2019): 497, https://doi.org/10.1038/s41576-019-0161-z.

45. See Gismondi, "How Indigenous Scientists Are Using Biomedical Research," for other examples, such as the proposed Northern Biobank Initiative to serve First Nations in northern British Columbia, Canada, and the Silent Genomes Project, also in Canada.

46. "Research for Natives, by Natives," NativeBioData.

47. See the SING consortium's mission statement, "Past Workshops," Summer internship for INdigenous peoples in Genomics, www.singconsortium.org/past -workshops (accessed October 18, 2022).

48. For a detailed discussion of how Haudenosaunee lands were defined through cultural and legal values on the part of white settlers, most prominently Thomas Jefferson, in order to formally displace and dispossess Haudenosaunee from their territories, see Palmer, *Rendering Settler Landscapes*, 2020.

49. Oscar Schwartz, "A Geneticist's Dilemma," *Washington Post*, November 23, 2020, www.washingtonpost.com/magazine/2020/11/23/many-scientists-believe -that-dna-holds-cure-disease-that-poses-problem-some-native-americans/.

50. For an in-depth investigative account, see Judy Pasternak, *Yellow Dirt: A Poisoned Land and the Betrayal of the Navajos* (New York: Free Press, 2011).

51. Antonella Spinazzola et al., "Lack of Founder Effect for an Identical mtDNA Depletion Syndrome (MDS)-Associated MPV17 Mutation Shared by Navajos and Italians," *Neuromuscular Disorders: NMD* 18, no. 4 (2008): 315–318, https://doi.org/10.1016/j.nmd.2007.12.007; also see Charalampos Karadimas et al., "Navajo Neurohepatopathy Is Caused by a Mutation in the MPV17 Gene," *American Journal of Human Genetics* 79, no. 3 (2006): 544–548, https://doi.org/10.1086/506913.

52. Schwartz, "Geneticist's Dilemma."

53. Pasternak, *Yellow Dirt.*

54. Susan E. Dawson and Gary E Madsen, "Psychosocial and Health Impacts of Uranium Mining and Milling on Navajo Lands," *Health Physics* 101, no. 5 (2011): 618–625, https://doi.org/10.1097/HP.0b013e3182243a7a.

55. Wise Uranium Project, "Compensation of Navajo Uranium Miners," last modified June 8, 2022, www.wise-uranium.org/ureca.html; see also "S.4119—RECA Extension Act of 2022," Congress.gov, www.congress.gov/bill/117th-congress/senate-bill/4119 (accessed November 25, 2022).

56. Joseph H. Hoover et al., "Exposure to Uranium and Co-Occurring Metals among Pregnant Navajo Women," *Environmental Research* 190, art. no. 109943 (2020): 1–12, https://doi.org/10.1016/j.envres.2020.109943.

57. "What Does It Mean to Give Away Our DNA?," *The Experiment* podcast of *The Atlantic*, published October 28, 2021, 31:22, www.theatlantic.com/podcasts/archive/2021/10/native-american-dna-navajo-ban-on-genetic-research/620510/.

58. There are several Indigenous data sovereignty networks in the United States, and internationally. See Stephanie Russo Carroll's chaired resource here, www.rd-alliance.org/groups/international-indigenous-data-sovereignty-ig; see a collaborative resource for researchers to connect on data sovereignty issues here, www.gida-global.org/. For the Māori network, see www.temanararaunga.maori.nz/. For the Indigenous Australian network, see www.maiamnayriwingara.org/.

59. Among many examples, see Stephanie Russo Carroll et al., "The CARE Principles for Indigenous Data Governance," *Data Science Journal* 19, art. no. 43 (2020): 1–12, http://doi.org/10.5334/dsj-2020-043; Megan Davis, "Data and the United Nations Declarations on the Rights of Indigenous Peoples," in *Indigenous Data Sovereignty: Toward an Agenda*, ed. Tahu Kukutai and John Taylor, 25–38 (Canberra: Australian National University Press, 2016); Nanibaa Garrison et al., "Genomic Research through an Indigenous Lens: Understanding the Expectations," *Annual Review of Genomics and Human Genetics* 20 (2019): 495–517, https://doi.org/10.1146/annurev-genom-083118-015434; C. Matthew

Snipp, "What Does Data Sovereignty Imply: What Does It Look Like?," in *Indigenous Data Sovereignty*, ed. Kukutai and Taylor, 39–55 (Canberra: Australian National University Press, 2016); and Rebecca Tsosie, "Tribal Data Governance and Informational Privacy: Constructing 'Indigenous Data Sovereignty," *Montana Law Review* 80, no. 2 (2019): 229–268.

60. "What Does It Mean to Give Away Our DNA?" *The Experiment* podcast.

61. Jessica Bardill, "Comparing Tribal Research and Specimens Policies: Models, Practices, and Principles," *International Indigenous Policy Journal* 8, no. 4 (2017): 1–22, https://doi.org/10.18584/iipj.2017.8.4.4; and Laura Arbour and Doris Cook, "DNA on Loan: Issues To Consider When Carrying out Genetic Research with Aboriginal Families and Communities," *Community Genetics* 9, no. 3 (2006): 153–160, https://doi:10.1159/000092651.

62. Kim TallBear, "Genomic Articulations of Indigeneity," *Social Studies of Science* 43, no. 4 (2013): 514 (emphasis added), https://doi.org/10.1177/03063127134838.

63. Tallbear, "Genomic Articulations of Indigeneity," 510.

64. For a detailed review of the latter, see Bardill, "Comparing Tribal Research and Specimens Policies."

CONCLUSION

1. Skloot, *Immortal Life of Henrietta Lacks*, 91.

2. Troy Duster has for years cautioned about how the rise of genetic science for health purposes is in lockstep with the forensic use of DNA and surveillance more broadly, see Troy Duster, *Backdoor to Eugenics* (UK: Routledge, 2003), 156–158. For forensic uses of DNA in other global contexts, see "Expansions of Police DNA Databases Worldwide Urgently Need Human Rights Safeguards," Press Release, GeneWatch UK Press Release, February 16, 2016, www.genewatch.org/article.shtml?als%5Bcid%5D=576198&als%5Bitemid%5D=576234; and Noah Tamarkin, "Forensics and Fortification in South African Self-captivity," *History and Anthropology* 30 no. 5 (2019): 521–526, 524–525,https://doi.org/10.1080/02757206.2019.1638774. Tamarkin writes: "As a security infrastructure, national criminal DNA databases contribute to the fantasy of the impenetrability of the built form. Through its association with safety and security, DNA has been rebranded in post-apartheid South Africa from an association with extraction for the benefit of researchers to the means to ascertain the truth, which can then facilitate justice. But the question remains: justice for whom?"

3. Geoffrey C. Bowker, *Memory Practices in the Sciences* (Cambridge, MA: MIT Press, 2008), 184.

4. Both Nancy Krieger and Alain Desrosières have argued that data within a societal field is not merely a set of "givens" (*les données*). In no way is it ever simply

"raw." Rather, it is "made" in cultural contexts and—to varying degrees—it contains the same politics as those of the society in which it is formed. See Alain Desrosières, *The Politics of Large Numbers: A History of Statistical Reasoning*, trans. Camille Naish (Cambridge, MA: Harvard University Press 2002), 103; and Nancy Krieger, "The Making of Public Health Data: Paradigms, Politics and Policy," *Journal of Public Health Policy* 13, no. 4 (1992): 412–413, https://doi.org/10.2307/3342531.

5. See Margaret Lock and Vinh-Kim Nguyen, *An Anthropology of Biomedicine* (London: Wiley, 2018), 329.

6. The ways that race itself works overtime is seen in the multiple uses of racialized DNA from one domain to another. One example of people halting some aspect of this overtime was in the withdrawals of the Edmonton Police Department's use of a Parabon Snapshot likeness of a Black male face that would unfairly target a whole community of Black men. See the CBC News, "Edmonton Police Issue Apology for Controversial Use of DNA Phenotyping," CBC, October 6, 2022, www.cbc.ca/news/canada/edmonton/edmonton-police-issue-apology -for-controversial-use-of-dna-phenotyping-1.6608457. We see a similar concern in the actions of the 1921 Tulsa Massacre descendants who backtracked on a plan to share their DNA with a Utah company that offered to try to match their samples to those of a mass grave of Black 1921 victims. Their decision changed when they learned that the all-Black database could be a resource for criminal surveillance of them, their families, and their community in the here and now. See Chris Polansky, "Massacre Descendants' Group: Don't Submit Your DNA to Tulsa Graves Investigators," Public Radio Tulsa, August 11, 2022, www .publicradiotulsa.org/local-regional/2022-08-11/massacre-descendants-group -dont-submit-your-dna-to-tulsa-graves-investigators.

7. Forrest Briscoe et al., "Evolving Public Views on the Value of One's DNA and Expectations for Genomic Database Governance: Results from a National Survey," *PLOS One* 15, no. 3 (2020): 7, figure 2, https://doi.org/10.1371/journal .pone.0229044.

8. See Megan Molteni, "23andMe's Pharma Deals Have Been the Plan All Along," *Wired*, August 3, 2018, www.wired.com/story/23andme-glaxosmithkline -pharma-deal/.

9. Ron Leuty, "23andMe Grabs Former Genentech Science Boss, Will Develop Drugs," *San Francisco Business Times*, March 12, 2015, www.bizjournals.com /sanfrancisco/blog/biotech/2015/03/23andme-genentech-richard-scheller.html. Biz Carson and Kathleen Chaykowsk, "Live Long and Prosper: How Anne Wojcicki's 23andMe Will Mine Its Giant DNA Database for Health and Wealth," *Forbes*, June 6, 2019 (emphasis added), www.forbes.com/sites /bizcarson/2019/06/06/23andme-dna-test-anne-wojcicki-prevention-plans-drug -development/?sh579f09853494d.

10. See Sandra Soo-Jin Lee, "Race, Risk, and Recreation an Personal Genomics: The Limits of Play," *Medical Anthropology Quarterly* 27, no. 4 (2013), 552,

https://doi.org10.1111/maq.12059. Soo-Jin Lee writes: "Citing as a core value individual control of genetic information, 23andMe states that it strives 'to help others take a bold, informed step toward self knowledge.' Framing it with the pleasures of hobbies and entertainment, 23andMe recasts the weighty enterprise of genetic testing for disease traditionally overseen by health care professionals into a private matter between the consumer and the company."

11. Fullwiley, "DNA and Our 21st Century Ancestors."

12. Erin Shaak, "Class Action Claims Ancestry.com Violated Genetic Privacy Law by Disclosing Data in Blackstone Acquisition," classaction.org, October 29, 2021.

13. As part of the auction marketing, the company writes: "The human genome provides an exciting opportunity for an NFT, as it is a unique encoding of an individual. Each human's genome is a non-changing representation of their most fundamental and personal data, which is inherently non-fungible." That said, people's genomes have been fungible in all sorts of transactions discussed in the previous pages. See "Genomic NFT: Professor George Church," Nebula Genomics, https://nebula.org/genomic-nft/#about (accessed November 27, 2022).

14. "Not knowing has social consequences that can also be useful for those in powerful positions." Sociologist Lisa McGoey writes: "Ignorance, I suggest, is often more valuable to institutions . . . the more pervasive it becomes." Lisa McGoey, "The Logic of Strategic Ignorance," *British Journal of Sociology* 63, no. 3 (September 2012): 556, https://doi.org/10.1111/j.1468-4446.2012.01424.x.

15. African Burial Ground, National Monument New York, "History and Culture," National Park Services, last modified November 23, 2021, www.nps.gov /afbg/learn/historyculture/index.htm.

16. For a detailed analysis concerning many aspects of this story, see Alondra Nelson, *The Social Life of DNA* (New York: Basic Books, 2016).

17. For an elegant in-depth survey of African history as well as the politics of history, and the ways that it gets told in the Americas, see Howard French, *Born in Blackness: Africa, Africans, and the Making of the Modern World, 1471 to the Second World War* (New York: Norton, Liveright, 2021).

18. See Sam Ford, "DNA and African American Family Genealogy Research," in *The SAGE Encyclopedia of African Cultural Heritage in North America*, ed. Mwalimu J. Shujaa and Kenya J. Shujaa, 365 (Thousand Oaks, CA: SAGE Publications, 2015), https://dx.doi.org/10.4135/9781483346373.

19. Deadria Farmer-Paellmann, "Dr. Rick Kittles for Nobel Peace Prize," Ipetitions: Your Voice Counts, www.ipetitions.com/petition/short_titlekittlesdna; also see Deadria Farmer-Paellmann, "Black History Month Effort Supports Young Doctor for Nobel Prize," PR Web, www.prweb.com/releases/2006/02 /prweb340767.htm (accessed November 21, 2022).

20. See Amy McGuire and Sara Beskow, "DNA Data Sharing: Research Participants' Perspectives," *Genetics in Medicine* 10, no. 1 (2008): 46–53, doi:10.1097/GIM.0b013e31815f1e00 p361.

21. Sui-Lee Wee, "China Uses DNA to Track Its People, with the Help of American Expertise," *New York Times*, February 21, 2019, www.nytimes.com /2019/02/21/business/china-xinjiang-uighur-dna-thermo-fisher.html. Among other details the article reads: "Dr. Kidd's data became part of China's DNA drive. In 2014, ministry researchers published a paper describing a way for scientists to tell one ethnic group from another. It cited, as an example, the ability to distinguish Uighurs from Indians. The authors said they used 40 DNA samples taken from Uighurs in China and samples from other ethnic groups from Dr. Kidd's Yale lab."

22. See Sui-Lee Wee, "China Uses DNA to Track Its People." During a trip to China, she writes: "Dr. Kidd met Li Caixia, the chief forensic physician of the ministry's Institute of Forensic Science. The relationship deepened. In December 2014, Dr. Li arrived at Dr. Kidd's lab for an 11-month stint. She took some DNA samples back to China."

23. See "DNAPrint Genomics," International Society of Genetic Genealogy Wiki (ISOGG Wiki), January 25, 2015, https://isogg.org/wiki/DNAPrint _Genomics.

24. See Duana Fullwiley, "The Song of Power: A Fugue of Dystopia and Racial Regeneration," *HAU Journal of Ethnographic Theory* 11, no. 2 (2021): 739–747, https://doi.org/10.1086/716919; and Duana Fullwiley, "The 'Contemporary Synthesis' When Politically Inclusive Genomic Science Relies on Biological Notions of Race," *ISIS* 105, no. 4 (2014): 803–814, https://doi.org/10.1086/679427.

25. The company has several more recent services that it offers, but the forensics aspect that claims to create facial sketches, or digital mugshots, from DNA was how it started. This product continues to be one of the features the company boasts about most. See "The Snapshot DNA Phenotyping Service," Parbon NanoLabs, https://snapshot.parabon-nanolabs.com/phenotyping (accessed November 19, 2022).

26. Jocelyn Kaiser, "New Federal Rules Limit Police Searches of Family Tree DNA Databases: Privacy Experts Welcome Policy on Use of Ancestry Sites," *Science*, September 25, 2019, https://doi.org/10.1126/science.aaz6336.

27. In an essay on formal equal opportunity and racial preferences, Patricia Williams wondered what a world without preference would look like. She writes: "Standards are nothing more than standard preferences. Preferential treatment isn't inherently dirty; seeing its ubiquity, within and without racial politics, is the key to the vaults of underground freedom locked up in the idea of who one likes." See Patricia J. Williams, *The Alchemy of Race and Rights* (Cambridge, MA: Harvard University Press, 1991), 103. On the one hand, using ancestry tests in the

classroom mentioned here might upend a sense that racialized DNA, whether "African" or "European," necessarily "belongs" to those with "Black" or "white" identities, especially given the enthusiasm for unexpected genetic origins. On the other hand, the fact of possessing "European" ancestry for Black students—given the continued biases against Blackness in the United States—did not usually translate to a "privilege."

In the *New York Times* coverage of this classroom's testing, a Black student recounted that the fact of his Blackness, regardless of his genetic test, still prevented him from easily getting a cab. Furthermore, that students' DNA was also collected for forensic applications produces a form of biosociality that requires more exploration as large sets of unrelated people's DNA are linked through technologies of relatedness to bring products into being that they might not agree with or that can curb their life chances for freedom as concerns future surveillance. On "biosociality" as a foundational concept, see Paul Rabinow, *Essays on the Anthropology of Reason* (Princeton, NJ: Princeton University Press, 2021), chapter 5, 91–111.

28. See Mary Ziegler, "What Is Race?: The New Constitutional Politics of Affirmative Action," *Connecticut Law Review* 50, no. 2 (2018): 314–315, https://opencommons.uconn.edu/law_review/401. For a larger discussion about how biology and genetics can be used for various, seemingly opposite ends, see Joseph Graves, "Science Can Be Used for Liberation or Oppression," *Socialist Worker*, April 11, 2018, https://socialistworker.org/2018/04/11/science-can-be-used-for-oppression-or-liberation.

29. Ziegler, "What Is Race?," 315.

30. Ziegler, "What Is Race?"

31. Joseph L. Graves, *The Emperor's New Clothes: Biological Theories of Race at The Millennium* (New Brunswick, NJ: Rutgers University Press, 2001), 11. Graves comes from a different political side of the spectrum when he writes: "I believe that the survival of the United States as a democracy depends on the dismantling of the race concept. This dismantling requires careful examination of the tortured history of race and racism."

32. For a clear example of this, see David E. Bernstein, *Classified: The Untold Story of Racial Classification in America* (New York: Bombardier Books, 2022). For a review of the legal cases to date, see Jonathan Kahn, "The Legal Weaponization of Racialized DNA: A New Genetic Politics of Affirmative Action," *Georgetown Journal of Law and Modern Critical Race Perspectives* 13, no. 2 (2021): 187–229.

33. See Lyndon B. Johnson, "To Fulfill These Rights, Commencement Address at Howard University." *Teaching American History*, June 4, 1965. Accessed December 5, 2022. https://teachingamericanhistory.org/document/to-fulfill-these-rights-commencement-address-at-howard-university-2/. On the "diversity" concept, see Charles R. Lawrence III, "Two Views of the River: A Critique of

the Liberal Defense of Affirmative Action," *Columbia Law Review* 101 (2001): 928–975, https://scholarship.law.georgetown.edu/facpub/340; Derrick Bell, "Diversity's Distractions," *Columbia Law Review* 103, no. 6 (2003): 1622–1633, https://doi.org/10.2307/3593396; and Rachel F. Moran, "Diversity's Distractions Revisited: The Case of Latinx in Higher Education," *UC Irvine School of Law*, Research Paper no. 20 (June 2022): 579–641.

34. For Frederick Douglass's own account of his struggle to illegally learn how to read and write, see *The Complete Autobiographies of Frederick Douglass* (Radford, VA: Wilder Publications, 2008), 120.

35. See Alexandra Minna Stern, *Eugenic Nation: Faults and Frontiers of Better Breeding in Modern America*, 2nd ed. (Oakland: University of California Press, 2016), chapter 3.

36. See Jon Reyhner and Jeanne Eder, *American Indian Education: A History*, 2nd ed. (Norman: University of Oklahoma Press, 2015), chapter 7; especially, for instance, p. 203.

37. See "Personal Genetics and Law Enforcement: Improving Public Safety, Ensuring Justice, and Balancing Civil Rights," 3rd congressional briefing, PG-ED: Personal Genomics Education Project, March 19, 2015, https://pged .org/march-2015-congressional-briefing/. Also see "Congresswoman Louise Slaughter and Senator Elizabeth Warren Host Panel on Genetics and Law Enforcement: Improving Public Safety, Ensuring Justice, and Balancing Civil Rights," PG-ED Press Release, PG-ED: Personal Genomics Education Project, https://pged.org/wp-content/uploads/2016/03/pged-PressRelease.pdf (accessed December 10, 2022).

38. Hartman writes: "To what end does one conjure the ghost of slavery, if not to incite the hopes of transforming the present? Given this, I refuse to believe that the slave's most capacious claims or wildest imaginings are for back wages or debt relief. There are too many lives at peril . . ." To the question "Are we yet free?" she imagines Black people on both sides of the Atlantic answering with a "resounding no." And within that "no" resides "a promise of affiliation better than that of brothers and sisters." See Saidiya Hartman, *Lose Your Mother: A Journey along the Atlantic Slave Route* (New York: Farrar, Straus and Giroux, 2008), 170–172.

Bibliography

Abu El-Haj, Nadia. *The Genealogical Science: The Search for Jewish Origins and the Politics of Epistemology.* Chicago: Chicago University Press, 2012.

African Ancestry. "African Ancestry—Trace Your DNA. Find Your Roots. Today." Accessed June 15, 2021. https://africanancestry.com/.

African Burial Ground, National Monument New York. "History and Culture." National Park Service. Last modified November 23, 2021. www.nps.gov/afbg/learn/historyculture/index.htm.

American Cancer Society. "American Cancer Society: Cancer Facts and Figures 2022." Accessed December 4, 2022. www.cancer.org/content/dam/cancer-org/research/cancer-facts-and-statistics/annual-cancer-facts-and-figures/2022/2022-cancer-facts-and-figures.pdf.

American Society of Human Genetics. "Facing Our History—Building an Equitable Future." Accessed February 4, 2023. www.ashg.org/wpcontent/uploads/2023/01/Facing_Our_History-Building_an_Equitable_Future_Final_Report_January_2023.pdf.

Angrist, Misha. *Here Is a Human Being: The Dawn of Personal Genomics.* New York: Harper Perennial, 2011.

Anti-Defamation League. "Profile: David Duke." Accessed June 10, 2022. www.adl.org/resources/profiles/david-duke.

Apata, Mario, Bernado Arriaza, Elena Llop, and Mauricio Moraga. "Human Adaptation to Arsenic in Andean Populations of the Atacama Desert."

American Journal of Physical Anthropology 163, no. 1 (2017): 192–199. https://doi.org/10.1002/ajpa.23193.

Arbour, Laura, and Doris Cook. "DNA on Loan: Issues To Consider When Carrying Out Genetic Research with Aboriginal Families and Communities." *Community Genetics* 9, no. 3 (2006): 153–160. https://doi.org/10.1159/000092651.

Aristotle. *On the Soul and Other Psychological Works*. Translated by Fred D. Miller Jr. Oxford: Oxford University Press, 2018.

Ayoz, Kareem, Erman Ayday, and A. Ercument Cicek. "Genome Reconstruction Attacks Against Genomic Data-Sharing Beacons, Proceedings on Privacy Enhancing Technologies." *Privacy Enhancing Technologies Symposium*, no. 3 (March 2021): 28–48. https://doi.org/10.2478/popets-2021-0036.

Baldwin, James. "Stranger in the Village." In *Notes of a Native Son*, 159–175. 1955; reprint, Boston: Beacon Press, 1984.

Ball, Madeline P., Jason Bobe, Michael F. Chou, Tom Clegg, Preston W. Estep, Jeantine E. Lunshof, Ward Vandewege, Alexander Wait Zaranek, and George M. Church. "Harvard Personal Genome Project: Lessons from Participatory Public Research." *Genome Medicine* 6, no. 10 (2014): 1–7. https://doi.org/10.1186/gm527.

Barad, Karen. *Meeting the Universe Halfway: Quantum Physics and the Entanglement of Matter and Meaning*. Durham, NC: Duke University Press, 2007.

Baraka, Amiri. "Ka Ba." In *I Am the Darker Brother: An Anthology of Modern Poems by African Americans*, edited by Arnold Adoff, 38–39. New York: Simon and Schuster, 1997.

Bardill, Jessica. "Comparing Tribal Research and Specimens Policies: Models, Practices, and Principles." *International Indigenous Policy Journal* 8, no. 4 (2017): 1–22. https://doi.org/10.18584/iipj.2017.8.4.4.

Bashir, Martin. "Confessions of a Sperm Donor." *ABC-News*, August 31, 2006. https://abcnews.go.com/Nightline/Health/story?id=1982328&page=1.

Beerli, Peter. "Effect of Unsampled Populations on the Estimation of Population Sizes and Migration Rates between Sampled Populations." *Molecular Ecology* 13, no. 4 (2004): 827–836. https://doi.org/10.1111/j.1365-294x.2004.02101.x.

Berlant, Lauren. *Cruel Optimism*. Durham, NC: Duke University Press, 2011.

Bell, Derrick. "Diversity's Distractions." *Columbia Law Review* 103, no. 6 (2003): 1622–1633. https://doi.org/10.2307/3593396.

Benjamin, Ruha. *Race after Technology: Abolitionist Tools for the New Jim Code*. Cambridge: Polity Press, 2019.

———. "Racial Fictions, Biological Facts: Expanding the Sociological Imagination through Speculative Methods." *Catalyst: Feminism, Theory, Technoscience* 2, no. 2 (2016): 1–28.

Benn Torres, Jada. "Race, Rare Genetic Variants, and the Science of Human Difference in the Post-Genomic Age." *Transforming Anthropology* 27 (2019): 37–49. https://doi-org.stanford.idm.oclc.org/10.1111/traa.12144.

Benn Torres, Jada, and Gabriel A. Torres Colón. *Genetic Ancestry: Our Stories, Our Pasts.* London: Routledge, 2021.

Bergström, A., S. A. McCarthy, R. Hui, M. A. Almarri, Q. Ayub, P. Danecek, Y. Chen, S. Felkel, P. Hallast, J. Kamm, H. Blanché, J. F. Deleuze, H. Cann, S. Mallick, D. Reich, M. S. Sandhu, P. Skoglund, A. Scally, Y. Xue, R. Durbin, and C. Tyler-Smith. "Insights into Human Genetic Variation and Population History from 929 Diverse Genomes." *Science* 367, no. 6484 (March 20, 2020): eaay5012. https://doi.org/10.1126/science.aay5012. PMID: 32193295; PMCID: PMC7115999.

Bernstein, David E. *Classified: The Untold Story of Racial Classification in America.* New York: Bombardier Books, an Imprint of Post Hill Press, 2022.

Black, Derek. "Transcript: Interview with Former White Nationalist Derek Black." Interview by Michael Barbaro. *The Daily* podcast, 34:49. *New York Times*, August 22, 2017. www.nytimes.com/2017/08/22/podcasts/the-daily -transcript-derek-black.html.

Bliss, Catherine. *Race Decoded: The Genomic Fight for Social Justice.* Stanford, CA: Stanford University Press, 2012.

Bock, Eric. "Housing Segregation a Central Cause of Racial Health Inequities." *NIH Record.* https://nihrecord.nih.gov/2021/03/05/housing-segregation -central-cause-racial-health-inequities.

Borrell, Luisa N., Jennifer R. Elhawary, Elena Fuentes-Afflick, Jonathan Witonsky, Nirav Bhakta, Alan H. B. Wu, Kirsten Bibbins-Domingo, et al. "Race and Genetic Ancestry in Medicine: A Time for Reckoning with Racism." *New England Journal of Medicine* 384, no. 5 (2021): 474–480. https://doi.org/10.1056/NEJMms2029562.

Bowker, Geoffrey C. *Memory Practices in the Sciences.* Cambridge, MA: MIT Press, 2008.

Braun, Lundy. *Breathing Race into the Machine: The Surprising Career of the Spirometer from Plantation to Genetics.* Minneapolis: University of Minnesota Press, 2014.

Briggs, Charles. "Futures of Writing in Medical Anthropology." Panel presented at "Raising our Voices" during the American Anthropological Association livestream. November 14, 2020.

———. *Unlearning: Rethinking Poetics, Pandemics, and the Politics of Knowledge.* Logan: Utah State University Press, 2021.

Briscoe, Forrest, Ifeoma Ajunwa, Allison Gaddis, and Jennifer McCormick. "Evolving Public Views on the Value of One's DNA and Expectations for Genomic Database Governance: Results from a National Survey." *PLOS One* 15, no. 3 (2020): 1–10. https://doi.org/10.1371/journal.pone.0229044.

Broad Cell Stem Research Center. "Induced Pluripotent Stem Cells (iPS)." Accessed November 24, 2022. https://stemcell.ucla.edu/induced -pluripotent-stem-cells.

Brown, Kristen V. "All Those 23andMe Spit Tests Were Part of a Bigger Plan." *Bloomberg*, November 4, 2021. www.bloomberg.com/news/features /2021-11-04/23andme-to-use-dna-tests-to-make-cancer-drugs.

Bruggeman, Leslie A., John F. O'Toole, and John R. Sedor. "Identifying the Intracellular Function of APOL1." *Journal of the American Society of Nephrology* 28, no. 4 (2017): 1008–1011. https://doi.org/10.1681/ASN.2016111262.

Buck, Laura T. "Homo heidelbergensis." In *Encyclopedia of Animal Cognition and Behavior,* edited by Jennifer Vonk and Tad Shackelford, 3187–3192. Switzerland: Springer Nature, 2020. https://doi.org/10.1007/978-3-319-47829-6 _1151-1.

Burchard González, Esteban, Luisa N. Borrell, Shweta Choudhry, Mariam Naqvi, Hui-Ju Tsai, Jose R. Rodriguez-Santana, Rocio Chapela, Scott D. Rogers, Rui Mei, William Rodriguez-Cintron, et al. "Latino Populations: A Unique Opportunity for the Study of Race, Genetics, and Social Environment in Epidemiological Research." *American Journal of Public Health* 95, no. 12 (2005): 2161–2168. https://doi.org/10.2105/AJPH.2005.068668.

Bustamante, Carlos D., Francisco M. De La Vega, and Esteban G. Burchard. "Genomics for the World." *Nature* 475 (2011): 163–165. https://doi.org/10 .1038/475163a.

Callier, Shawneequa L., Perry W. Payne Jr., Deborah Akinniyi, Kaitlyn McPartland, Terry L. Richardson, Mark A. Rothstein, and Charmaine D. M. Royal. "Cardiologists' Perspectives on BiDil and the Use of Race in Drug Prescribing." *Journal of Racial and Ethnic Health Disparities* 9, no. 6 (2022): 2146–2156. https://doi.org/10.1007/s40615-021-01153-x.

Carroll, Stephanie Russo, Ibrahim Garba, Oscar L. Figueroa-Rodríguez, Jarita Holbrook, Raymond Lovett, Simeon Materechera, Mark Parsons, et al. "The CARE Principles for Indigenous Data Governance." *Data Science Journal* 19, art. no: 43 (2020): 1–12. http://doi.org/10.5334/dsj-2020-043.

Carroll, Stephanie Russo, Maui Hudson, Maggie Walker, and Per Axelsson. "International Indigenous Data Sovereignty IG." Research Data Alliance. November 2, 2022. www.rd-alliance.org/groups/international-indigenous -data-sovereignty-ig.

Carson, Biz, and Kathleen Chaykowski. "Live Long and Prosper: How Anne Wojcicki's 23andme Will Mine Its Giant DNA Database for Health and Wealth." *Forbes*, June 6, 2019. www.forbes.com/sites/bizcarson/2019 /06/06/23andme-dna-test-anne-wojcicki-prevention-plans-drug -development/?sh579f09853494d.

CBC News. "Edmonton Police Issue Apology for Controversial Use of DNA Phenotyping." CBC, October 6, 2022. www.cbc.ca/news/canada/edmonton

/edmonton-police-issue-apology-for-controversial-use-of-dna-phenotyping
-1.6608457.

Cerdeña, Jessica P., Vanessa Grubbs, and Amy L. Non. "Genomic Supremacy:
The Harm of Conflating Genetic Ancestry and Race." *Human Genomics* 16,
no. 18 (May 19, 2022): 1–5. https://doi.org/10.1186/s40246-022-00391-2.

The Charlotte Lozier Institute. "About Us, James L. Sherley, M.D., Ph.D.
Associate Scholar." Accessed November 23, 2022. https://lozierinstitute.org
/team-member/james-l-sherley/.

Church, George. "The Personal Genome Project." *Molecular Systems Biology* 1,
no. 1 (2005): 1–3. https://doi.org/10.1038/msb4100040.

Clewell, Chu Beatriz, and Jomills Henry Braddock II. "Influences on Minority
Participation in Mathematics, Science and Engineering." In *Access Denied:
Race, Ethnicity and the Scientific Enterprise*, edited by George Campbell Jr.,
Ronni Denes, and Catherine Morrison, 89–136. Oxford: Oxford University
Press, 2006.

Coop, Graham. "Genetic Similarity and Genetic Ancestry Groups." July 23,
2022. Preprint. https://doi.org/10.48550/arXiv.2207.11595.

Cooper, Richard, Jay Kaufman, and Ryk Ward. "Race and Genomics." *New
England Journal of Medicine* 348, no. 12 (March 2003): 1166–1170. https://
doi.org/10.1056/NEJMsb022863.

Coriell Institute for Medical Research. "GM21846 LCL from B-Lymphocyte."
Accessed November 23, 2022. www.coriell.org/0/Sections/Search/Sample
_Detail.aspx?Ref=GM21846&Product=CC.

———. "MPG00021 Human Variation Panel." www.coriell.org/0/Sections/Search
/Panel_Detail.aspx?PgId=761&Ref=MGP00021.

———. "1000 Genomes Project." Accessed October 22, 2022. www.coriell
.org/0/Sections/Collections/NHGRI/1000genome.aspx?PgId=664&coll
=HG.

Corrales, Eloy Martín. "The Spain That Enslaves and Expels: Moriscos and
Muslim Captives (1492 to 1767–1791)." In *Muslims in Spain, 1492–1814*,
67–94. Leiden: Brill, 2020.

David Reich Lab. "Software." Harvard University. Accessed June 15, 2021.
https://reich.hms.harvard.edu/software.

Davis, Megan. "Data and the United Nations Declarations on the Rights of
Indigenous Peoples." In *Indigenous Data Sovereignty: Toward an Agenda*,
edited by Tahu Kukutai and John Taylor, 25–38. Canberra: Australian
National University Press, 2016.

Dawson, Susan E., and Gary E. Madsen. "Psychosocial and Health Impacts of
Uranium Mining and Milling on Navajo Lands." *Health Physics* 101, no. 5
(2011): 618–625. https://doi.org/10.1097/HP.0b013e3182243a7a.

De la Cadena, Marisol. *Indigenous Mestizos: The Politics of Race and Culture in
Cuzco, Peru, 1919–1991*. Durham, NC: Duke University Press, 2000.

de Léry, Jean. *History of a Voyage to the Land of Brazil, Otherwise Called America*. Berkley: University of California Press, 1578.

Delaney, Carol. "Columbus's Ultimate Goal: Jerusalem." *Comparative Studies in Society and History* 48, no. 2 (2006): 260–292. https://doi.org/10.1017/S0010417506000119.

Desrosières, Alain. *The Politics of Large Numbers: A History of Statistical Reasoning*, translated by Camille Naish. Cambridge, MA: Harvard University Press, 2002.

Doucet-Battle, James. "Ennobling the Neanderthal: Racialized Texts and Genomic Admixture." *Kalfou: A Journal of Comparative and Relational Ethnic Studies* 5, no. 1 (2018): 61–67.

Douglass, Frederick. *The Complete Autobiographies of Frederick Douglass (An African American Heritage Book)*. Radford, VA: Wilder Publications, 2008.

Duke, David. "Race and Medicine: A Reply from David Duke to a Quote by Dr. Esteban Burchard." March 10, 2007. https://davidduke.com/race-and-medicine-a-reply-from-david-duke-to-a-quote-of-dr-esteban-burchard/.

Duncan, David Ewing. "An Interview with Kari Steffanson, M.D." *The Believer*, July 1, 2003. www.thebeliever.net/an-interview-with-kari-steffanson-m-d/.

Durvasul, Arun, and Sriram Sankararaman. "Recovering Signals of Ghost Archaic Introgression in African Populations." *Science Advances* 6, no. 7 (February 2020): 1–9. https://doi.org/10.1126/sciadv.aax5097.

Duster, Troy. *Backdoor to Eugenics*. London: Routledge, 2003.

———. "A Post-Genomic Surprise." *British Journal of Sociology* 66, no. 1 (2015): 1–27. https://doi.org/10.1111/1468-4446.12118.

Edwards, A. W. F. "Human Genetic Diversity: Lewontin's Fallacy." *BioEssays* 25, no. 8 (2003): 798–801. https://doi.org/10.1002/bies.10315.

El-Haj, Nadia Abu. *The Genealogical Science: The Search for Jewish Origins and the Politics of Epistemology*. Chicago: Chicago University Press, 2012.

Ellison, Ralph. Introduction to *Invisible Man*, xvii–xxxiv. New York: The Modern Library, 1981.

Epstein, Steven. "Bodily Differences and Collective Identities: The Politics of Gender and Race in Biomedical Research in the United States." *Body & Society* 10, no. 2–3 (2004): 183–203. https://doi.org/10.1177/1357034X04042942.

———. "Inclusion, Diversity, and Biomedical Knowledge Making: The Multiple Politics of Representation." In *How Users Matter: The Co-construction of Users and Technologies*, edited by N. Oudshoorn and T. Pinch, 173–190. Cambridge: MIT Press, 2003.

———. *Inclusion: The Politics of Difference in Medical Research*. Chicago: Chicago University Press, 2007.

Enserink, Martin. "Physicians Wary of Scheme to Pool Icelanders' Genetic Data." *Science* 281, no. 5379 (1998): 890-891. https://doi.org/10.1126/science.281.5379.890.

Erlich, Yaniv, and Arvind Narayanan. "Routes for Breaching and Protecting Genetic Privacy." *Nature Reviews Genetics* 15, no. 6 (2014): 409–421. https://doi.org/10.1038/nrg3723.

Executive Office of the President, Office of Management and Budget (OMB). "Standards for Maintaining, Collecting, and Presenting Federal Data on Race and Ethnicity." *Federal Register* 81, no. 190 (September 30, 2016): 67398–67401. www.federalregister.gov/documents/2016/09/30/2016-23672 /standards-for-maintaining-collecting-and-presenting-federal-data-on -race-and-ethnicity.

The Experiment. "What Does It Mean to Give Away Our DNA?" *The Atlantic*, October 28, 2021. Podcast, 31:22. www.theatlantic.com/podcasts/archive /2021/10/native-american-dna-navajo-ban-on-genetic-research/620510/.

Fanon, Frantz. *Black Skin, White Masks*. London: Pluto Press, 2008.

Farmer-Paellmann, Deadria. "Black History Month Effort Supports Young Doctor for Nobel Prize." PR Web. Accessed November 21, 2022. www.prweb .com/releases/2006/02/prweb340767.htm.

———. "Dr. Rick Kittles for Nobel Peace Prize." Ipetitions: Your Voice Counts. www.ipetitions.com/petition/short_titlekittlesdna.

Federal Register. "Standards for Maintaining, Collecting, and Presenting Federal Data on Race and Ethnicity." *Federal Register*. Accessed June 16, 2022. www .federalregister.gov/documents/2016/09/30/2016-23672/standards-for -maintaining-collecting-and-presenting-federal-data-on-race-and-ethnicity.

Fernández-Morera, Darío. *The Myth of the Andalusian Paradise*. Wilmington, DE: Intercollegiate Studies Institute, 2016.

Figal, Sara. "The Caucasian Slave Race: Beautiful Circassians and the Hybrid Origin of European Identity." In *Reproduction, Race, and Gender in Philosophy and the Early Life Sciences*, edited by Susanne Lettow, 163–186. Albany: SUNY Press, 2014.

Fletcher, Richard. *Moorish Spain*. Berkeley: University of California Press, 2006.

Fontaine, Theodore. Foreword to *Colonial Genocide in Indigenous America*, edited by Andrew Woolford, Jeff Benvenuto, and Alexander Hinton, vii–x. Durham, NC: Duke University Press, 2014.

Ford, Richard. *Racial Culture: A Critique*. Princeton, NJ: Princeton University Press, 2004.

Ford, Sam. "DNA and African American Family Genealogy Research." In *The SAGE Encyclopedia of African Cultural Heritage in North America*, edited by Mwalimu J. Shujaa and Kenya J. Shujaa. Thousand Oaks, CA: SAGE Publications, 2015. https://dx.doi.org/10.4135/9781483346373.

Fortun, Michael. *Promising Genomics: Iceland and DeCODE Genetics in a World of Speculation*. Berkeley: University of California Press, 2008.

Foucault, Michel. "The Subject and Power." *Critical Inquiry* 8, no. 4 (1982): 777–795. www.jstor.org/stable/1343197.

Fournié-Martinez, Christine. "Contribution à l'étude de l'esclavage en Espange au Siècle d'or: Les esclaves devant l'inquisition." Thesis, Ecole Nationale des Chartes, Paris, 1988. https://bibnum.chartes.psl.eu/s/thenca/item/58599.

Fox, Keolu. "The Illusion of Inclusion—The 'All of Us' Research Program and Indigenous Peoples' DNA." *New England Journal of Medicine* 383, no. 5 (2020): 411–413. https://doi.org/10.1056/NEJMp1915987.

Freedman, Matthew L., Christopher A Haiman, Nick Patterson, Gavin J McDonald, Arti Tandon, Alicja Waliszewska, Kathryn Penney, et al. "Admixture Mapping Identifies 8q24 As a Prostate Cancer Risk Locus in African-American Men." *Proceedings of the National Academy of Sciences of the United States of America* 103, no. 38 (2006): 14068–14073. https://doi.org/10.1073/pnas.0605832103.

French, Howard. *Born in Blackness: Africa, Africans, and the Making of the Modern World, 1471 to the Second World War.* New York: Norton, Liveright, 2021.

Friedman, David J., and Martin R. Pollak. "APOL1 Nephropathy: From Genetics to Clinical Applications." *Clinical Journal of the American Society of Nephrology* 16, no. 2 (2021): 294–303. https://doi.org/10.2215/CJN.15161219.

Fujimura, J. H., and R. Rajagopalan. "Different Differences: The Use of 'Genetic Ancestry' versus Race in Biomedical Human Genetic Research." *Social Studies of Science* 41, no. 1 (2011): 5–30. https://doi.org/10.1177/0306312710379170.

Fullwiley, Duana. "The Biologistical Construction of Race: 'Admixture' Technology and the New Genetic Medicine." *Social Studies of Science* 38, no. 5 (October 2008): 695–735. https://doi.org/10.1177/0306312708090796.

———. "The 'Contemporary Synthesis' When Politically Inclusive Genomic Science Relies on Biological Notions of Race." *ISIS* 105, no. 4 (2014): 803–814. https://doi.org/10.1086/679427.

———. "DNA and Our Twenty-first-century Ancestors." *Boston Review,* February 4, 2021. www.bostonreview.net/articles/duana-fullwiley-dna-and-our-twenty-first-century-ancestors/.

———. *The Encultured Gene: Sickle Cell Health Politics and Biological Difference in West Africa.* Princeton, NJ: Princeton University Press, 2011.

———. "Futures of Writing in Medical Anthropology: Voice and Practice (year 2)." Panel presented at the American Anthropological Association (Zoom), November 18, 2021, Maryland, MD.

———. "The Molecularization of Race: Institutionalizing Human Difference in Pharmacogenetics Practice." *Science as Culture* 16, no. 1 (2007): 1–30. https://doi.org/10.1080/09505430601180847.

———. "The Song of Power: A Fugue of Dystopia and Racial Regeneration." *HAU Journal of Ethnographic Theory* 11, no. 2 (2021): 739–747. https://doi.org/10.1086/716919.

Galanter, Joshua M., Juan Carlos Fernandez-Lopez, Christopher R. Gignoux, Jill Barnholtz-Sloan, Ceres Fernandez-Rozadilla, Marc Via, et al. "Development of a Panel of Genome-Wide Ancestry Informative Markers to Study Admixture Throughout the Americas." *PLOS Genetics* 8, no. 3 (2012): 1–16. https://doi.org/10.1371/journal.pgen.1002554.

Ganguly, Prabarna, and Rachael Zisk. "Researchers Generate the First Complete, Gapless Sequence of a Human Genome." National Human Genome Research Institute. March 31, 2022. www.genome.gov/news/news-release/researchers -generate-the-first-complete-gapless-sequence-of-a-human-genome.

Garofalo, Leo J. "The Shape of a Diaspora: The Movement of Afro-Iberians to Colonial Spanish America." In *Africans to Spanish America: Expanding the Diaspora*, edited by Sherwin K. Bryant and Rachel Sarah O'Toole, 27–49. Champaign: University of Illinois Press, 2012.

Garrison, Nanibaa A., Māui Hudson, Leah L. Ballantyne, Ibrahim Garba, Andrew Martinez, Maile Taualii, Laura Arbour, Nadine R. Caron, and Stephanie Carroll Rainie. "Genomic Research through an Indigenous Lens: Understanding the Expectations." *Annual Review of Genomics and Human Genetics* 20 (2019): 495–517. https://doi.org/10.1146/annurev-genom -083118-015434.

GeneWatch UK. "Expansions of Police DNA Databases Worldwide Urgently Need Human Rights Safeguards." Press release. February 16, 2016. www .genewatch.org/article.shtml?als%5Bcid%5D=576198&als%5Bitemid %5D=576234.

Genovese, Giulio, David J. Friedman, Michael D. Ross, Laurence Lecordier, Pierrick Uzureau, Barry I. Freedman, Donald W. Bowden, et al. "Association of Trypanolytic ApoL1 Variants with Kidney Disease in African Americans." *Science* 329, no. 5993 (2010): 841–845. https://doi.org/10.1126/science .1193032.

Gharala, Norah L. A. *Taxing Blackness: Free Afromexican Tribute in Bourbon New Spain*. Tuscaloosa: University of Alabama Press, 2019.

Gibbon, Sahra. "Translating Population Difference: The Use and Re-Use of Genetic Ancestry in Brazilian Cancer Genetics." *Medical Anthropology* 35, no. 1 (2016): 58–72. https://doi.org/10.1080/01459740.2015.1091818.

Gismondi, Melissa. "How Indigenous Scientists Are Using Biomedical Research To Seek 'Genomic Justice." *CBC Radio*, September 28, 2021. www.cbc.ca /radio/ideas/how-indigenous-scientists-are-using-biomedical-research-to -seek-genomic-justice-1.6190855.

Goetz, Thomas. "The Gene Collector." *Wired*, July 26, 2008. www.wired.com /2008/07/ff-church/.

Goodman, Alan. "Toward Genetics in an Era of Anthropology." *American Ethnologist* 34 (2007): 227–229. https://doi-org.stanford.idm.oclc.org /10.1525/ae.2007.34.2.227.

Graves, Joseph L. *The Emperor's New Clothes: Biological Theories of Race at the Millennium.* New Brunswick, NJ: Rutgers University Press, 2001.

———. "Science Can Be Used for Liberation or Oppression." *Socialist Worker,* April 11, 2018. https://socialistworker.org/2018/04/11/science-can-be-used-for-oppression-or-liberation.

Green, Eric D., Chris Gunter, Leslie G. Biesecker, Valentina Di Francesco, Carla L. Easter, Elise A. Feingold, Adam L. Felsenfeld, et al. "Strategic Vision for Improving Human Health at the Forefront of Genomics." *Nature* 586, no. 7831 (2020): 683–692. https://doi.org/10.1038/s41586-020-2817-4.

Gudmundsson, Julius, Patrick Sulem, Andrei Manolescu, Laufey T. Amundadottir, Daniel Gudbjartsson, Agnar Helgason, Thorunn Rafnar, et al. "Genome-wide Association Study Identifies a Second Prostate Cancer Susceptibility Variant at 8q24." *Nature Genetics* 39, no. 5 (2007): 631–637. https://doi.org/10.1038/ng1999.

GWAS Diversity Monitor. "Total GWAS Participants Diversity." Accessed December 3, 2022. https://gwasdiversitymonitor.com/.

Gymrek, Melissa, Amy L. McGuire, David Golan, Eran Halperin, and Yaniv Erlich. "Identifying Personal Genomes by Surname Inference." *Science* 339, no. 6117 (2013): 321–324. https://doi.org/10.1126/science.1229566.

Hammer, Michael F., August E. Woerner, Fernando L. Mendez, Joseph C. Watkins, and Jeffrey D. Wall. "Genetic Evidence for Archaic Admixture in Africa." *Proceedings of the National Academy of Sciences of the United States of America* 108, no. 37 (2011): 15123–15128. https://doi.org/10.1073/pnas.1109300108.

Haraway, Donna. "Situated Knowledges: The Science Question in Feminism and the Privilege of Partial Perspective." *Feminist Studies* 14, no. 3 (1988): 575–599.

Harmon, Amy. "Why White Supremacists Are Chugging Milk (and Why Geneticists Are Alarmed)." *New York Times,* October 17, 2018. www.nytimes.com/2018/10/17/us/white-supremacists-science-dna.html.

Harry, Debra. "The Human Genome Diversity Project and Its Implications for Indigenous Peoples." Indigenous People's Council on Biocolonialism. Accessed 22, 2022. www.ipcb.org/publications/briefing_papers/files/hgdp.html.

Hartman, Saidiya. *Lose Your Mother: A Journey along the Atlantic Slave Route.* New York: Farrar, Straus and Giroux, 2008.

Hawks, John. "Neanderthals and Covid-19, beyond the Hype: What a Gene Discovery Means To Understanding the Biology of This Pandemic." *Medium,* July 5, 2020. https://medium.johnhawks.net/neanderthals-and-covid-19-beyond-the-hype-c258dc8bc2c9.

Hayden, Ericka C. "Privacy Protections: The Genome Hacker." *Nature* 497 (May 2013): 172–174. https://doi.org/10.1038/497172a.

Heath, Deborah, Rayna Rapp, and Karen-Sue Taussig. "Genetic Citizenship." In *A Companion to the Anthropology of Politics*, edited by David Nugent and Joan Vincent, 152–167. Hoboken: Blackwell Publishing Ltd., 2007. https://doi.org/10.1002/9780470693681.ch10.

"Heterocyclic Amine." Wikipedia. Accessed April 24, 2021. https://en.wikipedia.org/wiki/Heterocyclic_amine.

Homer, Nils, Szabolcs Szelinger, Margot Redman, David Duggan, Waibhav Tembe, Jill Muehling, John V. Pearson, Dietrich A. Stephan, Stanley F. Nelson, and David W. Craig. "Resolving Individuals Contributing Trace Amounts of DNA to Highly Complex Mixtures Using High-Density SNP Genotyping Microarrays." *PloS Genetics* 4, no. 8 (August 2008): 1–9. https://doi.org/10.1371/journal.pgen.1000167.

Hoover, Joseph H., Esther Erdei, David Begay, Melissa Gonzales, NBCS Study Team, Jeffery M. Jarrett, Po-Yung Cheng, and Johnnye Lewis. "Exposure to Uranium and Co-occurring Metals among Pregnant Navajo Women." *Environmental Research* 190, art. no: 109943 (2020): 1–12. https://doi.org/10.1016/j.envres.2020.109943.

Hoppe, Travis A., Aviva Litovitz, Kristine A. Willis, Rebecca A. Meseroll, Matthew J. Perkins, Ian Hutchins, Alison F. Davis, et al. "Topic Choice Contributes to the Lower Rate of NIH Awards to African-American/Black Scientists." *Science Advances* 5, no. 10 (October 9, 2019): 1–12. https://doi.org/10.1126/sciadv.aaw7238.

Howie, Bryan, Jonathan Marchini, and Matthew Stephens. "Genotype Imputation with Thousands of Genomes." *G3 Genes/Genomes/Genetics* 1, no. 6 (2011): 457–470. https://doi.org/10.1534/g3.111.001198.

Hsieh, PingHsun, August E. Woerner, Jeffrey D. Wall, Joseph Lachance, Sarah A. Tishkoff, Ryan N. Gutenkunst, and Michael F. Hammer. "Model-based Analyses of Whole-Genome Data Reveal a Complex Evolutionary History Involving Archaic Introgression in Central African Pygmies." *Genome Research* 26, no. 3 (2016): 291–300. https://doi.org/10.1101/gr.196634.115.

HUGO. *The Human Genome Diversity (HGD) Project.* Summary document. December 31, 1993. UNT Digital Library. https://digital.library.unt.edu/ark:/67531/metadc692124/.

International HapMap Consortium. "A Haplotype Map of the Human Genome." *Nature* 437, no. 27. (2005): 1299–1320. https://doi.org/10.1038/nature04226.

———. "Integrating Ethics and Science in the International HapMap Project." *Nature Reviews Genetics* 5, no. 6 (June 2004): 467–475. https://doi.org/10.1038/nrg1351.

International Society of Genetic Genealogy Wiki (ISOGG Wiki). "DNAPrint Genomics." January 25, 2015. https://isogg.org/wiki/DNAPrint_Genomics.

"Introgression." Wikipedia. Accessed October 22, 2022. https://en.wikipedia
.org/wiki/Introgression.

Jablonski, G. Nina, and George Chaplin. "Hemispheric Difference in Human
Skin Color." *American Journal of Physical Anthropology* 107, no. 2 (1998):
221–223; discussion 223–224. https://doi.org/10.1002/
(SICI)1096-8644(199810)107:2221::AID-AJPA83.0.CO;2-X.

Jagadeesan, Anuradha, Ellen D. Gunnarsdóttir, Sunna Ebenesersdóttir, Valdis
B. Guðmundsdóttire, Elisabet Linda Thordardottir, Margrét S. Einarsdóttir,
Hákon Jónsson, et al. "Reconstructing an African Haploid Genome from the
18th Century." *Nature Genetics* 50, no. 2 (2018): 199–205. https://doi.org
/10.1038/s41588-017-0031-6.

Jain, S. Lochlann. "The WetNet: What the Oral Polio Vaccine Hypothesis
Exposes about Globalized Interspecies Fluid Bonds." *Medical Anthropology
Quarterly* 34 (2020): 504–524. https://doi.org/10.1111/maq.12587.

Johnson, Lyndon B. "To Fulfill These Rights, Commencement Address at
Howard University." *Teaching American History*, June 4, 1965. Accessed
December 5, 2022. https://teachingamericanhistory.org/document/to
-fulfill-these-rights-commencement-address-at-howard-university-2/.

Joyner, Michael J., Nigel Paneth, and John P. A. Ioannidis. "What Happens
When Underperforming Big Ideas in Research Become Entrenched?" *JAMA*
316, no. 13 (2016): 1355–1356. https://doi.org/10.1001/jama.2016.11076.

Kabo, Raymond. "Les esclaves Africains face à l'inquisition espanole: Les procès
de sorcellerie et de magie." Thesis, Université de Montpellier, 1984. www
.sudoc.fr/041209532.

Kahn, Jonathan. "Getting the Numbers Right: Statistical Mischief and Racial
Profiling in Heart Failure Research." *Perspectives in Biology and Medicine*
46, no. 4 (2003): 473–483. https://doi.org/10.1353/pbm.2003.0087.

———. "The Legal Weaponization of Racialized DNA: A New Genetic Politics of
Affirmative Action." *Georgetown Journal of Law and Modern Critical Race
Perspectives* 13, no. 2 (2021): 187–229. www.law.georgetown.edu/mcrp
-journal/in-print/volume-13-issue-2-fall-2021-2/519-2/.

———. "Mandating Race: How The USPTO Is Forcing Race into Biotech
Patents." *Nature Biotechnology* 29, no.5 (2011): 401–403. https://doi.org/10
.1038/nbt.1864.

———. "Patenting Race." *Nature Biotechnology* 24 (2006): 1349–1351. https://
doi.org/10.1038/nbt1106-1349.

———. "Revisiting Racial Patents in an Era of Precision Medicine." *Case
Western Reserve Law Review* 67, no. 4 (June 7, 2017): 1153–1169. https://
ssrn.com/abstract=2982539.

Kaiser, Jocelyn. "New Federal Rules Limit Police Searches of Family Tree DNA
Databases: Privacy Experts Welcome Policy on Use of Ancestry Sites."
Science, September 25, 2019. https://doi.org/10.1126/science.aaz6336.

Kampourakis, Kostas. *Ancestry Re-imagined: Dismantling the Myth of Genetic Ethnicities*. New York: Oxford University Press, 2023.

Kampourakis, Kostas, and Erik L Peterson. "The Racist Origins, Racialist Connotations, and Purity Assumptions of the Concept of 'Admixture' in Human Evolutionary Genetics." *Genetics* 223, no. 3 (March 2023): 1–7. https://doi.org/10.1093/genetics/iyad002.

Karadimas, Charalampos L., Tuan H. Vu, Stephen A. Holve, Penelope Chronopoulou, Catarina Quinzii, Stanley D. Johnsen, Janice Kurth, et al. "Navajo Neurohepatopathy Is Caused by a Mutation in the MPV17 Gene." *American Journal of Human Genetics* 79, no. 3 (2006): 544–548. https://doi.org/10.1086/506913.

Keel, Terence. *Divine Variations: How Christian Thought Became Racial Science*. Stanford, CA: Stanford University Press, 2018.

Kittles, Rick A., and Kenneth M. Weiss. "Race, Ancestry, and Genes: Implications for Defining Disease Risk." *Annual Review of Genomics and Human Genetics* 4, no. 1 (2003): 33–67. https://doi.org/10.1146/annurev.genom.4.070802.110356.

Knorr Cetina, Karen. *Epistemic Cultures*. Cambridge, MA: Harvard University Press, 1999.

Kolata, Gina. "Targeting the Uneven Burden of Kidney Disease on Black Americans." *New York Times*, May 17, 2022. www.nytimes.com/2022/05/17/health/kidney-disease-black-americans.html.

Krieger, Nancy. "The Making of Public Health Data: Paradigms, Politics and Policy." *Journal of Public Health Policy* 13, no. 4 (1992): 412–4217. https://doi.org/10.2307/3342531.

———. "Stormy Weather: Race, Gene Expression, and the Science of Health Disparities." *American Journal of Public Health* 95 (2005): 2155–2160. https://doi.org/10.2105/AJPH.2005.067108.

———. "Structural Racism, Health Inequities, and the Two-Edged Sword of Data: Structural Problems Require Structural Solutions." *Frontiers in Public Health* 9, art. no. 655447 (April 2021): 1–10. https://doi.org/10.3389/fpubh.2021.655447.

Krieger, Nancy, Jaquelyn L. Jahn, and Pamela D. Waterman. "Jim Crow and Estrogen-Receptor-Negative Breast Cancer: US-born Black and White Non-Hispanic Women, 1992–2012." *Cancer Causes & Control* 28, no. 1 (2017): 49–59. https://doi.org/10.1007/s10552-016-0834-2.

Kroll-Zaidi, Rafil. "Your DNA Test Could Send a Relative to Jail." *New York Times*, December 27, 2021. www.nytimes.com/2021/12/27/magazine/dna-test-crime-identification-genome.html.

Lachance, Joseph, Benjamin Vernot, Clara C. Elbers, Bart Ferwerda, Alain Froment, Jean-Marie Bodo, Godfrey Lema, et al. "Evolutionary History and Adaptation from High-coverage Whole-Genome Sequences of Diverse

African Hunter-Gatherers." *Cell* 150, no. 3 (2012): 457–469. https://doi.org /10.1016/j.cell.2012.07.009.

Lamberth, Royce C. "Memorandum Opinion." Civ. No. 1:09-cv-1575 (RCL), United States District Court for the District of Columbia, August 23, 2010. https://ecf.dcd.uscourts.gov/cgi-bin/show_public_doc?2009cv1575-44.

Langefeld, Carl D., Mary E. Comeau, Maggie C. Y. Ng, Meijian Guan, Latchezar Dimitrov, Poorva Mudgal, Mitzie H. Spainhour, et al. "Genome-wide Association Studies Suggest That APOL1-Environment Interactions More Likely Trigger Kidney Disease in African Americans with Nondiabetic Nephropathy Than Strong APOL1-Second Gene Interactions." *Kidney International* 94, no. 3 (2018): 599–607. https://doi.org/10.1016/j.kint.2018.03.017.

The Late Show with Stephen Colbert. "White Supremacists, You Won't Like Your DNA Results." Published on October 17, 2017. YouTube video, 05:27. https://youtu.be/yvS2gjMMXBQ.

Latour, Bruno. *We Have Never Been Modern*. Cambridge, MA: Harvard University Press, 1993.

Lawrence, Charles R., III. "Two Views of the River: A Critique of the Liberal Defense of Affirmative Action." *Columbia Law Review* 101 (2001): 928–975. https://scholarship.law.georgetown.edu/facpub/340.

Lee, Sandra Soo-Jin. "Race, Risk, and Recreation in Personal Genomics: The Limits of Play." *Medical Anthropology Quarterly* 27, no. 4 (2013): 550–569. https://doi.org10.1111/maq.12059.

Leuty, Ron. "23andMe Grabs Former Genentech Science Boss, Will Develop Drugs." *San Francisco Business Times*, March 12, 2015. www.bizjournals .com/sanfrancisco/blog/biotech/2015/03/23andme-genentech-richard -scheller.html.

Lewontin, C. Richard. "The Apportionment of Human Diversity." In *Evolutionary Biology*, edited by T. Dobzhansky, M. K. Hecht, and W. C. Steere, 381–398. New York: Springer, 1972. https://doi.org/10.1007/978-1-4684 -9063-3_14.

Lewontin, Richard. *Biology as Ideology*. New York: Harper Perennial, 1993.

Lock, Margaret. "The Alienation of Body Tissue and the Biopolitics of Immortalized Cell Lines." *Body & Society* 7, no. 2–3 (2001) 63–91. https://doi.org /10.1177/1357034X0100700204.

———. "The Epigenome and Nature/Nurture Reunification: A Challenge for Anthropology." *Medical Anthropology* 32, no. 4 (2013): 291–308. https:// doi.org/10.1080/01459740.2012.746973.

Lock, Margaret, and Vinh-Kim Nguyen. *An Anthropology of Biomedicine*. London: Wiley, 2018.

Lock, Susannah F. "Meet my Genome: 10 People Release Their DNA on the Web." *Scientific American*, October 21, 2008. https://blogs.scientificamerican .com/news-blog/meet-my-genome-10-people-release-th-2008-10-21/.

Locke, John. *The Works, vol. 1 An Essay concerning Human Understanding Part 1*. London: Rivington, 1689.

López-Durán, Fabiola. *Eugenics in the Garden: Transatlantic Architecture and the Crafting of Modernity*. Austin: University of Texas Press, 2018.

Lowney, Chris. *A Vanished World: Muslims, Christians and Jews in Medieval Spain*. Oxford: Oxford University Press, 2005.

Lu, Chen, Aaron B. Wolf, Wenqing Fu, Liming Li, and Joshua M. Akey. "Identifying and Interpreting Apparent Neanderthal Ancestry in African Individuals." *Cell* 180, no. 4 (January 2020): 677–687.e16. https://doi.org/10.1016/j.cell.2020.01.012.

Lunshof, E. Jeantine, Ruth Chadwick, Daniel B. Vorhaus, and George M. Church. "From Genetic Privacy to Open Consent." *Nature Reviews Genetics* 9, no. 5 (2008): 406–411. https://doi.org/10.1038/nrg2360.

Lunshof, Jeantine E., Jason Bobe, John Aach, Misha Angrist, Joseph V. Thakuria, Daniel B. Vorhaus, Margret R. Hoehe, and George M. Church. "Personal Genomes in Progress: From the Human Genome Project to the Personal Genome Project." *Dialogues in Clinical Neuroscience* 12, no. 1 (2010): 47–60. https://doi.org/10.31887/DCNS.2010.12.1/jlunshof.

M'charek, Amade. *The Human Genome Diversity Project: An Ethnography of Scientific Practice*. Cambridge, UK: Cambridge University Press, 2005.

Marks, Jonathan M. "Ten Facts about Human Variation." In *Human Evolutionary Biology*, edited by M. P. Muehlenbein, 265–276. Cambridge, UK: Cambridge University Press, 2010.

Martin, Emily. *Experiments of the Mind: From the Cognitive Psychology Lab to the World of Facebook and Twitter*. Princeton, NJ: Princeton University Press, 2022.

Martinez, María Elena. *Genealogical Fictions: Limpieza de Sangre, Religion and Gender in Colonial Mexico*. Stanford, CA: Stanford University Press, 2008.

Marx, Karl. "The German Ideology." In *The Marx-Engels Reader*, edited by Robert C. Tucker, 146–200. New York: Norton, 1978.

Mason, Katherine A. *Infectious Change: Reinventing Chinese Public Health after an Epidemic*. Stanford, CA: Stanford University Press, 2016.

Mathieson, Iain, and Aylwyn Scally. "What Is Ancestry?" *PLoS Genetics* 16, no. 3 (March 9, 2020): e1008624. https://doi.org/10.1371/journal.pgen.1008624.

Matory, J. Lorand. "The English Professors of Brazil: On the Diasporic Roots of the Yoruba Nation." *Comparative Studies in Society and History* 41, no. 1 (1999): 72–103. www.jstor.org/stable/179249.

Max-Planck-Gesellschaft. "Neandertal Mother, Denisovan Father!" August 22, 2018. www.mpg.de/12208106/neandertals-denisovans-daughter.

Mayo Clinic. "Prostate Cancer." Accessed July 2, 2022. www.mayoclinic.org/diseases-conditions/prostate-cancer/symptoms-causes/syc-20353087.

———. "Saint John's Wort." Accessed June 21, 2022. www.mayoclinic.org/drugs -supplements-st-johns-wort/art-20362212.

McGoey, Lisa. "The Logic of Strategic Ignorance." *British Journal of Sociology* 63, no. 3 (September 2012): 533–576. https://doi.org/10.1111/j.1468-4446 .2012.01424.x.

McGuire, Amy L., and Laura M. Beskow. "Informed Consent in Genomics and Genetic Research." *Annual Review of Genomics and Human Genetics* 11 (2010): 361–381. https://doi.org/10.1146/annurev-genom-082509-141711.

Menocal, María Rosa. *Ornament of the World: How Muslims, Jews and Christians Created a Culture of Tolerance in Medieval Spain.* New York: Back Bay Books, Little Brown and Co., 2002.

Meyer, Matthias, Martin Kircher, Marie-Theres Gansauge, Heng Li, Fernando Racimo, Swapan Mallick, Joshua G. Schraiber, et al. "A High-Coverage Genome Sequence from an Archaic Denisovan Individual." *Science* 338, no. 6104 (August 2012): 222–226. https://doi.org/10.1126/science.1224344.

Mills, Melinda C., and Charles Rahal. "The GWAS Diversity Monitor Tracks Diversity by Disease in Real Time." *Nature Genetics* 52 (2020): 242–243. https://doi.org/10.1038/s41588-020-0580-y.

———. "A Scientometric Review of Genome-Wide Association Studies." *Communications Biology* 2, art. no. 9 (2019): 1–11. https://doi.org/10.1038 /s42003-018-0261-x.

Molteni, Megan. "Buffalo Shooting Ignites a Debate over the Role of Genetics Researchers in White Supremacist Ideology." *Stat News*, May 23, 2022. www.statnews.com/2022/05/23/buffalo-shooting-ignites-debate-genetics -researchers-in-white-supremacist-ideology/.

———. "23andMe's Pharma Deals Have Been the Plan All Along." *Wired*, August 3, 2018. www.wired.com/story/23andme-glaxosmithkline-pharma-deal/.

Montoya, Michael. *Making the Mexican Diabetic: Race, Science, and the Genetics of Inequality.* Berkeley: University of California Press, 2011.

Moran, Rachel F. "Diversity's Distractions Revisited: The Case of Latinx in Higher Education." *UC Irvine School of Law*, no. 20 (June 2022): 579–641. https://ssrn.com/abstract=4137816.

Muhammad, Elijah. *Message to the Blackman in America.* Phoenix, AZ: Secretarius MEMPS, 1973.

Museo Nacional del Virreinato. "Cuadro de Castas." Accessed May 30, 2023. https://lugares.inah.gob.mx/es/museos-inah/museo/museo-piezas/8409 -8409-10-241348-cuadro-de-castas.html?lugar_id=475.

NASEM Health and Medicine. "Welcome&Session 1." Uploaded on April 14, 2022. YouTube video, 2:01:35. www.youtube.com/watch?v=ObtrFNlydSg& list=PLGTMA6QkejfjFG77TzlbT_-ieI3E7BwPH&index=1.

Nash, Catherine. *Genetic Geographies: The Trouble with Ancestry.* Minneapolis: University of Minnesota Press, 2015.

National Academies. "Committee on Use of Race, Ethnicity, and Ancestry as Population Descriptors in Genomics Research (Meeting 2 and Public Workshop)." Accessed April 10, 2022. www.nationalacademies.org/event/04-04 -2022/committee-on-use-of-race-ethnicity-and-ancestry-as-population -descriptors-in-genomics-research-meeting-2-and-public-workshop.

National Academies of Sciences, Engineering, and Medicine. 2023. *Using Population Descriptors in Genetics and Genomics Research: A New Framework for an Evolving Field.* Washington, DC: National Academies Press. https://doi.org/10.17226/26902.

National Archives and Records. "Executive Order 13505—Removing Barriers to Responsible Scientific Research Involving Human Stem Cells, Memorandum of March 9, 2009." *Federal Register* 74, no. 46 (2009): 10667–10668.

National Human Genome Research Institute. "Complete Neanderthal Genome Sequenced." Accessed October 23, 2022. www.genome. gov/27539119/2010-release-complete-neanderthal-genome-sequenced.

———. "Human Heredity and Health in Africa Announced in London." Accessed February 20, 2012. www.genome.gov/27539880/human -heredity-and-health-in-africa-announced-in-london.

———. "Whose DNA Was Sequenced?" The Genome Project. Accessed April 25, 2022. www.genome.gov/human-genome-project/Completion-FAQ.

NativeBioData. "Research for Natives, by Natives." Accessed April 25, 2022. https://nativebio.org/.

Nebula Genomics. "Genomic NFT: Professor George Church." Accessed April 25, 2022. https://nebula.org/genomic-nft/#about.

Nelson, Alondra. *The Social Life of DNA.* New York: Basic Books, 2016.

Ness, Marilyn. "GENOME: The Future Is Now WEBISODE 1." Uploaded on July 29,2009. YouTube video, 07:21. www.youtube.com/watch?v=mVZI7NBgcWM.

———. "GENOME: The Future Is Now WEBISODE 2." Uploaded on August 6, 2009. YouTube video, 06:06. www.youtube.com/watch?v=2r9DpthvNKM.

———. "GENOME: The Future Is Now WEBISODE 3." Uploaded on August 13, 2009. YouTube video, 06:53. www.youtube.com/watch?v=mgXAO8pv-X4.

NIH *All of Us* Research Program. "*All of Us* Research Program Tribal Consultation Final Report." March 2021. https://allofus.nih.gov/all-us-research -program-tribal-consultation-final-report.

———. "*All of Us* Tribal Leader and Urban Indian Organization Letter." May 24, 2019. https://allofus.nih.gov/about/tribal-engagement/all-us-tribal-leader -and-urban-indian-organization-letter.

NIH Grants and Funding. "Policy and Guidelines on The Inclusion of Women and Minorities as Subjects in Clinical Research." Accessed June 10, 2022. https:// grants.nih.gov/policy/inclusion/women-and-minorities/guidelines.htm.

Nurk, Sergey, Sergey Koren, Arang Rhie, Mikko Rautiainen, Andrey V. Bzikadze, Alla Mikheenko, Mitchell R. Vollger, Nicolas Altemose, Lev Uralsky, and

Ariel Gershmana. "The Complete Sequence of a Human Genome." *Science* 376, no. 6588 (March 2022): 44–53. https://doi.org/10.1126/science .abj6987.

Ogundudiran, Akinwumi. *The Yorúbà: A New History*. Bloomington: Indiana University Press, 2020.

Ong, Aihwa. *Fungible Life: Experiment in the Asian City of Life*. Durham, NC: Duke University Press, 2016.

Oni-Orisan, Akinyemi, Yusuph Mavura, Yambazi Banda, Timothy A. Thornton, and Ronnie Sebro. "Embracing Genetic Diversity to Improve Black Health." *New England Journal of Medicine* 384, no. 12 (February 2021): 1163–1167. https://doi.org/10.1056/NEJMms2031080.

Online Ethics Center for Engineering and Science. "Case: Big Data & Genetic Privacy: Re-identification of Anonymized Data." Accessed November 29, 2022. https://onlineethics.org/cases/big-data-life-sciences-collection /case-big-data-genetic-privacy-re-identification-anonymized.

Otele, Olivette. *African Europeans: An Untold History*. New York: Basic Books, 2021.

Palmer, Meredith. "Rendering Settler Landscapes: Race and Property in the Empire State." *Environment and Planning D: Society and Space* 38, no. 5 (2020): 793–810.

Pálsson, Gísli. *The Man Who Stole Himself: The Slave Odyssey of Hans Jonathan*. Chicago: University of Chicago Press, 2016.

Panofsky, Aaron, and Joan Donovan. "Genetic Ancestry Testing among White Nationalists: From Identity Repair to Citizen Science." *Social Studies of Science* 49, no. 5 (2019): 653–681. https://doi.org/10.1177/0306312719861434.

Parbon NanoLabs. "The Snapshot DNA Phenotyping Service." Accessed November 19, 2022. https://snapshot.parabon-nanolabs.com/phenotyping.

Pasternak, Judy. *Yellow Dirt: A Poisoned Land and the Betrayal of the Navajos*. New York: Free Press, 2011.

PBS. "Faces of America | A Piece of the Pie." *Faces of America*. Video, 4:18. https://ca.pbslearningmedia.org/resource/foa10.sci.living.gen.piecepie/faces -of-america-a-piece-of-the-pie/.

———. *The Ornament of the World*. Aired December 17, 2019. Video, 1:55:31. www.pbs.org/show/ornament-world/.

Personal Genome Project. "Participant Profiles." Accessed November 23, 2022. https://my.pgp-hms.org/users.

———. "Public Profile—hu604D39." Accessed November 23, 2022. https://my .pgp-hms.org/profile_public?hex=hu604D39.

Pfaff, Carrie L., Jill Barnholtz-Sloan, Jennifer K. Wagner, and Jeffrey C. Long. "Information on Ancestry from Genetic Markers." *Genetic Epidemiology* 26 (2004): 305–315. https://doi.org/10.1002/gepi.10319.

PG-ED: Personal Genomics Education Project. "Congresswoman Louise Slaughter and Senator Elizabeth Warren Host Panel on Genetics and Law Enforcement: Improving Public Safety, Ensuring Justice, and Balancing Civil Right." Press release. Accessed December 10, 2022. https://pged.org/wp-content/uploads/2016/03/pged-PressRelease.pdf.

———. "Personal Genetics and Law Enforcement: Improving Public Safety, Ensuring Justice, and Balancing Civil Rights." 3rd Congressional Briefing. March 19, 2015. https://pged.org/march-2015-congressional-briefing/.

———. "What We Do." Accessed November 21, 2022. www.transvection.org/pged.

Picture, Bill. "First: Do No Harm?" Asianweek.com. Accessed August 1, 2008. http://news.asianweek.com/news/view_article.html?article_id=1352b889890ad56828d852d98d0d50e4.

Pinker, Steven. "My Genome, My Self." *New York Times*, January 7, 2009. www.nytimes.com/2009/01/11/magazine/11Genome-t.html.

Polansky, Chris. "Massacre Descendants' Group: Don't Submit Your DNA to Tulsa Graves Investigators." *Public Radio Tulsa*, August 11, 2022. www.publicradiotulsa.org/local-regional/2022-08-11/massacre-descendants-group-dont-submit-your-dna-to-tulsa-graves-investigators.

Povinelli, Elizabeth A. *Between Gaia and Ground: Four Axioms of Existence and the Ancestral Catastrophe of Late Liberalism*. Durham, NC: Duke University Press, 2021.

Powell, Kendall. "The Broken Promise That Undermines Human Genome Research." *Nature* 590, no. 7845 (2021): 198–220. https://doi.org/10.1038/d41586-021-00331-5.

Rabinow, Paul. *Essays on the Anthropology of Reason*. Princeton, NJ: Princeton University Press, 2021.

Rajan, Kaushik Sunder. *Biocapital: The Constitution of Postgenomic Life*. Durham, NC: Duke University Press, 2006.

Ramya, Rajagopalan, and Joan H. Fujimura. "Variations on a Chip: Technologies of Difference in Human Genetics Research." *Journal of the History of Biology* 51, no. 4 (2018): 841–873. https://doi.org/10.1007/s10739-018-9543-x.

Rawla, Prashanth. "Epidemiology of Prostate Cancer." *World Journal of Oncology* 10, no. 2 (2019): 63–89. https://doi.org/10.14740/wjon1191.

Reardon, Jenny. *The Postgenomic Condition: Ethics, Justice and Knowledge after the Genome*. Chicago: Chicago University Press, 2017.

———. *Race to the Finish: Identity and Governance in an Age of Genomics*. Princeton, NJ: Princeton University Press, 2005.

Relethford, John H., and Deborah Ann Bolnick. *Reflections of Our Past: How Human History is Revealed in Our Genes*. New York: Routledge, 2018. Kindle edition.

Research and Markets. "China Omeprazole Market Report, 2018–2022: Sales, Competition, Manufacturers, Prices & Market Prospects." *PR Newswire.* Accessed June 21, 2022. www.prnewswire.com/news-releases/china-omeprazole-market-report-2018-2022-sales-competition-manufacturers-prices—market-prospects-300743001.html.

Reyhner, Jon, and Jeanne Eder. *American Indian Education: A History.* 2nd ed. Norman: University of Oklahoma Press, 2015.

Richard, Green E., Johannes Krause, Adrian W. Briggs, Tomislav Maricic, Udo Stenzel, Martin Kircher, Nick Patterson, Heng Li, Weiwei Zhai, and Markus Hsi-Yang Fritz. "A Draft Sequence of the Neandertal Genome." *Science* 328, no. 5979 (2010): 710–722. https://doi.org/10.1126/science.1188021.

Risch, Neil, Esteban Burchard, Elad Ziv, and Hua Tang. "Categorization of Humans in Biomedical Research: Genes, Race and Disease." *Genome Biology* 3, no. 7 (2002): 1–12. https://doi.org/10.1186/gb-2002-3-7-comment2007.

Robbins, Christiane, Jada Benn Torres, Stanley Hooker, Carolina Bonilla, Wenndy Hernandez, Angela Candreva, Chiledum Ahaghotu, Rick Kittles, and John Carpten. "Confirmation Study of Prostate Cancer Risk Variants at 8q24 in African Americans Identifies a Novel Risk Locus." *Genome Research* 17, no. 12 (2007): 1717–1722. https://doi.org/10.1101/gr.6782707.

Roberts, Dorothy. *Fatal Invention: How Science, Politics and Big Business Re-create Race in the Twenty-first Century.* New York: New Press, 2012.

Schoofs, Mark. "Advance in Quest for HIV Vaccine." *Wall Street Journal,* July 9, 2010. www.wsj.com/articles/SB10001424052748703609004575355072271264394.

Schumacher, Fredrick R., Heather Spencer Feigelson, David G. Cox, Christopher A. Haiman, Demetrius Albanes, Julie Buring, Eugenia E. Calle, et al. "A Common 8q24 Variant in Prostate and Breast Cancer from a Large Nested Case-Control Study." *Cancer Research* 67, no. 7 (2007): 2951–2956. https://doi.org/10.1158/0008-5472.CAN-06-3591.

Schwartz, Oscar. "A Geneticist's Dilemma." *Washington Post,* November, 23, 2020. www.washingtonpost.com/magazine/2020/11/23/many-scientists-believe-that-dna-holds-cure-disease-that-poses-problem-some-native-americans/.

Shaak, Erin. "Class Action Claims Ancestry.com Violated Genetic Privacy Law by Disclosing Data in Blackstone Acquisition." Classaction.org. October 29, 2021. www.classaction.org/news/class-action-claims-ancestry.com-violated-genetic-privacy-law-by-disclosing-data-in-blackstone-acquisition.

Sherley, James L. "Including the Excluded: Concepts for Successful Integration of the U.S. Academy." In *Racism: Global Perspectives, Coping Strategies and Social Implications,* edited by Tracey Lowell, 37–46. Hauppauge, NY: Nova

Science Publishers. https://novapublishers.com/wp-content/uploads/2019/04/Concepts-for-Successful-Integration-of-the-US-Academy.pdf.

Shreeve, James. *The Genome War: How Craig Venter Tried to Capture the Code of Life and Save the World*. New York: Knopf, 2004.

Simpson, Audra. "The Ruse of Consent and the Anatomy of 'Refusal': Cases from Indigenous North America and Australia." *Postcolonial Studies* 20, no. 1 (2017): 18–33. https://doi.org/10.1080/13688790.2017.1334283.

SING consortium. "Past Workshops." Summer Internship for INdigenous Peoples in Genomics. Accessed September 23, 2023. www.singconsortium.org/past-workshops/.

Singer, Emily. "The Genome Pioneers: Early Adopters of Personal Genome Sequencing Gather to Reflect on What We Need to Do to Move the Field Forward." *MIT Technology Review*, April 30, 2010. www.technologyreview.com/2010/04/30/204008/the-genome-pioneers/.

Skloot, Rebecca. *The Immortal Life of Henrietta Lacks*. New York: Random House, 2010.

Slon, Viviane, Fabrizio Mafessoni, Benjamin Vernot, Cesare de Filippo, Steffi Grote, Bence Viola, Mateja Hajdinjak, Stéphane Peyrégne, Sarah Nagel, and Samantha Brown. "The Genome of the Offspring of a Neanderthal Mother and a Denisovan Father." *Nature* 561, no. 7721 (2018): 113–116. https://doi.org/10.1038/s41586-018-0455-x.

Smedley, Audrey. "'Race' and the Construction of Human Identity." *American Anthropologist* 100, no. 3 (1998): 690–702. www.jstor.org/stable/682047.

———. *Race in North America: Origin and Evolution of a Worldview*. 4th ed. Boulder: Routledge, 2011. https://search-ebscohostcom.stanford.idm.oclc.org/login.aspx?direct=true&db=nlebk&AN=421158&site=ehost-live.

Snipp, Matthew C. "What Does Data Sovereignty Imply: What Does It Look Like?" In *Indigenous Data Sovereignty: Toward an Agenda*, edited by Tahu Kukutai and John Taylor, 39–55. Canberra: Australian National University Press, 2016.

Spillers, Hortense. *Black, White and Color: Essays on American Literature and Culture*. Chicago: University of Chicago Press, 2003.

Spinazzola, Antonella, Valeria Massa, Michio Hirano, and Massimo Zeviani. "Lack of Founder Effect for an Identical mtDNA Depletion Syndrome (MDS)-Associated MPV17 Mutation Shared by Navajos and Italians." *Neuromuscular Disorders: NMD* 18, no. 4 (2008): 315–318. https://doi.org/10.1016/j.nmd.2007.12.007.

Starkman, Evan. "What Are HeLa Cells?." WebMD. January 22, 2022. www.webmd.com/cancer/cervical-cancer/hela-cells-cervical-cancer.

Stepan, Nancy Leys. *The Hour of Eugenics: Race, Gender, and Nation in Latin America.* Ithaca, NY: Cornell University Press, 1996.

Stern, Alexandra Minna. *Eugenic Nation: Faults and Frontiers of Better Breeding in Modern America.* 2nd ed. Oakland: University of California Press, 2016.

Sweeney, Latanya, Akua Abu, and Julia Winn. "Identifying Participants in the Personal Genome Project by Name." Harvard University. 2013. https://privacytools.seas.harvard.edu/files/privacytools/files/1021-1.pdf.

TallBear, Kim. "Genomic Articulations of Indigeneity." *Social Studies of Science* 43, no. 4 (2013): 509–533. https://doi.org/10.1177/0306312713483893.

———. *Native American DNA: Tribal Belonging and the False Promise of Genetic Science.* Minneapolis: University of Minnesota Press, 2013.

Tamarkin, Noah. "Forensics and Fortification in South African Self-captivity." *History and Anthropology* 30, no. 5 (2019): 521–526. https://doi.org/10.1080/02757206.2019.1638774.

———. *Genetic Afterlives: Black Jewish Indigeneity in South Africa.* Durham, NC: Duke University Press, 2020.

Tanner, Adam. "Harvard Professor Re-Identifies Anonymous Volunteers in DNA Study." *Forbes,* April 25, 2013. www.forbes.com/sites/adamtanner/2013/04/25/harvard-professor-re-identifies-anonymous-volunteers-in-dna-study/?sh=383d932092c.

Tate, Sarah K., and David B. Goldstein. "Will Tomorrow's Medicines Work for Everyone?" *Nature Genetics* 36, Suppl. 11 (2004): S34–42. https://doi.org/10.1038/ng1437.

Tian, Chao, David Hinds, Russel Shigeta, Rick Kittles, Denise Ballinger, and Michael F. Seldin. "A Genomewide Single-Nucleotide-Polymorphism Panel with High Ancestry Information for African American Admixture Mapping." *American Journal of Human Genetics* 79, no. 4 (2006): 640–649. https://doi.org/10.1086/507954.

Tripp, Simon, and Martin Grueber. "The Economic Impact and Functional Applications of Human Genetics and Genomics." TEConomy Partners LLC. May 2021. www.ashg.org/wp-content/uploads/2021/05/ASHG-TEConomy-Impact-Report-Final.pdf.

Tsosie, Krystal S., Joe Yracheta, and Donna Dickenson. "Overvaluing Individual Consent Ignores Risks to Tribal Participants." *Nature Reviews Genetics* 20, no. 9 (2019): 497–498. https://doi.org/10.1038/s41576-019-0161-z.

Tsosie, Krystal S., Keolu Fox, and Joseph M. Yracheta. "Genomics Data: The Broken Promise Is to Indigenous People." *Nature* 591 (2021): 529. https://doi.org/10.1038/d41586-021-00758-w.

Tsosie, Rebecca. "Tribal Data Governance and Informational Privacy: Constructing 'Indigenous Data Sovereignty.'" *Montana Law Review* 80, no. 2 (2019): 229–268.

23andMe Blog. "Celebrate Your Ancient DNA with a New Neanderthal Report." Accessed October 22, 2022. https://blog.23andme.com/ancestry-reports /new-neanderthal-report/.

Twinam, Ann. *Purchasing Whiteness: Pardos, Mulattos, and the Quest for Social Mobility in the Spanish Indies.* Stanford, CA: Stanford University Press, 2015.

US Congress. 117th Congress. "S.4119—RECA Extension Act of 2022." Accessed November 25, 2022. www.congress.gov/bill/117th-congress /senate-bill/4119.

US Food and Drug Administration. "Grapefruit Juice and Some Drugs Don't Mix." Accessed June 21, 2022. www.fda.gov/consumers/consumer-updates /grapefruit-juice-and-some-drugs-dont-mix.

UC San Diego. "Keolu Fox." Department of Anthropology. Accessed November 16, 2022. https://anthropology.ucsd.edu/people/faculty/faculty-profiles /Keolu%20Fox.html.

University of California TV. "Sriram Sankararaman: Recovering Signals of Ghost Archaic Introgression in African Populations." Recorded on February 21, 2020. YouTube video, 20:09. www.youtube.com/watch?v=vXOyVG-LNoA.

University of Hawaii Cancer Center. "Composition of the Cohort." The Multiethnic Cohort Study. Accessed December 2, 2021. www.uhcancercenter.org /for-researchers/mec-cohort-composition.

US National Library of Medicine. "Study Of Pharmacogenetics in Ethnically Diverse Populations (SOPHIE Study) (SOPHIE)." ClinicalTrials.gov . Accessed June 21, 2022. https://clinicaltrials.gov/ct2/show/NCT00187668.

Vaughn, Emily. "What's Behind the Research Funding Gap For Black Scientists?" NPR, Accessed May 30, 2022. www.npr.org/sections/health-shots /2019/10/18/768690216/whats-behind-the-research-funding-gap-for-black -scientists.

Vespucci, Amerigo. *Mundus Novus: Letter to Lorenzo Pietro di Medici* [1504]. Translated by George Tyler Northrup. Princeton, NJ: Princeton University Press, 1916.

Vinson III, Ben. *Before Mestizaje: The Frontiers of Race and Caste in Colonial Mexico.* New York: Cambridge University Press, 2018.

Vuong, Ocean. *On Earth We're Briefly Gorgeous.* New York: Penguin, 2019.

Wade, Nicholas. "A Genomic Treasure Hunt May Be Striking Gold." *New York Times*, June 8, 2002. www.nytimes.com/2002/06/18/science/a-genomic -treasure-hunt-may-be-striking-gold.html.

Wade, Peter. *Mestizo Genomics: Race Mixture, Nation, and Science in Latin America.* Durham, NC: Duke University Press, 2014.

Wadman, Meredith. "High Court Ensures Continued US Funding of Human Embryonic-Stem-Cell Research." *Nature* (2013). https://doi.org/10.1038/nature.2013.12171.

Walsh, Stephen, and Michal Seaman. "Broadly Neutralizing Antibodies for HIV-1 Prevention." *Frontiers in Immunology* 12 (2021), Sec. Vaccines and Molecular Therapeutics. https://doi.org/10.3389/fimmu.2021.712122.

Weber, Max. *The Protestant Ethic and the Spirit of Capitalism*. Oxford: Oxford University Press, 2010.

Wee, Sui-Lee. "China Uses DNA to Track Its People, with the Help of American Expertise." *New York Times*, February 21, 2019. www.nytimes.com/2019/02/21/business/china-xinjiang-uighur-dna-thermo-fisher.html.

Weiss, Kenneth M., and Jeffrey C. Long. "Non-Darwinian Estimation: My Ancestors, My Genes' Ancestors." *Genome Research* 19 (2009): 703–710. https://doi.org/10.1101/gr.076539.108.

Wen-Wei Liao, Mobin Asri, Jana Ebler, et al. "A Draft Human Pangenome Reference." *Nature* vol 617 (2023): 312–324. https://doi.org/10.1038/s41586-023-05896-x.

Williams, David, and Chiquita Collins. "Racial Residential Segregation: A Fundamental Cause of Racial Disparities in Health." *Public Health Reports* 116, no. 5 (2001): 404–416. https://doi.org10.1093/phr/116.5.404.

Williams, Patricia J. *The Alchemy of Race and Rights*. Cambridge, MA: Harvard University Press, 1991.

———. *Seeing a Color-Blind Future: The Paradox of Race*. New York: Farrar, Straus and Giroux, 2016.

Wilson, Reginald. "Barriers to Minorities Success in College Science, Mathematics and Engineering Programs." In *Access Denied: Race, Ethnicity and the Scientific Enterprise*, edited by George Campbell Jr., Ronni Denes, and Catherine Morrison, 193–205. Oxford: Oxford University Press, 2006.

Winstein, Keith J. "Harvard's Gates Refines Genetic-Ancestry Searches for Blacks: Scholar Founds a Firm after DNA Tracer Put Forebear in Wrong Place." *Wall Street Journal*, November 17, 2007. www.wsj.com/articles/SB119509026198193566.

Winther, R. G. "The Genetic Reification of 'Race'? A Story of Two Mathematical Methods." In *Phylogenetic Inference, Selection Theory, and History of Science: Selected Papers of AWF Edwards with Commentaries*, edited by R. G. Winther, 488–508. Cambridge, UK: Cambridge University Press, 2018.

Wise Uranium Project. "Compensation of Navajo Uranium Miners." Last modified June 8, 2022. www.wise-uranium.org/ureca.html.

Wondermondo. "Denisova Cave: The Only Find of Denisovan Humans." Wonders of the World. Accessed October 19, 2022. www.wondermondo.com/denisova-cave/.

Wu, June Q. "Harvard Prof's Personal Genome Project Reveals DNA Secrets: 'PGP-10' Volunteers Release Their Medical Records and Personal Genome Sequences Online." *Harvard Crimson*, October 23, 2008. www.thecrimson .com/article/2008/10/23/harvard-profs-personal-genome-project-reveals/.

Zeberg, Hugo, and Svante Pääbo. "The Major Genetic Risk Factor for Severe COVID-19 Is Inherited from Neanderthals." *Nature* 587, no. 7835 (2020): 610–612. https://doi.org/10.1038/s41586-020-2818-3.

Ziegler, Mary. "What Is Race?: The New Constitutional Politics of Affirmative Action." *Connecticut Law Review* 50, no. 2 (2018): 279–338. https:// opencommons.uconn.edu/law_review/525.

Þorgilsson, Ari. *Islendingabok, Kristnisaga: The Book of the Icelanders, the Story of the Conversion* [1130]. Translated by Siân Grønlie. Viking Society for Northern Research, University of College London. Exeter: Short Run Press Limited, 2006.

Index

'Abd al-Rahman III, 73
Abrabanel, Isaac, 75
adaptive traits, 9, 31, 59–62, 123
ADHD, 202–203
admixture: ancestry inference, 68, 223–226, 283n4, 284n5; assumptions about, 9, 12, 14, 142, 147, 208, 284n5; applied to African Americans, 138–139, 143; applied to Latinos, 67, 104–105, 222–225; and circularity, 142–143; and confounding, 140; Denisovan and modern humans, 214–215; mapping, 136, 137–138, 221, 222, 224, 278n9, 296n11; and neanderthal and modern humans, 212–214, 305n5; term, 8, 277n3; under appreciated for European populations, 71–73, 209–210
affirmative action, 261, 262, 300n20
Africa, sub-Saharan, xxii–xxiii, 72, 78, 139, 208, 307n17
African Americans: admixture maps of, 136–139, 222; asthma of, 67; in Chicago, 112; in DNA databases, xx, xxi–xxii, 257, 259; 8q24 locus and, 137–138, 140; and forensic uses of, 259–261; fear and mistrust of experimentation, xx, 146; genetic diversity of, xx; health of, xiii–xiv, 120–121, 130; Lower Manhattan Burial

Ground, xxii, 255, 257; mainstream history of, 256; marginalization of in science, 35, 120, 122, 127, 130–132; policing of, xvi, 259; prostate cancer in, 118, 130, 134–135, 137–138, 140–142, 299n15; scarce numbers in genetics, 17, 118, 166; recruitment into studies, 103–105, 152–153, 155–159; reference populations for, 139, 223, 290n19; risk for *APOL1*-linked kidney disease, 59–60, 61; as a "special" population, 127; 23andMe campaign for, xxii. *See also* Black men; Black women; health disparities
African ancestry: construction of panel, 290n19; coveted for genetic research, xxii–xxiii; diversity of, xx, 168; Duffy-null marker and, 49; 8q24 locus and, 137, 138, 140–144; genetic risk and, 59–61, 126–130, 132, 135–142, 222, 288n34; health disparities and, xvi–xvii, 108, 122, 288n34; Mexicans having, 67; *mulatos* with, 37; Old vs. New World terminology and, 78; prostate cancer and, 154–155, 168, 223, 224, 255–257; in Puerto Ricans, 162. See also *APOL1*; 8q24 locus
African Ancestry (company), xx, xxi, xxii, 110, 257

Founded in 1893,
UNIVERSITY OF CALIFORNIA PRESS
publishes bold, progressive books and journals
on topics in the arts, humanities, social sciences,
and natural sciences—with a focus on social
justice issues—that inspire thought and action
among readers worldwide.

The UC PRESS FOUNDATION
raises funds to uphold the press's vital role
as an independent, nonprofit publisher, and
receives philanthropic support from a wide
range of individuals and institutions—and from
committed readers like you. To learn more, visit
ucpress.edu/supportus.